Zulu Bird Names and Bird Lore

The Common Ostrich (Zulu **intshe**). The traditional use of its feathers is discussed in Chapter 11. © Adrian Koopman

Zulu Bird Names and Bird Lore

Adrian Koopman

UNIVERSITY OF KwaZulu-Natal Press

Published in 2019 by University of KwaZulu-Natal Press
Private Bag X01
Scottsville, 3209
Pietermaritzburg
South Africa
Email: books@ukzn.ac.za
Website: www.ukznpress.co.za

ISBN: 978 1 86914 425 8
e-ISBN: 978 1 86914 426 5

Managing editor: Sally Hines
Editor: Catherine Munro
Proofreader: Judith Shier
Typesetter: Patricia Comrie
Cover design: Marise Bauer, M Design
Cover art: Red Bishop by Adrian Koopman

Typeset in Times New Roman 10.5pt

Contents

Colour plates fall between pages 238 and 239

Preface   xiii

Acknowledgements   xvii

1 Introduction   1

1.1   The names of the White-eye   1

1.2   Proper names vs common nouns   3

1.3   A preliminary typology of bird names   5

1.4   Human and bird interaction   10

1.5   Rise of interest in birds and birding among Zulu speakers   11

1.6   Review of source material   12

        1.6.1   Primary source material   12

        1.6.2   Secondary source material   12

1.7   Overview of the chapters   15

2 Names and identity   20

2.1   Introduction   20

2.2   Exploring the notion 'eagle'   21

2.3   The issue of generic and specific names   25

        2.3.1   The 'problem' of bird names in African languages   26

        2.3.2   Nguni names for the Bateleur   29

2.4   Folk-taxonomic systems   32

2.5   How do we handle 'Maclean's problem'?   38

2.6   Dove and pigeon names as an example of 'polytypy'   41

    2.6.1   Introducing the notion of polytypy   41

    2.6.2   Extracts from Doke and Vilakazi's 1958 *Zulu-English Dictionary* relating to doves and pigeons   43

    2.6.3   Doves and pigeons of KwaZulu-Natal: a comparison of the scientific taxonomy with Zulu taxonomy   44

3   The meaning of African bird names   49

    3.1   Introduction: Semantics – the theory of meaning   49

       3.1.1   The notion of 'meaning'   49

       3.1.2   Secondary meanings and secondary functions   51

    3.2   An overview of meaning in African bird names   54

       3.2.1   Typologies of bird names based on semantics   55

       3.2.2   Song   59

       3.2.3   Appearance   66

       3.2.4   Habitat   72

       3.2.5   Diet   74

       3.2.6   General behaviour   79

       3.2.7   Motion   81

       3.2.8   Belief names   84

       3.2.9   Nesting   86

       3.2.10   Other   87

4   The meaning of the new Zulu bird names   90

    4.1   Introduction   90

    4.2   Comparative semantic profiles of African bird names, old Zulu bird names and the new Zulu bird names   92

    4.3   A semantic profile of the coined Zulu names   94

       4.3.1   The semantic category 'Song'   94

       4.3.2   The semantic category 'Appearance'   96

       4.3.3   The semantic category 'Habitat'   99

       4.3.4   The semantic category 'Diet'   100

       4.3.5   The semantic category 'General behaviour'   102

       4.3.6   The semantic category 'Motion'   103

4.3.7	The semantic category 'Belief'	103
4.3.8	The semantic category 'Nest'	104
4.3.9	The semantic category 'Other'	104

5	**Zulu bird names used for other purposes**	**106**
5.1	Bird names as other lexical items	107
5.1.1	Polysemy as illustrated by Zulu bird names	108
5.1.2	Bird names and cattle colours	113
5.1.3	From cattle colour patterns linked to birds to general colours linked to birds	117
5.2	Bird names in secondary lexemes	121
5.2.1	Onymisation	121
5.2.2	Names of people	123
5.2.3	Names of geographical features	125
5.2.4	Street names	128
5.2.5	Brand names and logos	130

6	**The morphology of bird names**	**132**
6.1	Introduction	132
6.2	Classification of noun stems	134
6.3	Names with simple stems	134
6.4	Names with reduplicated stems	136
6.5	Names with complex stems	137
6.5.1	The use of the prefix *–no–*	139
6.5.2	The use of the prefix *–ma–*	140
6.5.3	The use of the prefix *–so–*	141
6.5.4	The use of the prefix *–sa–*	141
6.5.5	The use of the suffixes *–ana* and *–ane*	142
6.5.6	The use of the suffixes *–azi* and *–kazi*	144
6.5.7	Various other suffixes	145
6.6	Names with compound stems	146
6.6.1	Noun + noun	146
6.6.2	Noun + adjective	148
6.6.3	Noun + possessive locative	151
6.6.4	Noun + verb	152

	6.6.5	Verb + noun	152
	6.6.6	Other verb-based names	154
6.7		The value of a derivational system	154

7 Verbalisation of birdsong in Zulu   157

7.1		Introduction	157
	7.1.1	Preamble	157
	7.1.2	Birdsong renderings in field guides for birdwatchers	158
	7.1.3	Birdsong references in bird names	161
7.2		Descriptive names, including the use of metaphor and simile	163
7.3		Imitative names: the use of onomatopoeia	165
	7.3.1	Onomatopoeia	165
	7.3.2	The call and the names of the Hadeda Ibis	166
	7.3.3	Previously recorded Zulu onomatopoeic names	168
	7.3.4	Zulu workshop-coined onomatopoeic names	171
	7.3.5	Onomatopoeia that becomes generic	171
7.4		Interpreted calls: what the bird is really saying	175
	7.4.1	The notion of interpretation	175
	7.4.2	Contributors to the interpretations of Zulu bird calls	177
	7.4.3	Selected Zulu interpretations of bird calls	179

8 Birds in Zulu praise poetry   185

8.1		Introduction	185
	8.1.1	Different genres of praise poetry	186
	8.1.2	Historical figures in Zulu praise poetry	187
	8.1.3	The copying of bird-related memes	188
	8.1.4	Birds in Zulu praise poetry	190
8.2		*Ulwandle kaluwelwa, luwelwa yizinkonjane*: swallows over the sea	190
8.3		The 'red birds': **ungqwashi** and **igwalagwala**	195
8.4		The raptors	201
8.5		**Ujojo** and other finches	205
	8.5.1	The **ujojo** finch	207

8.6		Miscellaneous other birds	210
	8.6.1	Domestic hen	211
	8.6.2	The owl	211
	8.6.3	The Hamerkop	212
	8.6.4	The wagtail	213
	8.6.5	Mousebird and egret	214
	8.6.6	Honeysucker	214
	8.6.7	The dove	215
	8.6.8	Waterfowl and vulture	215
	8.6.9	The lark **ucilo**	216
	8.6.10	Glossy starlings (perhaps)	216
	8.6.11	The **intendele** partridge	217
	8.6.12	The plover	218
	8.6.13	The Southern Ground Hornbill	218
8.7		*Izinyoni nje* 'Just birds'	220

9		**Praises, proverbs and riddles**	**224**
9.1		The praises of birds	224
	9.1.1	Introduction: praise names to praise poetry	224
	9.1.2	The praises of selected birds in Zulu	229
	9.1.3	The praises of the longclaw **inqomfi**	234
9.2		Birds in proverbs, idiomatic expressions and sayings	246
	9.2.1	Introduction to proverbs	246
	9.2.2	Proverbs relating to the hunting and trapping of birds	248
	9.2.3	Other aspects of birds leading to proverbs	251
	9.2.4	Cognitive links between the three elements of a 'bird proverb'	259
	9.2.5	Proverbs and folk tales	260
9.3		Birds in riddles and children's games	261

10		**Bird beliefs: Portents and heralds**	**270**
10.1		Introduction	270
	10.1.1	Explanation of terminology	271
	10.1.2	Sources of material	272

10.2   Omens, portents, taboos and charms   274
    10.2.1   Manifestation of omens and portents   274
    10.2.2   Omens of death and illness, witchcraft and evil   277
    10.2.3   Omens of war   282
    10.2.4   Omens of luck, bad and good   283
    10.2.5   Taboos against killing, eating and imitating birds   287
    10.2.6   Birds used as charms   293
10.3   Harbingers and heralds: birds of weather, seasons and times
of day   295
    10.3.1   Weather forecasters: rain birds, storm birds and
lightning birds   295
    10.3.2   The heralds: birds of spring and summer   303
    10.3.3   Announcers of dawn and dusk   308

11 Feathers, food and fancies   312
11.1   Use of feathers for decoration and ornament   313
    11.1.1   The Blue Crane (*Grus paradisea*), Z. **indwa**
(also frequently **indwe**)   313
    11.1.2   The Common Ostrich (*Struthio camelus*),
Z. **intshe**   317
    11.1.3   The Long-tailed Widowbird (*Euplectes progne*),
Z. **isakabuli**   321
    11.1.4   Purple-crested Turaco (*Tauraco porphyreolophus*),
Z. **igwalagwala** and mousebirds (**indlazi** and
umtshivovo)   323
    11.1.5   Paradise Flycatcher (*Terpsiphone viridis*),
Z. **u(lu)ve** or **inzwece**   324
    11.1.6   The Kori Bustard (*Ardeotis kori*),
Z. **umngqithi**   325
    11.1.7   The dove (**ihobhe**, **ijuba**), the eagle (**ukhozi**)
and the vulture (**inqe**)   326
    11.1.8   General plumage (unspecified birds)   327
    11.1.9   A lexicon of feather terms   328
11.2   Use of birds for food   329
11.3   Miscellaneous beliefs and uses   332
    11.3.1   Humans, cattle and birds   333
    11.3.2   Where birds winter   336

	11.3.3	Curious miscellanea	337
	11.3.4	Practical uses with some modern implications	339
11.4	Conclusion: attitudes towards traditional beliefs		340

12 New names and new identities — 346

12.1	Introduction		346
12.2	Background to the workshops		347
12.3	Preparations for the first workshop		349
	12.3.1	Logistics of the workshop	349
	12.3.2	Collating of existing material	350
12.4	Modus operandi of the first workshop		350
12.5	A case study of shore birds and waders		351
12.6	A summary of the linguistic strategies and processes used		357
	12.6.1	Confirmation	358
	12.6.2	Selection and relegation	358
	12.6.3	Redirection	364
	12.6.4	Assignment	365
	12.6.5	Coinage	366
12.7	Interrogation of the results of the 2013–2015 workshops		372
12.8	The second round of workshops		374
12.9	Final list of Zulu bird names: the process of acceptance		378
12.10	Appendix: The kingfisher's tale		381
	12.10.1	Nine birds need nine names	381
	12.10.2	Nine names for nine birds	387

13 Change is in the air — 389

13.1	Yesterday, today and tomorrow		389
13.2	Zulu bird names in their historical context		391
	13.2.1	Oral contributions and their transcription	391
	13.2.2	Earlier contributors to the Zulu ornithonomasticon	395
13.3	The modern Zulu bird namers		401
	13.3.1	Those involved in the 2013–2017 Zulu bird name workshops	401
	13.3.2	Sakhamuzi Mhlongo	404

	13.3.3	Themba Mthembu	406
	13.3.4	Junior Gabela	409
13.4	Birds and education		410
	13.4.1	Birds and conservation or environmental education	411
	13.4.2	Bird clubs and the Women's Leadership and Training Programme (WLTP)	413
	13.4.3	Writing of children's readers	415
	13.4.4	Bird guide training	415
	13.4.5	Community bird guides	418
13.5	Birding and avitourism		420
13.6	The evolution of memes: changing dynamics in traditional beliefs		423
	13.6.1	The dying out of a meme	424
	13.6.2	The (virtually) unchanged survival of a meme	424
	13.6.3	The adaptation of a meme: the Southern Ground Hornbill	425
	13.6.4	The birth of a new meme: the Southern Ground Hornbill (again)	426
	13.6.5	The deliberate forced reversal of memes: owls and vultures	427
13.7	Bird names: challenges for the future		429
	13.7.1	Bird names in South African Bantu languages	429
	13.7.2	Species-specific names in Swahili	432
	13.7 3	Names of birds in Seychelles Creole	437

Bibliography	442
General index	450
Bird name index	457

Preface

The major focus in this book concerns Zulu[1] bird names. These names come from different time periods, which for convenience could be labelled 'then' and 'now'. 'Then' refers to all Zulu bird names that were recorded in print up to 2013. This refers to all the Zulu bird names to be found in existing Zulu dictionaries, as well as the bird names found in bird guides such as the *Roberts* series. In 2013 Professor Noleen Turner of the University of KwaZulu-Natal organised a Zulu bird name workshop, the first of a series of annual workshops that ran until 2017. These workshops, which brought together a total of twenty Zulu-speaking bird experts, had as their ultimate goal a unique name for every species of bird found in KwaZulu-Natal (KZN). After the last workshop in 2017, names had been found for every species, including Zulu generic names playing the same role as 'babbler' or 'white-eye' in English. The process is fully described in the penultimate chapter of this book – Chapter 12. These names are the 'now' names I mentioned earlier.

Both 'then' and 'now' names are described in this book, and wherever possible a distinction is made. Indeed, very often the difference between the 'then' situation and the 'now' situation is a subject itself for discussion. For

---

1. I specifically do not use the Zulu word *isiZulu* (Zulu language) in this book. For full details on why I do not, see Koopman (2011b). But briefly: using 'isiZulu' in English discourse is linguistically and stylistically inconsistent in a number of different ways. For example, to be consistent, writers in English who use 'isiZulu' for 'Zulu' should also use *Française* instead of *French*, *Deutsch* instead of *German,* and *Suomi* instead of *Finnish*. Writers in Zulu should use *English* instead of *isiNgisi*, *Française* instead of *isiFulentshi* and *Deutsch* instead of *isiJalimane*. In addition, writers in English should use *amaZulu* instead of 'Zulu people', *kwaZulu* instead of 'Zululand' and *ubuZulu* instead of 'Zulu culture'. The Zulu word *isiZulu* is a noun, not an adjective, so it should not be used adjectivally as in 'isiZulu bird names', which is a linguistic horror. And so on, and so on.

example, in Chapter 2, where the notions 'generic' and 'specific' are discussed with regards to bird names, we see that a lack of species-specific names for birds in Zulu was seen by an ornithological writer as a problem. In chapters 3 and 4 where the underlying meanings of bird names are discussed for African languages in general, and the Zulu language in particular, it can be seen that the new Zulu bird names fit perfectly into the semantic profiles of previously existing Zulu bird names and of African language bird names. In Chapter 5, where the morphology (grammar) of Zulu bird names is discussed, we see again that the new Zulu names conform to the same morphological structure as the old names.

Chapters 6 through 11 deal with various aspects of how Zulu society has integrated birds and their names into their oral literature and their traditional beliefs, so clearly the new Zulu bird names have no role to play in these chapters. However, the last chapter ('Change is in the Air') includes discussion on modern avitourism (where local and international birders visit KZN to 'tick off' individual species of birds), and on the use of birds in the education of Zulu-speaking children. In these domains identification of discrete species of bird by unique names is required, and so the development of the new bird names becomes an essential part of this.

I was privileged to be a part of the stimulating – indeed, exciting – Zulu bird name workshops since the very first one in 2013, and have tried to interweave the creative thought-patterns that emanated from these workshops into the traditional thought-patterns about birds over the years.

Conventions used in writing about birds and their names

In this book, I write about bird names, mainly about Zulu names, but at the same time I also use examples from a number of other African languages: Afrikaans, Xhosa and Sotho from South Africa, and languages such as Chewa, Nyabongo and Nyanja from other parts of sub-Saharan Africa, as well as Malagasy from Madagascar. At the same time I continually use English names to identify the bird whose name I am writing about. Occasionally it is necessary to use the Latin-based scientific name as well. To keep these all separate, I use the following fonts throughout:

bold for Zulu bird names: **ititihoye, isakabuli, uthekwane**

Roman with Upper Case for English names: Blue Crane, Yellow-billed Stork, Rufous-naped Lark

Roman with lower case for English names when talking of generic groups: swallow(s), owl(s), duck(s)

Italics for scientific names: *Struthius camelus*, *Pternistis swainsonii*, *Morus capensis*

Italics for all other languages: *fiandrivoditatatra, ndege chai, mmamasionoke, kanyamarhaza, kakelaar, bosmusikant, spookvoêl.*

When writing about bird names, it is obviously essential to also identify which bird the name refers to. In other words bird names are used in two ways in this book: to identify birds, which is the normal and primary function of any name in any language, but also as an object of linguistic inquiry. Look at the following two sentences:

1) I usually use a dishcloth as do most people: for drying dishes.
2) The English word *dishcloth* is a compound word where the first element *dish* qualifies the second element *cloth*.

Statement (1) is the kind of statement any person would use; statement (2) is that typically uttered by a linguist when talking about language.

We can set up the same sort of contrast when talking about birds and about bird names:

3) The Hamerkop builds a very large nest of twigs and other plant material, into which it incorporates all sorts of discarded rubbish.
4) The English name *hamerkop* is derived from the Afrikaans or Dutch word meaning literally 'hammer head' and refers to the shape of the bird's head.

As sentence (3) talks about the bird and sentence (4) talks about the name, it is important to distinguish between them. As we see above, italics are usually used when talking about a name or other linguistic item, but as I am using italics in the book when giving scientific names and other languages, I need another typological device. So in this book, single inverted commas will be used with an English name in uppercase Roman font as in the following:

5) The Hamerkop builds a very large nest of twigs and other plant material, into which it incorporates all sorts of discarded rubbish.

6) The English name 'Hamerkop' is derived from the Afrikaans or Dutch word meaning literally 'hammer head' and refers to the shape of the bird's head.

There is no need to make similar distinctions between use of a name as a referent to a species and use of a name as a subject of linguistic investigation for any of the other languages as only English bird names are used for referential purposes in this book.

The sign < indicates 'derived from'. Thus the name **umehlwane** (white-eye) < *amehlo* (eyes).

The sign > means 'give rise to'. The noun *igazi* > the name **ugazini** for the Red-headed Finch.

An asterisk in front of a word, phrase or sentence indicates that what follows is considered ungrammatical or substandard: *Red Bishop occurs in open grassland. It should be the Red Bishop or Red Bishops.

Acknowledgements

This book is one of the products of the 2013–2018 Zulu Bird Name Project, which had at its core the 2013–2017 Zulu bird names workshops. These workshops were initiated and administered by Professor Noleen Turner, a long-time colleague and friend from the University of KwaZulu-Natal. I am grateful to her for her role in this, for without the Zulu bird name workshops, this book would never have come into being. Retired ecologist Roger Porter joined the third workshop in 2017 and has been present in all subsequent workshops and discussions. I thank him for his frequent and useful suggestions that have helped link the avian and the onomastic, enabling me to see links between bird name and bird behaviour that I had not seen before.

These workshops were all funded by research grants awarded to Turner and to myself by the National Research Foundation (NRF) – grants that are awarded to NRF 'rated researchers'. We are both grateful to the NRF for these grants.

The Zulu Bird Name Project, and its core workshops, could also not have come into being without the traditional knowledge held by the Zulu-speaking bird guides and bird experts who took part. I am grateful, therefore, to Theo Bukhosi, Siya Dlamini, Abednigo Dube, Junior Gabela, Thabile Khuzwayo, Jethro Mdlalose, Bheki Mhlongo, Sakhamuzi Mhlongo, Bongani Mthembu, Sakhile Mthembu, Themba Mthembu, Sakhile Mthenjwa, Daluxolo Ngcobo, Benson Ngubane, Vusi Ngwenya, Phindile Ntshangase, Bheki Nyandeni, Bheki Sithole and Nontuthuko Xaba. Of these, I am particularly grateful to Junior Gabela, Sakhamuzi Mhlongo and Themba Mthembu, who sent me detailed information about their paths to becoming successful bird guides, and details of how they integrated bird guiding and conservation education into their lives.

In the earlier research stages Professor Colleen Downs, of the University of KwaZulu-Natal in Pietermaritzburg, was instrumental in helping me find access to earlier copies of the British journal *Ibis*, and the South African journals *Ostrich* and *Bokmakierie*. Over the next two or three years she continued to send me links to useful articles on various aspects of ornithology and bird-naming. I am very grateful for her help. Thanks are also due to American historian and author of the 2016 *Birders of Africa*, Professor Nancy Jacobs, who on a visit to Pietermaritzburg in 2017 made suggestions about the ordering of the chapters. She later sent me a number of useful links. Professor John Wright, another erstwhile colleague of the former University of Natal, has been working on editing Volume 7 of the *James Stuart Archive*, which contains praises left out of volumes 1–4. He very kindly sent me previously unpublished Zulu praises of several birds, and my thanks go to him for that. Dr Johan Meyer of Pretoria, currently working on Zulu bird names in Northern Sotho and Venda, has sent me useful information on bird names from a number of South Africa languages. His information on Southern Sotho bird names has been particularly helpful, and I am very grateful to him for his contributions.

In the latter stages of putting the book together, when illustrations were being sought, renowned bird artist Ingrid Weiersbye allowed me access to her extensive library of bird photographs. I am most grateful for this. Her archive of bird photographs contain a number taken by Hugh Chittenden, and I thank him as well for his permission to use these. Each photograph in the colour plate sections has been individually acknowledged as from either Weiersbye or Chittenden. These colour photographs have been integrated with watercolour paintings of birds that I have painted myself. I thank the late Alan Turton, for many years art teacher extraordinaire at Durban High School, who in my years under his guidance in the 1960s inculcated in me a love of watercolour painting that would last a lifetime.

And then in the final production stages of the book, thanks must go to all contributing members of the University of KwaZulu-Natal Press team: managing editor Sally Hines, editor Catherine Munro, proofreader Judith Shier, typesetter Patricia Comrie, and cover designer Marise Bauer. It is their professional skills that have turned a manuscript into this book.

And last but not least, I thank my wife Jewel for her help in compiling the General Index for this book and for providing the space needed for writing this book by uncomplainingly taking over my share of the domestic duties while I grappled with notions of folk taxonomy, knitted together various narrative strands about the Southern Ground Hornbill, or decided on which out of nine Zulu names to choose for the Rufus-naped Lark.

1 Introduction

1.1 THE NAMES OF THE WHITE-EYE

The Reverend James Sibree was a missionary in Madagascar in the late nineteenth century. Like so many other reverend gentlemen whose names will appear in this book, Reverend Sibree was also a 'naturalist' with a deep and abiding interest in living things. Birds are what held his attention most, and his birdwatching records, which first appeared in the *Antananarivo Journal* in 1890, were reprinted later in the British ornithological journal *The Ibis*. The following brief extract is taken from an *Ibis* article of 1891:

> As might have been expected, the White-eyes (*Zosterops*) have several names referring to the prominent white ring round their eyes; e.g. *Tsàramàso*, "Beautiful-eyes," *Sipàromàso*, *Paríamàso*, and also *Ramanjèreky*, from a root meaning "to be conspicuous", "to be obvious to the sight" . . .
>   . . . The White-eye (*Zosterops madagascariensis*) builds a very pretty open nest on the end of some hanging branch. Its eggs are very pale blue (1891a: 426).

A number of points can be made relating to this extract, brief as it is. We note that Sibree talks about two different aspects of the birds belonging to the genus *Zosterops*: one is an aspect of the bird itself – its nidification (nest shape, eggs and their colour). The other aspect is not of the bird itself, but of a label humans have assigned to it, i.e. its name. Or rather, names in this case, as Sibree recognises that the genus as a whole shares more than one name. It is clear that Sibree has a scientific ornithological background, as he uses a genus name for White-eyes generally, but a genus name with a specific epithet for a particular type of White-eye. But Sibree has also acquainted himself with the Malagasy language, giving four different Malagasy names for this group of birds, and identifying the underlying meaning of each.

In this short extract, Sibree shows how three different 'notional systems' interact:

- *Birds*: Birds live in various habitats where they fly, eat, mate, nest and produce young. Humans occasionally enter this avian world and hunt, eat, observe, identify and name these birds.
- *Language*: In this brief extract Sibree has identified names in three different languages: Malagasy, English and Latin. He understands that these names have the primary function of identification, but also that the names have underlying meanings that refer directly to some aspect of the bird, in this case the prominent white markings around the eye.
- *Humans*: Language is a human construct. In six short lines Sibree has identified three different groups of humans: (i) the indigenous population of Madagascar whose language marks the white eyes of this group of birds; (ii) the unknown taxonomists of museums in Europe and America who have coined the word *Zosterops* from the Greek *zoster* 'girdle' and *ops* 'eye' for this group of White-eyes, and have added the locative suffix *–ensis* to 'Madagascar' to form the specific epithet *madagascarensis*, and (iii) the travelling, exploring writer and naturalist, often a military man, a colonial official, a doctor, or, as in this case, a missionary.

Sibree was thus not just a keen 'birder', in modern terms, but one who wrote about the wonder of birds, tried to learn and interpret their 'native' names, and one who tried to integrate birds and language. In another extract from the same article where he wrote about White-eyes, Sibree explains the underlying meaning of the name *Goàika* for a crow, and then shows how the Malagasy keep this bird as a pet and construct proverbs, songs and riddles around its name.

The Reverend James Sibree, as can be seen from this brief introduction, was a missionary, a naturalist and a writer. As such, he is one among many such clerics. We think of Bishop John Colenso – missionary, and renowned Zulu grammarian and lexicographer; the missionary brothers the Reverends Richard and John Woodward – naturalists, collectors and writers; the Reverend Robert Godfrey – missionary, naturalist, writer and Xhosa lexicographer; Father Arthur Bryant – Catholic priest, Zulu lexicographer, Zulu historian and Zulu ethnographer; and the Catholic priest Father Jacob Gerstner who was a naturalist and specifically a botanist.

Much of the material I have used in this book about Zulu bird names and bird lore has come from these recorders of the traditional knowledge of the Zulu and related peoples. Most of these people were multidisciplinary in their approach, interested not only in birds, but in the Zulu language, and in the people who spoke the language, and the Zulu names that they gave to the birds which were so much part of their lives.

1.2   PROPER NAMES VS COMMON NOUNS

In the introduction above, using a brief extract from the 1891 writing of a British missionary in Madagascar, I identified three different types of names: the Latin-based scientific name *Zosterops madagascariensis*, the English vernacular name 'White-eye' and the Malagasy vernacular names *Tsàramàso, Sipàromàso, Paríamàso*, and *Ramanjèreky*. To the onomastician (names scholar), however, none of these are true proper names. The term 'proper name' is usually applied to the names of people (given names, nicknames, baptismal names, surnames, etc.), to the names of places (rivers, cities, mountains, countries, etc.) and to the names of a few other entities, such as brands and characters in literature.

The difference between the two types of names can clearly be seen when an ornithologically minded person writes about birds and includes anecdotes about pet birds he or she has known. Earlier I mentioned the Woodward brothers. In their 1899 *Natal Birds* they regularly give four different names for each species listed: a scientific name, an English 'book name', a 'popular' name (functioning as a nickname) and the local 'native' name. Thus for the bird '*Estrilda astrilda*' they use the 'book name' 'Common Waxbill', tell the reader that "red-billed Waxbills are generally known by the Dutch name 'Rooibekje' or Redbill" (Woodward and Woodward, 1899: 74), and give the 'native' name as 'Intiyane', correctly pointing out that this Zulu word is used for all waxbills.

In their entry for the Wattled Crane, they give the scientific name *Bugeranus carunculatus*, and the Zulu name as 'Unohemu'.[1] But in this particular entry they add a different kind of name altogether when they tell a lengthy anecdote about the Wattled Crane that they reared from the nest and which became a tame pet bird. The following is just a brief extract from the story:

---

1. The Woodwards are usually quite accurate in their Zulu names. Here though they are mistaken: **unohhemu** is the Zulu name of the Crowned Crane; the Wattled Crane is **ubhamukwe**.

It used to come to its name of 'Mick', and follow its master about like a dog for long distances; and on his return from being away, would run and dance to meet him, showing unmistakable signs of pleasure. 'Mick' did not understand the art of bathing . . . (1899: 173).

'Mick' here is a true proper name. It passes all linguistic tests for 'properhood', for example:

- Like any other proper name it can function as the subject of a sentence without needing 'a' or 'the' as in 'Mick did not understand . . .', where one could not say *'Wattled Crane did not understand . . .'[2] or *'*Bugeranus carunculatus* did not understand . . .'
- Like all proper names, 'Mick' here is the name of an <u>individual</u> Wattled Crane, whereas <u>every single member</u> of the species *Bugeranus carunculatus* may be identified by the labels '*Bugeranus carunculatus*', 'Wattled Crane' and **ubhamukwe**.

In another example, the Woodwards identify "one of the smallest of our larks" as the Short-tailed Pipit (*Anthus brachyurus*). There is no popular nickname for this entry, nor is there a Zulu name. But there is a story of Mr Layard's "little lark which he kept in an aviary for six years" and the brothers happily recount all the little tricks and personal traits of 'Brownie' – "Brownie was the first to sing in the morning . . .", and so on.[3]

This phenomenon of giving personal names to wild animals, or 'wild' birds in this case, has been noted and described in an earlier article of mine.[4] In this article I suggest that this can happen when an animal or bird ceases to be an undifferentiated member of a group and becomes an individual through interaction with humans. In this article, and in the database on which the article was based, we can find examples of a song-thrush named Breac, a ptarmigan named Jim, a Hadeda Ibis named Harriet, penguin chicks named Kola and Kelp, and Martha, the last surviving passenger pigeon in the world, who died in Cincinnati Zoo in 1914. And there are more Wattled Cranes, to keep the Woodwards' Mick company: the cranes Elvi (♀) and Amanzi (♂) and their

---

2. In linguistic discourse an asterisk placed in front of a word or an utterance indicates that it is substandard and/or ungrammatical.
3. The story is in Layard and Sharpe (1867: 540).
4. Koopman (2014a).

two chicks Blake and Trinity, all resident in 2012 at the Hlatikulu Crane and Wetland Sanctuary in the KwaZulu-Natal (KZN) Midlands.

Mark Cocker's 2014 *Birds and People* is also full of stories about birds, which for one reason or another have been given personal names.

These are all true proper names, whereas 'Short-tailed Pipit', '*Anthus brachyurus*' and '**ubhamukwe**' are all what an onomastician would call 'appellatives, i.e. not 'proper names'. The study of these proper names of birds is in itself intriguing, especially to onomasticians. Why are so many birds given human names like Jim, Harriet, and Martha?[5] Are pairs of chicks often named alliteratively like Kola and Kelp?[6] Is there a story behind the names Blake and Trinity?[7] Fascinating those questions may be, these are not the kind of names that are going to be looked at in this book. Instead I shall be using the word 'name' in the same way as any writer of books about birds does – to refer to designations for different species of birds. In so doing we will be referring to four different types of bird names:

1.3   A PRELIMINARY TYPOLOGY OF BIRD NAMES

Below are entries for two birds, the Yellow-bellied Bulbul and the Rock Pigeon, from two different books. The entry for the bulbul is from page 325 of Clancey's 1964 *The Birds of Natal and Zululand*, and the entry for the pigeon is from page 303 of Maclean's 1984 *Roberts' Birds of Southern Africa*.

---

389   *ANDROPADUS FLAVIVENTRIS* (Smith)   YELLOW-BELLIED BULBUL

Andropadus flaviventris flaviventris (Smith)

Trichophorus flaviventris A. Smith, *S. Afr. Quart. Journ.*, 2nd series, No. 2, 1834, p. 143: near Port Natal, *i.e.*, Durban, Natal.

---

349 (311)   **Rock Pigeon** (Speckled Pigeon)   **Plate 33**
Kransduif (Bosduif)

Columba guinea

Haifoko (K), Leeba, Lehoboi (SS), Leeba, Leeba-rope, Letseba (Tw), Ivukuthu (X), iJuba, iVuku-thu (Z)

---

5. The phenomenon is not restricted to birds. Koopman noted (2014a) how many wild animals that are individually named get human personal names, and one only has to think of pet names in one's own society.

6. Yes – this is a common practice.

7. Yes – this has to do with the funders of the crane-breeding project. 'Trinity', for example, refers to funding received from students at Trinity College in Cambridge.

In comparing these two entries from different authors, a number of things are immediately apparent:

- Clancey gives far more prominence to the scientific name than Maclean does: the name *Andropadus flaviventris* is the same sized font as the name Yellow-bellied Bulbul, but takes the prime position on the left. It is then repeated again below with the Natal/Zululand sub-species – *Andropadus flaviventris flaviventris*. The original name given by Dr Andrew Smith – *Trichophorus flaviventris*, is also given, with the details of where Smith discovered the type specimen "near Port Natal, *i.e.*, Durban", as well as where and when he published the original description and name "*S.Afr. Quart. Journ.*, 2nd series, No. 2, 1834, p. 143". Maclean simply gives the genus name (*Columba*) and the specific epithet (*guinea*), and this is in a much smaller font than the English and Afrikaans vernacular names.
- Besides the detail on the scientific name, Clancey offers one vernacular name: the English name 'Yellow-bellied Bulbul'. That is it. Maclean on the other hand gives an English vernacular name, an Afrikaans vernacular name and names in Kwangali, Southern Sotho, Tswana, Xhosa and Zulu.
- Maclean acknowledges that names have alternatives, and that the book names may change from time to time. As regards the English vernacular, *Roberts* (McLachlan and Liversidge 1957, 1978) only have 'Rock Pigeon'. Maclean suggests the alternative name 'Speckled Pigeon', which by the time of Chittenden, Davies and Weiersbye in 2016 has become the main book name with "Alt name Rock Pigeon" at the end of the entry for this bird. As regards the Afrikaans name, *Roberts* (1957, 1978) only have *bosduif*, but Maclean gives *bosduif* as an alternative to *kransduif*. In Chittenden, Davies and Weiersbye (2016), the only Afrikaans name given is *Kransduif*.

In the two extracts above three different types of names are illustrated, and I enlarge on these below, as well as adding a fourth type. This, then, is my suggested typology of names of birds, and the categories I will be referring to are:

- *The scientific name*: The Latin-based binomial name that designates a species is too well known to need further discussion. They are usually in the form of a binomial (the genus name and the specific epithet), but occasionally authors use a trinomial and include a sub-species name, as in the case of Clancey's *Andropadus flaviventris flaviventris* shown above. These names are mainly used purely for identificative purposes in this book, and will not be specifically discussed unless the underlying meaning is of relevance to a point made about a Zulu bird name.

- *The artificially constructed 'book' name*: By this I mean the phrasal names such as 'White-fronted Plover', 'Buff-spotted Flufftail', 'Fiery-necked Nightjar' and the like. Clancey's 'Yellow-bellied Bulbul' is a perfect example. These specialist 'vernacular' names are mostly used by serious 'birders' and not by the man in the street. I give English examples here, but such official book names may be in any language, and we see that Maclean has included names in six southern African languages other than English. The processes involved in constructing such phrasal names, especially the processes of doing the same for Zulu, are discussed in Chapter 12 of this book.

- *The folk name*: By 'folk name' we mean the names used by the vast majority of the population who are neither birders nor ornithologists, i.e. the 'common folk'. The man in the street sees a 'hawk' in the sky, not a 'Dark Chanting Goshawk', and hears an 'owl' in the night, not a 'Pearl-spotted Owlet'. So by 'folk names' we mean names such as 'owl', 'eagle', 'stork', 'lark' and 'duck', and their equivalents in Afrikaans, Zulu, Sotho, Malagasy, and any other language. In the extracts above the names 'pigeon' and '*duif*' are folk names, and become 'vernacular book names' when 'rock' or 'speckled' or '*krans*' or '*bos*' are added to them.

    These are names that are found in dictionaries as well as in in bird books, and Zulu dictionaries have provided much of the information that I have used in my discussion of Zulu names. Discussion of various aspects of Zulu 'folk names' for birds constitutes the major part of this book, and as these names are the names that are still used by Zulu-speaking people today, I have tried to get as much input as possible from knowledgeable Zulu speakers.

- *The 'popular' nickname*: Above I noted how the Woodward brothers gave the Dutch '*rooibekje*' as a popular name used by English-speaking colonists for various species of waxbills. In the same way such colonists used 'turkey buzzard' to refer to the Southern Ground Hornbill and the Zulu-based '**sakaboola**' for the Long-tailed Widowbird. Zulus, Xhosa and other peoples of southern Africa have had 'praise names' for birds in addition to the everyday name used for identification and reference.

The popular nicknames used by English and Afrikaans speakers are not discussed in this book, but the Zulu and Xhosa praise names will appear in Chapter 8 when birds in praise poetry are discussed.

In Table 1.1 I set out a matrix with the four suggested name types running along one axis, and a number of aspects of each name type running along the other axis. These are Denotation (in other words what avian identities are referenced by the name type), Structure (the syntax of the name type) and Language (Latin, English, French, Zulu, etc.). Very brief examples are given in the table of each suggested type:

Table 1.1   A proposed typology of bird names.

Type	A. Scientific binomial	B. Vernacular book name	C. Folk generic names	D. Alternative names, nicknames, earlier names, praise names
Denotation	genus and species	species	folk-generic	both species and folk-generics
Structure	two-word phrase	almost always phrasal, often with elements hyphenated	usually single words	varied structures
Language	Latin, often derived from Greek	Any language but Latin	Any language but Latin	Any language but Latin
Examples	*Columba guinea*; *Streptopelia capicola*; *Aplopelia larvata*	Rock Pigeon/ *kransduif*; Cape Turtle Dove/ *gewone tortelduif*; Lemon Dove/ *kaneelduifie*/ **isagqukwe**†	dove, pigeon, **ijuba**, **ihobhe**	Speckled Pigeon/ *bosduif*; White-breasted Dove; Too-zoo,‡ Cooscot,‡ Timmer Doo,‡ Quisty‡

† A species-specific name for the Lemon Dove (*Aplopelia larvata*) in Zulu.

‡ These are four of the seventeen names given by Greenoak (1997: 118) for the Wood Pigeon (*Columba palumbus*) – not a South African bird – as regional 'nicknames'.

Column four is a difficult category to pin down. In an earlier draft of this typology, I had this category marked only for 'nicknames' – the informal names so well known to South African English speakers, like 'jacky hangman' for the Southern Fiscal (formerly the Fiscal Shrike), 'turkey buzzard' for the Southern Ground Hornbill, and 'toppie' for the Black-eyed Bulbul. Then I found myself adding in alternative names often used as book names by other writers, earlier names used as vernacular book names, and a very large number of regional names. Greenoak's 1997 book *British Birds: their Folklore, Names and Literature* introduced me to an alternative world of bird naming, where the Wren (*Troglodytes troglodytes*) has no less than 36 alternative names in Great Britain and Ireland. None of these fit into categories A, B or C in the table above. What makes this category so difficult to pin down is that the names are hovering in a number of intersecting continua: the continuum between formal and informal, between oral and written, between yesterday and today, and between highly localised to more general distribution.

Zulu has such alternative names as well, and whether they are names with a wider earlier currency, or specifically regional names, or names only used in specific circumstances, must remain the subject of a separate study. For example, consider the Zulu names of the Bateleur, the Southern Ground Hornbill, the Black-headed Oriole and the Hamerkop.

In the study of the naming of the Bateleur eagle (see chapter 2), we see the 'praise name' **indlamadoda** 'eater of men' for a bird whose denotative name is usually given as **ingqungqulu**.

The Southern Ground Hornbill is usually referred to as **insingizi**, but when it is heard calling, with its booming 'du du du' sound, it is called **ingududu**.

The Black-headed Oriole (*Oriolus larvatus*) has been recorded with the following names: **umqoqongo**, **umbhicongo** and **umgongolozi**. The name **umqoqongo** appears in the form **umqaqongo** as well, either an error in the earlier recording of this name – yet another factor to be taken into consideration – or a regional variation. Yet despite all these recorded names for the oriole, the Zulu-speaking birders at the 2013–2017 Zulu bird name workshops said it is best-known in KZN as **usibó**, with the stress on the final syllable. Which of these is the 'default' name, which are 'nicknames', and which are regional names?

When the name **uthekwane** for the Hamerkop appears in the form **uThekwane kaZiluba** – 'Mr Hamerkop, son of Mr Plumes' – then this is at least clearly a praise name.

Specialised onomastic terminology

While still on the topic of types of bird names, here is a brief rundown of some specialised onomastic terminology:

The sum total of words in any language, together with their meanings, is known as a lexicon. Lexicons exist for languages as a whole, and each individual in a society holds at least one individual lexicon in his or her head, that of the mother tongue. In the discipline of onomastics, there are, as in other disciplines, specialist terms.[8] The word 'onomastics' itself may be combined with 'lexicon' to form 'onomasticon', meaning a list of <u>names</u> in any languages. The terms 'anthroponym' for a name of a person and 'toponym' for a place name may be familiar. The terms 'anthroponomasticon' and 'toponomasticon' are probably not so familiar. Here the bases *anthropos* (human) and *topos* (place) are combined with 'onomasticon' to form terms meaning 'the total list of personal names in any language' and 'the total list of place names in any language'. In my 2015 *Zulu Plant Names* I used the same idea to coin the term 'botonomasticon' (from bot[any] + onomasticon 'list of names') to mean a list of plant names in any language. Similarly, I coin here the term 'ornithonomasticon' to mean a list of bird names in any language, and I will be using it in phrases like 'the traditional Zulu ornithonomasticon' and 'an extended Zulu ornithonomasticon'.

1.4   HUMAN AND BIRD INTERACTION

Humans and birds interact in a number of different ways. Four broad categories of such interaction can be identified:

- *Birds in 'indigenous birder' communities*: Here we think of communities like the Zulu community, like other African communities, and other 'indigenous', (i.e. non-colonial) societies who have traditionally hunted birds for food, used their feathers for decoration and ornamentation, used them as symbols and metaphors in song, dance and poetry, have seen them as harbingers of seasons of the year or as signs of approaching weather, and have constructed belief systems around birds that include various kinds of taboos and the use of birds as omens and portents. All of these are dealt with in chapters 8 to 11 of this book.

---

8. The International Council of Onomastic Sciences (ICOS) maintains an official terminology list, which is regularly updated. Author Adrian Koopman is a member of the Terminology Board of ICOS.

- *Birds and language*: This is very definitely the main topic of this book: The majority of the chapters of this book deal with bird names (their morphology, typology, semantics, syntax, etc.); together with bird vocalisations and human verbalisations of these, with overlaps with both names and oral poetry; and birds in literature, where we will look at sayings, proverbs, riddles, songs, stories and poetry.
- *Birds in ornithology*: The way birds are approached by ornithological scientists from universities and museums is not an area covered in this book. I have no interest in the diet, nidification, migration, etc., of birds, unless these relate specifically to the underlying meaning of a name. Nor is this book interested in taxonomy and systematics unless in the most general terms in contrast to how traditional Zulu society perceives clusters of birds. This is the topic of the next chapter.
- *'Birdwatching' or 'birding'*: This aspect of human-bird interaction comes in when we talk about the issue of naming and identification, and how a rise in both South African avitourism and the rise of interest in birding among African communities has led to a need for 'new' Zulu names for birds. This will be discussed later in chapters 12 and 13.

1.5   RISE OF INTEREST IN BIRDS AND BIRDING AMONG ZULU SPEAKERS

Throughout this book I will be referring to the 2013–2017 Zulu bird name workshops, and in Chapter 12 these workshops and their aims, goals and results are described in detail. These workshops basically came about as a result of a perceived need in the early years of this century for species-specific Zulu names for discrete bird species found in KZN. Birdwatching or 'birding' as a leisure activity has grown exponentially in the last several decades, and this is certainly so for South Africa in general and KZN in particular. Birders require unique names for individual species so they can record what they have seen at specific times in specific places. Together with the rise of interest in birding, has been the rise of interest in conservation issues and in environmental awareness. This is particularly the case with the education of schoolchildren, for Zulu-speaking schoolchildren no less than any others. Conservation planning, like recreational birding, requires the exact identification of birds, meaning, again, that species-specific names are needed. When Zulu-speaking children are involved in the teaching of environmental awareness and conservation planning, and when

Zulu-speaking bird guides are involved in recreational birding, the need for species-specific names for birds in Zulu becomes equally obvious.

In the final chapter of this book, details will be given of a wide variety of educational programmes involving birds and bird-related outings. I will be introducing specific Zulu-speaking bird guides and looking at how they are involved in educational initiatives as well as their guiding activities. I will also be looking at how birdwatching clubs are being developed as part of self-help developmental initiatives among the Zulu-speaking population of KZN.

1.6   REVIEW OF SOURCE MATERIAL

1.6.1   Primary source material

I have twice been involved in collecting primary material about Zulu bird names, and the two periods were more than 30 years apart. My interest in Zulu names goes back to 1975 when I was doing research for my Honours degree and chose Zulu personal names for my research project. My interest in birds came about when I moved from Durban to Pietermaritzburg in 1978 and bought a property in the paperbark acacia country on the eastern outskirts of town where there was a great variety of birdlife. Soon after my arrival in Pietermaritzburg I was involved in fieldwork on Zulu place names in the Drakensberg, and from there it was a short step to researching Zulu bird names. The then Natal Parks Board allowed me access to a number of their older Zulu game guards and I had a number of interviews with these knowledgeable men in various locations in the Drakensberg and in Zululand.

More than thirty years later, in 2013, my long-time colleague and friend Professor Noleen Turner invited me to the first of what would be a number of workshops on Zulu bird names where again Zulu-speaking bird experts from various parts of KZN were willing to share their considerable knowledge about birds and their Zulu names. These workshops are described in detail in Chapter 12 of this book.

1.6.2   Secondary source material

The secondary research material can be roughly divided into three areas: (i) Zulu and other dictionaries; (ii) bird guides and general 'bird books'; and (iii) a wide variety of published articles in folkloric, anthropological, ornithological, linguistic and onomastic journals.

Zulu dictionaries

The historical development of Zulu bird names in dictionaries and other published sources is covered in detail in Koopman (2018b, 2019), but briefly here:

- Bishop John Colenso's *Zulu-English Dictionary* was first published in 1861, with revised editions appearing in 1878, 1884 and 1905, the last being revised by Colenso's daughter Harriette Colenso. In the 1884 third edition I used, the entries for 'bird' in the most part were unable to specify a particular species of bird, (e.g. the entry for **idada** is simply 'duck'). The fourth edition, however, contains a far great number of bird names, with a far greater link to species-specific birds.

- The Rev. A.T. Bryant's 1905 dictionary is a great improvement in terms of the number of names and overall identification achieved, with much lore added.

- R.C.A. Samuelson's 1928 *King Cetywayo Zulu Dictionary* is simply a poor version of Bryant.

- C.M. Doke and B.W. Vilakazi's 1948 dictionary contains all Bryant's material, with reduced detail in the separate entries but with more bird names.

- C.L.S. Nyembezi's 1996 *isiChazamazwi* has a very limited number of names, and species identification is poor to non-existent – the whole dictionary is almost a Zulu version of Colenso from 130 years earlier.

Bird guides

- Woodward and Woodward's 1899 *Natal Birds* was the first to use Zulu names for most of the species recorded; Stark and Sclater's 1900–1906 four volumes of *The Birds of South Africa* uses Zulu names, but all or most of these are the Woodwards' names.

- Austin Roberts' 1948 *Birds of South Africa* gives names for birds in all South African languages – the Zulu names have many lacunae and errors. Many of the names appear to have been collected by Bell-Marley in the 1920s and early 1930s. These are duplicated without change in the McLachlan and Liversidge editions of *Roberts* in 1957, 1973 and 1978. There is an improvement in Maclean's 1984 edition of *Roberts'* but his Zulu names remain unchanged in the sixth edition of 1993, and are left out altogether in the 2005 seventh edition. Maclean's Zulu names return in Hugh Chittenden's 2007 *Roberts Bird Guide*, although only in an index, not in the text of the book next to each bird, thus ensuring that you cannot find out the Zulu name of any bird unless you already know it! By the time Chittenden, Davies and Weiersbye had produced the 2016 second edition of the *Roberts Bird Guide*, the Zulu names had disappeared again.

- Philip Clancey's 1964 *The Birds of Natal and Zululand* also has no Zulu names at all.
- O.P.M. Prozesky's 1970 *Birds of Southern Africa* has some African names in it, but even fewer than the 1957 edition of *Roberts* and what it does have are no different from the *Roberts* names of that time. I have mainly used Prozesky for comparative material on the bird calls.
- I have not used any of the editions of Kenneth Newman's *Birds of Southern Africa,* nor any editions of Sinclair's *SASOL Birds of Southern Africa.*

Overall, the different editions of the bird guides were useful in tracking changes and development in bird names in both Zulu and in the scientific, English and Afrikaans names.

Miscellaneous bird books

Other books on birds – not bird guides – have proved useful. Some stand out: Jeremy Mynott's 2009 *Birdscapes* was particularly useful in providing theoretical background to human-bird interactions, and especially to bird naming issues. Nancy Jacobs' 2016 *Birders of Africa* provided further suggestions about human-bird interactions, especially in Africa. Edward Armstrong's 1958 *The Folklore of Birds* provided contextual material for the chapter on bird lore, as did Philippa Waring's 1978 *A Dictionary of Omens and Superstitions*. Mark Cocker's 2013 *Birds and People* has only a few snippets of information about Zulu bird lore, but is otherwise a huge collection of fascinating stories about the human-bird interface, and also has much interesting and thought-provoking material on bird names and how they come about.

Without any question, the most useful book was the Rev. Robert Godfrey's 1941 *The Bird-Lore of the Eastern Cape Province*. Although it is on Xhosa bird names rather than Zulu, this publication is a veritable treasure trove on Nguni bird names and bird lore.

Articles in journals

As mentioned above, a variety of journals, including folkloric, anthropological, ornithological, linguistic and onomastic ones, were checked for relevant articles. Professor Colleen Downs of the Department of Zoology at the Pietermaritzburg campus of the University of KwaZulu-Natal was particularly helpful in facilitating access to electronic back copies of the British journal *Ibis*, and the South African journals *Ostrich* and *Bokmakierie*. These journals provided material on taxonomic issues, general naming issues (for example, on the debate between the use of scientific versus vernacular names, and on the wisdom of continually changing species names), and the older issues of *Ibis* contained much material on bird naming in indigenous African societies, including in the Congo (Hendrickx), Tanzania (Moreau), Madagascar (Sibree), and Zululand (Woodward and Woodward).

1.7   OVERVIEW OF THE CHAPTERS

The aim of **Chapter 2, Names and identity**, is to explore the notions of 'genus' and 'species', both in science-based ornithology and 'folk taxonomy'. Previously recorded names have mainly referred to 'clusters' of birds, an example being the word **idada** which has been used for several species of duck.

The chapter uses 'eagle identity' to explore our understanding of different 'groups' or 'clusters' or 'families' of birds in traditional Zulu thought, and how these can be compared to scientific notions of family, genus and species. The Latin-based scientific names, the English and Afrikaans names, and the Zulu names (both old and new) for all the different eagles found in KZN are explored with a view to linking name and identity. The chapter then goes on to focus on only one species of eagle – the Bateleur – and look at all the regional variations of the names of this bird in the Nguni language cluster (mainly Xhosa and Zulu).

Once the identity 'problems' have been identified, the chapter ends with an appendix comparing the 'clustering' of names for doves and pigeons, showing how traditional Zulu taxonomy on the one hand and science-based taxonomy on the other have identified different clusters of these birds.

Chapter 3, The meaning of African bird names, provides a background to the semantics of Zulu bird names. The chapter begins with a layman's overview of different types and layers of meaning, and the relationship between the different meanings and their different functions. The chapter then goes on to look at different meaning-based typologies of bird names put forward by various authors, and proposes an integrated typology for African bird names.

The various semantic categories in this integrated typology – such as song-based names, appearance-based names, and behaviour-based names – are then explored in depth, using examples from a variety of different languages from southern, central and eastern Africa, as well as bird names from Madagascar.

Chapter 4, The meaning of the new Zulu bird names, uses the same integrated typology and applies it to Zulu bird names, both the old and the new. A statistical analysis of African bird names in general, the older Zulu bird names, and the new Zulu bird names, shows that Zulu bird naming is typical of African bird naming generally, and that the thought processes underlying the new Zulu bird names have produced names that are equally typical.

The chapter then follows the pattern of the previous chapter: dividing Zulu bird names both old and new into various semantic categories and exemplifying these fully.

Chapter 5, Zulu bird names used for other purposes, is about bird names that refer to entities other than birds. After an introduction, which explains the notion of 'polysemy', i.e. 'multiple meanings', the chapter goes on to look at how bird names are used as Zulu cattle colour terms, as terms for colours in an artist's palette, and as a variety of different types of proper names: the names of people, the names of geographical features such as mountains and rivers, the names of streets, and the names of brands. Under the last heading, the two Zulu names **inkwazi** 'African Fish Eagle' and **uthekwane** 'Hamerkop' are given close scrutiny.

Chapter 6, The morphology of bird names, looks at the underlying grammar of Zulu bird names, both old and new. As Zulu bird names are all nouns, this is essentially about the grammar of nouns: noun classes, singulars and plurals, noun prefixes and noun stems, and the different types of noun stems: simple, reduplicated, complex and compound.

The Zulu language has a wide range of derivational prefixes, allowing for an extensive variety of creative processes for forming new nouns. By using both old and new bird names in this chapter, it can be seen that the Zulu-speaking participants of the workshops used the creative structures of their own mother tongue to the full in the adaptation and extension of old names and the coining of new names for birds.

Chapter 7, Verbalisation of birdsong in Zulu, begins by looking at how avian vocalisation is recast as human language generally. The way is which bird vocalisation is articulated in human speech is looked at here, as well as how bird names themselves may reflect the song of a particular species.

In examining Zulu bird names specifically, the chapter makes a distinction between names which are imitative, i.e. onomatopoeic, and those which are descriptive, which includes the use of metaphor and simile.

The chapter then goes on to look at the interpretation of bird vocalisation in traditional Zulu societies, in other words 'folk beliefs' about what the bird is really saying. For example, the authors of modern field guides to South African birds describe the vocalisation of the Emerald-spotted Wood Dove as 'mournful' or 'plaintive'. The Zulu interpretation of the call explains that having lost all members of its family, it is now being robbed of its eggs.

Similar analyses are made of the 'folk interpretations' of a number of species of birds.

Chapter 8, Birds in Zulu praise poetry, and **Chapter 9, Praises, proverbs and riddles**, together look at the roles that birds have played in Zulu traditional oral literature. Chapter 8 is dedicated to birds in oral praises, including the praises of kings and chiefs, the praises of commoners and clan praises (*izithakazelo*). Chapter 9 begins with birds that have their own praise poems, and then compares the praises of the Yellow-throated Longclaw (**inqomfi**) with a poem written by the Zulu poet B.W. Vilakazi titled "Inqomfi".

The chapter then explores Zulu proverbs that feature birds, riddles that do the same, and then finally the children's bird-name game *Bhula 'Ntsentse Bo!*

Chapter 10, Bird beliefs: portents and heralds, deals with traditional beliefs about birds. As the chapter title suggests, half of the chapter is devoted to omens, portents, taboos and charms. In this section are found birds that are linked to witchcraft, birds that portend death and illness, birds that prophesy good luck, and those whose behaviour indicates bad luck. Linked with these beliefs are various taboos directly related to birds: birds that are taboo to kill, those that are taboo to eat, and even those whose calls may not be imitated.

The second half of the chapter is about birds that are harbingers and heralds: birds that herald approaching weather, both welcome rains, and unwelcome thunder and lightning. Then there are the birds that herald the spring or the summer, as well as those that mark the break of day and those whose calls mark nightfall.

Chapter 11, Feathers, food and fancies, is a chapter about the practical uses of birds. At least half of the chapter deals with the use of bird plumage as ornamentation and decoration, not least as insignia. There is a wealth of documentation about this, so it is possible to link the Ostrich, the Turaco,[9] Long-tailed Whydahs and various other birds to specific regiments that served under the Zulu kings of the nineteenth century.

Conversely, there is little documentation about birds used as food, most probably because as in other parts of Africa indigenous bird species have never played a major role in the diets of indigenous cultures. There is likewise little material on the medicinal use of birds, for the same reason.

The chapter ends with various traditional beliefs: birds as cattle herders, the Yellow-billed Kite as a 'tooth fairy' and other miscellanea.

Chapter 12, New names and new identities, deals specifically with the processes adopted during the five Zulu bird name workshops held from 2013 to 2017. The contributors, the logistics, the linguistic processes and most of all the results of these workshops are looked at here in detail. While clearly it would take an entire book just on its own to detail how every new name was arrived at, this chapter at least gives case studies involving a cluster of shorebirds and waders, together with a detailed appendix on kingfisher names.[10]

The linguistic processes of adaptation, extension, relegation and most of all, coinage, are described and exemplified with the names of a wide variety of KZN birds.

Chapter 13, Change is in the air, begins with an overview of the diachronic recording of Zulu bird names. Beginning with the difficulties in converting orally-held Zulu knowledge into written English knowledge, the chapter goes on to look at a number of the major recorders of Zulu bird names over the second half of the nineteenth century and the first half of the twentieth century. The diachronic overview then jumps into the first part of the twenty-first century by giving details of all the Zulu-speaking bird namers who contributed to the 2013–2017 Zulu bird name workshops. Three of these in particular are placed within the steadily growing development of birding (birdwatching) and the exponential growth of avitourism in this country.

The role of birds in education is investigated from two points of view: the educating of young Zulu-speaking children about environmental and conservation issues by using birds (with their Zulu names and the associated

---

9. Turaco is the current name for the Lourie (also spelt Loerie); spellings in sources are varied.
10. Professors Turner and Koopman are in fact planning a book with the full list of new names for all the birds that occur in KwaZulu-Natal.

traditional beliefs) and the educating of Zulu-speaking bird guides to enable them to acquire the necessary recognised qualifications.

Traditional Zulu beliefs about birds are also re-examined within the modern context, and using a basic form of meme theory, various changes of bird beliefs are examined, such as loss of a belief, metamorphosis of a belief and the rise of new beliefs to suit modern circumstances.

The chapter continues with suggestions of how the use of Zulu bird names can play various and diverse economic roles, ranging from the development of bird-linked brand names and their associated logos, to children's readers about traditional beliefs.

The chapter, and the book, ends with challenges to scholars of other southern African languages to develop bird nomenclature, showing in passing how this has been done with Swahili and with Seychelles Creole.

2   Names and identity

2.1 INTRODUCTION

In this chapter the link between name and identity is examined. An important theme in the chapter is the difference between genus and species, in both scientific taxonomy and in the 'folk taxonomy' of non-scientific environments.

The ethnobiologist Brent Berlin writes:

> Before human beings can utilize the biological resources of a local environment, they must first be classified . . . People must be able to recognize, categorize, and identify examples of one species, group similar species together, differentiate them from others, and be capable of communicating this knowledge to others (1992: 5).

There are a number of points in this brief quote that are relevant to this chapter:

- Birds (which are part of the biological resources of an environment) can be "utilized" by humans;
- To do so, humans must be able to recognise, categorise and identify them;
- In grouping them into categories, they must be able to identify which features birds have in common, and which set them apart from others; and
- Having done this categorisation, the knowledge must be able to be passed on, i.e. the different groups must be named.

Berlin goes on to ask (1992: 5) "Why do human societies classify nature in the ways they do?", and part of the answer to this question is found three pages later under the heading "The bases of ethnobiological classification":

One of the main claims in this book is that human beings everywhere are constrained in essentially the same ways – by nature's basic plan – in their conceptual recognition of the biological diversity of their natural environments. In contrast, social organization, ritual, religious beliefs, notions of beauty – perhaps most of the aspects of social and cultural reality that anthropologists have devoted their lives to studying – are *constructed* by human societies . . . When human beings function as ethnobiologists, however, they do not construct order, they discern it (1992: 8).

Berlin's point is an interesting one: social organisation, rituals, religious beliefs, notions of beauty – in short, the order of human society – are all constructed by humans, but they do not construct the order found in the <u>natural</u> world. Instead, they discern it, categorise it and then name the categories. Names, as will be pointed out a number of times in this book, are linguistic items, and language is a human construct.[1]

*   *   *

This chapter is, then, about the link between name and identity, between the linguistic item and its <u>referent</u> – the entity the word refers to in the 'real' world. The essential question in this chapter could be phrased in a number of ways, but they all mean the same:
- What entity does a particular bird name refer to?
- What birds are included in the meaning of a bird name?
- What are the members of a category with such and such a label?

The working through of these questions, and the seeking for answers, could be done with a number of bird genera and/or species. I have chosen to use the 'eagle'.

2.2   EXPLORING THE NOTION 'EAGLE'

The questions above are illustrated below with different ways of looking at the word 'eagle' and its equivalents in Latin, Zulu and Afrikaans. The questions could now be recast as:

---

1. I am deliberating leaving out here the notion of communication between animals, birds, insects, etc., as being 'language'. Language, in the sense of a communication device with features such as phonemes, words, parts of speech, etc., is decidedly a human construct.

- 'What entity or entities is/are referred to by the word "eagle"?'
- 'Does the word "eagle" have the same reference as its Latin equivalent *aquila*?'
- 'An eagle that is black is a black eagle. Is this the same as "a Black Eagle"?'

In exploring these and other questions, I will also look at extended forms in various languages, such as Southern Banded Snake Eagle and *Witkruisarend*, and we will look at semantic equivalents in various languages, such as *bateleur*, *berghaan* and *ingqungqulu*, and whether these words refer to 'eagle' in its most general sense, or to specific species of eagle.

We begin by exploring the notions of inclusivity and exclusivity in the following paradigmatic sets:

Table 2.1   Paradigmatic sets of eagle names.

Set one	Set two	Set three	Set four
eagle	Martial Eagle	*bateleur*	*Aquila verreauxii*
aquila	Wahlberg's Eagle	*berghaan*	*witkruisarend*
arend	Long-crested Eagle	*stompstertarend*	*dassievanger*
ukhozi	Crowned Eagle	*indlamadoda*	Black Eagle
		ingqungqulu	Verreaux's Eagle
			*ukhozolumnyama

All nineteen of the names above refer to the same referent as the first name, i.e. 'eagle'. In other words, all the birds designated by the nineteen names above are eagles. But which eagles? And what is an 'eagle' anyway? (And anyway, who decides what an 'eagle' is – the man in the street? The scientifically trained ornithologist? The author of a field guide for birdwatching?)

Again, these are all questions which are tackled in this chapter.

In Set one above, the words 'eagle', '*arend*' and '**ukhozi**' are what I call 'folk names' or 'folk-generics'. They are used by the man in the street, who is neither a professionally trained ornithologist, nor an expert birder. In the three languages, English, Afrikaans and Zulu, these words mean 'big bird with hooked beak and big talons', or, more succinctly, 'large raptor'. The Latin word *aquila* also means 'eagle' in this generic sense, but it is used scientifically as a genus name for only five distinct species of eagle occurring in South Africa (but not for all the seventeen species of eagle found in southern Africa). The word *aquila* is thus the odd one out in this set: it is exclusive, rather than inclusive like the other three words.

In Set two, all four words are English names for specific species of eagle. Each name excludes the other three names, indeed, each name excludes all other known eagles, in South Africa and elsewhere in the world. The meanings are thus far more narrow (or specific) than the meanings of the words in Set one.

In Set three, there are five 'vernacular', (i.e. 'non-scientific') names for the same bird. The word *bateleur*, meaning 'acrobat' or 'tumbler', is a French word given by the explorer/ornithologist Le Vaillant, but now adopted as an English 'book name' in field guides (see Maclean, 1984: 130). The Afrikaans word *berghaan* literally means 'mountain cock', but it is the 'book name' that Maclean gives for the same bird. In brackets after *berghaan*, Maclean has *stompstertarend* (stump-tailed eagle), either an alternative to *berghaan*, or an earlier book name of other authors. Maclean gives, correctly, the Zulu name **ingqungqulu** for this species of eagle; the name **indlamadoda** (what eats men) is a well-known Zulu alternative that does not appear in Maclean. So here we have an inclusive set: each of the five names includes all the others, although the names themselves do not have identical 'official' status in the field guide selected.

In Set four, there are similar dynamics as in Set three: all six names refer to the same species of bird. *Aquila verreauxii* is the Latin-based binomial with *Aquila* referring to the genus, and *verreauxii* the specific epithet that distinguishes this species from the other four eagles in the genus *Aquila*. Maclean (1984: 117) has chosen Black Eagle as his English book name for this bird, but has included Verreaux's Eagle in brackets after 'Black Eagle', again suggesting either an alternative name, or a previously used book name for this bird.[2] His Afrikaans book name for this bird is *witkruisarend* (white cross eagle). He does not include the name *dassievanger* (catcher of rock rabbits), which Muir (1940: 4) records as an Afrikaans 'folk name' for this bird in the Western Cape. Maclean gives the folk-generic **ukhozi** (eagle) as the Zulu name for this bird. The Zulu species-specific name ***ukhozolumnyama** (black eagle) was

---

2. It is now the 'official' South African English book name again (see Chittenden, Davies and Weiersbye 2016: 166).

coined in the 2013–2017 Zulu bird name workshops, and at the time of writing (2017) has not yet achieved any 'official' status.[3]

We see then, in the four sets of names for 'eagle' above, a number of different dynamics: scientific names as opposed to vernacular names; inclusive sets as opposed to exclusive sets; generic names (both in scientific terms as well as 'folk' terms) as opposed to species-specific names; current book names as chosen by a particular author as opposed to earlier book names, or more informal names for the same bird. In one set we see a newly coined species-specific name that has not yet received any official recognition, as opposed to names for the same species of bird that have long been accepted in 'official' publications.

The basic theme, though, of this chapter, is how the notions 'genus' and 'generic name' on the one hand, and 'species' and 'specific name' (or 'species-specific name') on the other hand, are handled in both scientific ornithology and in what has been termed variously 'ethno-ornithology' (or more loosely 'ethnobiology'), 'folk taxonomy' and the like. Put differently: how do Western-trained scientists perceive the notions 'genus' and 'species' and in what ways can these distinctions be compared to identification and categorisation of living entities in non-Western traditional societies around the world? These, obviously, as with all the questions asked rhetorically so far in this chapter, are related to the chapter title "Names and Identity".

Before getting into the issue of generic and specific names, it may be useful to be reminded of where these two levels of identification fit into the scientific hierarchy of birds, using the Bateleur Eagle (*Terathopius ecaudatus, berghaan,* **ingqungqulu**) and starting at the taxonomic level of species:

→ The Bateleur (*Terathopius ecaudatus*) is the only species in the genus *Terathopius*.[4]

→ This is one of several genera in the family *Accipitridae*, which includes all the eagles, hawks, kites, buzzards and so on.

---

3. I have used, perhaps incorrectly, the linguistic convention of a preceding asterisk to suggest this word's non-official status. In linguistics, a preceding asterisk usually indicates a grammatically incorrect word or utterance.

4. The genus name *Terathopius* is based on Greek roots meaning 'acrobatic juggler', which is roughly what the French name *bateleur* means. The specific epithet *ecaudatus* means 'without a tail'.

→ The family *Accipitridae* is one of four families in the order *Falconiformes*, which includes the Secretarybird, the osprey and all falcons and kestrels as well.

→ The order *Falconiformes* is one of 27 orders (in southern Africa) in the class *Aves*, which contains all known birds.

There are no sub-species of the Bateleur recorded for its distribution range. To sum up, in Western ornithology, the Bateleur fits into the following hierarchy, from the widest (most general) level to the narrowest (most specific):

Class:	*Aves*
Order:	*Falconiformes*
Family:	*Accipitridae*
Genus:	*Terathopius*
Species:	*ecaudatus*
Sub-species:	none recorded

With few exceptions, the levels 'order' and 'family' are not filled with identified categories (and therefore corresponding names) in 'folk taxonomies' – the human-perceived categories in traditional societies. When it comes to birds the English language has terms such as 'nightfowl', 'waterfowl', 'raptors' and 'gamebirds', which refer to bigger categories than generic categories reflected in names such as 'owl', 'finch', 'lark' and 'eagle'. Zulu, however, has no such equivalents, and all names (apart, of course, from *inyoni* 'bird') occur at the generic or specific level.[5]

2.3   THE ISSUE OF GENERIC AND SPECIFIC NAMES

Under this heading we look generally at the so-called 'problem' of African language bird names, look at four sets of names for birds that have the word 'eagle' as part of their English vernacular book names, and then go on to a much more specific and detailed analysis of Nguni names for the Bateleur Eagle.

---

5. One exception here may be the word *intaka*, which means 'bird' in Xhosa, but more specifically 'finch' in Zulu. The word is used particularly for females of various species of weavers, bishops, widows and whydahs.

2.3.1   The 'problem' of bird names in African languages

Of all the bird guides mentioned on pages 13–14 in Chapter 1 above, Maclean's two editions of *Roberts' Birds* (1984, 1993) are the only ones to discuss the issue of 'African' names for birds.[6]

In the Introduction to the 1984 edition, he states, in what we will refer to several times below as his 'core quote' or the 'Maclean problem':[7]

> **The 'Maclean problem'**
>
> Bird names in the African languages present far more problems than in the European-derived languages. Many of them are generic (*i.e.* all species of sparrows may have the same name), others are regionally limited in application, one name may be applied to two or more different birds, some well-known birds may have more than one name in a single language, and so on. Most bird species have no African names at all (Maclean 1984: xxix).

He then goes on to explain how he handled the 'African' names, by "soliciting help from experts in the different language groups" and how he integrated their input into the names recorded in previous editions of *Roberts' Birds*.

Maclean's statement above could be illustrated with a number of different groups of birds, but I will continue with the 'eagle group' to show what Maclean is talking about. Table 2.2 gives information from Maclean's own 1984 edition of *Roberts*, so it reflects a situation for Zulu at the time of the publication of this edition, i.e. after "soliciting help from experts".[8] The table has four columns, each one reflecting a different overall naming pattern.

---

6. Maclean uses the term 'African languages' to refer to the southern African languages from the Bantu language family: Zulu, Xhosa, Sotho, Swazi, Tsonga, etc.
7. I have put the quote in a text box so as to make it easier to find, as I come back to it frequently in the course of this chapter.
8. As the Zulu names for birds (and the names in all other South African languages) have remained officially unchanged up to the time of writing (2017), Maclean's names, and his thoughts on these names, remain valid for the period covering the last thirty years and more.

Table 2.2   The naming of eagles in three languages.

Aquila	Black (Verreaux's) Eagle	**ukhozi**	**ukhozolumnyama** 'black eagle'
Aquila	Tawny Eagle	**ukhozi**	**ukhozolunsundu** 'brown eagle'
Aquila	Steppe Eagle		**ukhozimuhlwa** 'termite eagle'
Aquila	Lesser Spotted Eagle		**ukhozolumabala** 'spotted eagle'
Aquila	Wahlberg's Eagle		**ukhozolusisila** 'tailed eagle'
Hieraaetus	Booted Eagle		**ukhozolumadladla** 'shaggy eagle'
Hieraaetus	African Hawk Eagle		**ukhozolumidwayidwa** 'streaked eagle'
Hieraaetus	Ayres' Eagle		**ukhozolumabalabala** 'many-spotted eagle'
Lophaetus	Long-crested Eagle	**isiphungumangathi**	**isiphungumangathi** 'long-crested eagle'
Polemaetus	Martial Eagle	**ukhozi, isihuhwa**	**inkosiyezinkozi** 'king of the eagles'
Stephanoaetus	Crowned Eagle	**isihuhwa**	**isihuhwa** 'crowned eagle'
Circaetus	Brown Snake Eagle		**indlanyokensundu** 'brown snake-eater'
Circaetus	Black-breasted Snake Eagle	**ukhozi**	**indlanyokemnyama** 'black snake-eater'
Circaetus	Southern Banded Snake Eagle		**indlanyokephuzi** 'yellow snake-eater'
Terathopius	Bateleur Eagle	**ingqungqulu**[†]	**ingqungqulu** 'bateleur'
Haliaeetus	African Fish Eagle	**inkwazi**	**inkwazi** 'fish-eagle'
Gypaetus	Bearded Vulture	**ukhozilwentshebe**	**ukhozilwentshebe** 'bearded eagle'

† The name **ingqungqulu** is highlighted here as it is the subject of further detailed discussion below.

As can be seen from the column of scientific genus names, the taxonomic position in 1984 reflected that although each of seventeen species of birds bears the English vernacular name 'eagle',[9] they have been divided by ornithological scientists into no less than nine genera: *Aquila*, with five species of eagle; *Hieraaetus*, with three; *Lophaetus*, *Polemaetus* and *Stephanoaetus* with one each; *Circaetus* with three species; and *Terathopius*, *Haliaeetus* and *Gypaaetus* with one each. Maclean apparently sees nothing disturbing about nine scientific genera sharing one English generic name: his 'problem' is with the third column, where eight species of eagle have no Zulu names at all, and where **ukhozi** is used for four species of eagle, one of which is also carrying the name **isihuhwa**, which it shares with yet another eagle. Four other eagles have different names.

The fourth column contains revised and newly coined Zulu eagle names from the 2013–2017 Zulu bird name workshops, (discussed fully in Chapter 12). For the moment it is sufficient to note that this column contains separate and distinct names for all the eagles.

Column three, then, represents, as far as could be accurately established in the early 1980s when Professor Gordon Maclean was putting together his fifth edition of the *Roberts' Birds* series, the naming of eagles in traditional Zulu thought-patterns. We can note some of the features that he refers to in the 'problem' quote above: many eagle species have no Zulu names; one name (**ukhozi**) is used for many eagles; one species has two names.

To understand the onomastic profile of 'eagle' in traditional Zulu thinking, we need to examine both the names and the birds a little more closely, with especial attention to the role the various species play in Zulu society. Firstly, it is clear that **ukhozi** is a generic with the indigenous meaning of 'very large raptor'.[10] Even though there are eight species of eagle in the list above that appear to have no Zulu names, point to any one of these unnamed eagles circling in the sky above and ask the average Zulu speaker to identify what it is, and nine times out of ten the answer will be "*Wukhozi*" ('It is an eagle'). Note that the Bearded Vulture is also **ukhozi** in Zulu, with an extra *lwentshebe* (of the beard) tacked onto it in exactly the same manner as the extensions tacked on to **ukhozi** in the workshop names in the fourth column. And the name for

---

9. The Bearded Vulture is included here as it is an 'eagle' in both Latin and Zulu. The genus name is derived from Gk *gups* 'vulture' + *aetos* 'eagle'.
10. In the final 2017 Zulu bird name workshop, the name **ukhozi** was formally declared a generic term, not to be used for any of the distinct species unless further extended.

the African Fish Eagle is also basically **ukhozi**, with the vowel /a/ inserted between /o/ and /z/, and the stem moved into noun class 9.[11]

The name **isihuhwa** can be seen as a praise name for the two largest eagles. Although Maclean records only this name for the Crowned Eagle, there is no doubt this bird is also an '**ukhozi**' like the other eagles. Effectively, then, there are only two eagles that definitely have different names, and for each of these, there is a socio-cultural reason:

- The Bateleur (**ingqungqulu**) is seen in Nguni cultures generally as a portent of battle and death. Godfrey (1941: 33–6) devotes much space to this portentous role among the Xhosa people, and other writers say the same of the bird in Zulu society, where the bird frequently occurs as a metaphor in praise poetry. Chapters 8 to 11 in this book, on the folklore of birds and their role in traditional oral poetry, have specific and extensive details on this.
- The Long-crested Eagle (**isiphungumangathi**), on the other hand, is a more 'peaceful' bird, and is called upon by those who have lost cattle to indicate where they might be found. The details of this, which the bird shares with certain kinds of chrysalis, are also described later in Chapter 11 (pages 334–5).

In the 2013–2017 workshop processes, it was found necessary to create only one new 'generic' name, the name **indlanyoka** (snake-eater), to represent the three species in the genus *Circaetus* (the snake-eagles), thus bringing Zulu into onomastic line with Latin and English.

2.3.2   Nguni names for the Bateleur

The issue of naming eagles in Nguni[12] is in fact far more complex than the above brief discussion suggests. There are regional differences in naming, different regional nicknames, and even variations in forms of all of these according to

---

11. Inserting /a/ into the noun stem produces 'kho-a-zi', which is pronounced and written as 'khwazi'. Moving this into class 9, with the nasal prefix *in*– causes the /h/ to fall away, giving **in-kwazi**. Noun classes, noun stems and noun prefixes are discussed in detail in Chapter 6 below.
12. The term 'Nguni' usually applies to the southern African languages Xhosa, Swazi, Zulu and Zimbabwean Ndebele. I use it here to cover the languages (many regard them as dialects) spoken in the Eastern Cape and KZN: Gcaleka, Gaika (or Ngqika), Thembu, Hlubi, Pondo, Mpondomise and Zulu.

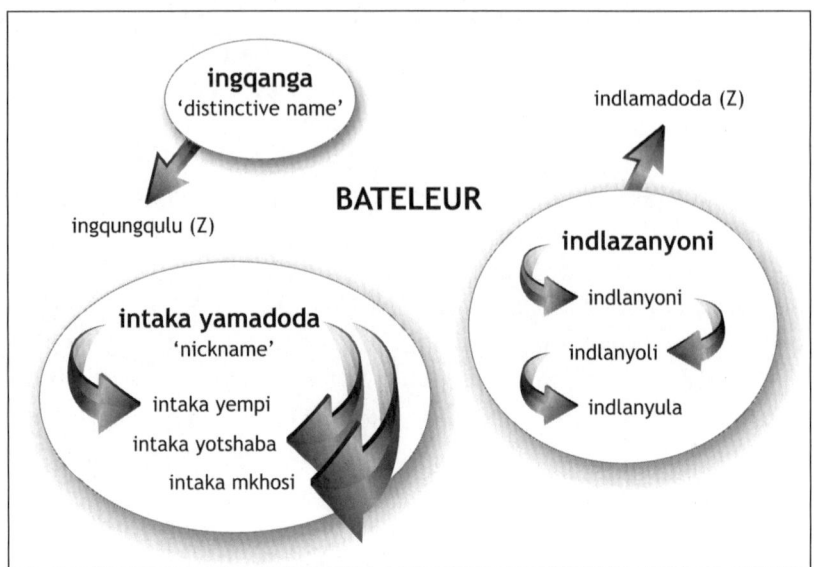

The names of the Bateleur Eagle.

the region. Taking the Eastern Cape and KwaZulu-Natal (KZN) as representing the linguistic region 'Nguni' we look below at the onomastic profile of the Bateleur Eagle in Xhosa and Zulu.

The diagram above illustrates the Xhosa and related Zulu names of the Bateleur Eagle.[13] The three rings show that the Bateleur basically has three names, of which two have a number of variants. Godfrey refers to the name *ingqanga* as the "distinctive" Xhosa name for the bird. The term 'distinctive' here apparently means the 'real' name, or the 'basic' name, or the 'official' name (the 'vernacular book name', in the typology suggested on pages 5–9 above). The Zulu name **ingqungqulu** is phonetically similar and is clearly a cognate. The literature reflects that some Zulu speakers have seen the repeated 'ngqu' syllables as representing the sound made by the eagle clapping its wings.

Godfrey refers to *intaka yamadoda* as a "nickname", with the meaning "bird of the warriors". Not directly derived from this name, but semantically

---

13. The names are all Xhosa names from Godfrey (1941: 33) except for the two marked (Z).

similar, are *intaka yempi* (bird of the army), *intaka yotshaba* (bird of the enemy) and *intaka mkhosi* (raiser of the war-cry). The Zulu 'nickname' **indlamadoda** (eater of the men) is also semantically linked to the notion of this eagle being involved in war, but the diagram above links it to *indlazanyoni* because of the use of the verb *dla*.

Godfrey's third name for the Bateleur is *indlazanyoni* (eater of birds). It is not clear whether he considers this to be another 'nickname' or another 'distinctive name'. I would personally regard its Zulu equivalent **indlamadoda** as a 'praise name'. The remaining three Xhosa names in this group are all derived morphologically or phonetically one from the other. *Indlanyoni* is simply an abbreviated form of *indlazanyoni* (morphological change), *indlanyoli* exhanges the final alveolar /n/ of *indlanyoni* for another alveolar /l/ and *indlanyula* has the same consonantal shape as *indlanyoli*, but the final two vowels have been changed.

Not indicated in this diagram, but recorded by Godfrey, is how these name variants are frequently regionally based. The name *ingqanga*, for example ranges from the erstwhile Ciskei to Flagstaff, while the name *indlazanyoni* (with its variants) runs eastwards from Flagstaff into KZN.

Maclean (1984: 130) gives only *ingqanga* and **ingqungqulu** as the Xhosa and Zulu names respectively of the Bateleur Eagle. Clearly the Nguni onomastic profile of this bird is considerably more complex. This raises the question of how much of this detail should go into a book such as a bird guide. I recall working with Maclean in the early 1980s when he was preparing the manuscript of the fifth edition of *Roberts.* When it came to the African names, he was keen to have only one name for each language, but would accept two if both were equally well known. Only under the most exceptional circumstances would he accept three. And here we see the Bateleur, having three Xhosa names without counting the variations, and with two Zulu names. This situation is by no means unusual for both Zulu and Xhosa. Bryant's 1905 dictionary, for example, records for the Black-headed Oriole (*Oriolus larvatus*) the names **umbhicongo**, **umgongolozi** and **umqoqongo**, with the last name having the variants **umqaqongo** and **ungoqongo**. The 2013–2017 workshops, aiming for a single name for each species, eventually, after much discussion, settled for **umqoqongo** as being the most widely spread name, and then added yet another name – **usibó** – saying that this was the name used today by most people they knew.

It is easy to see here how Maclean's comment quoted above, namely that "[b]ird names in the African languages present far more problems than in the European-derived languages" only scratches the surface.

Before going on to see how Maclean's 'problem' can be resolved when it comes to African language names in today's bird guides, it may be useful to look at the Zulu names within the context of 'indigenous' species naming generally in classification systems usually known as 'folk taxonomy' within the discipline of 'ethnobiology'.

2.4   FOLK-TAXONOMIC SYSTEMS

This section is based on Brent Berlin's 1992 *Ethnobiological classification: Principles of Categorization of Plants and Animals in Traditional Societies*. Berlin is used first to establish the more general notions about 'genus' and 'species' without going into specialised taxonomy. Berlin's principles of classification do not relate only to birds, and the traditional societies he discusses are mostly not African. To get a more focused idea of how Berlin's ideas relate to the distinction between 'genus' and 'species' in the thinking about birds in traditional southern Africa societies, I have relied heavily on Louis Louwrens' 2004 article on the generic nature of Northern Sotho bird names. What follows, then, is a synopsis and summary of the principal ideas espoused by both Berlin and Louwrens, as applied to Zulu bird names.

Early in his 1992 book, Berlin makes the key observation:

> Generic taxa are the basic building blocks of any folk taxonomy, are the most salient psychologically, and are likely to be among the first taxa learnt by the child (1992: 15).

The issue of 'salience' will be looked at shortly, but first we should consider the comment that "generic taxa are the basic building blocks". In the earliest works that detail the links between bird names and bird categories, namely Woodward and Woodward's 1899 *Natal Birds*, and A.T. Bryant's 1905 *Zulu-English Dictionary*, we find several acknowledgements that some bird names are generic in nature.

The entries in Woodward and Woodward are by English vernacular name:

> Lesser Double-collared Sunbird: "Native name, 'Incwincwi,' which is the general name for all the honey-suckers."
> European Swallow: "Native name, 'Inkonjane,' which is the general name for all the swallows."
> Common Waxbill: "Native name, 'Intiyane,' which is the general name for all the waxbills."

African Pied Wagtail: "Native name, 'Umvemve'."
Cape Wagtail: "Native name, same as preceding."
Green Starling: "Native name, 'Ikwinsi'."
Black-bellied Starling: Native name, same as the preceding species."

In Bryant's dictionary, the entries are by the Zulu name:

isi-Kova: "Hooting owl, of which there are several varieties" ['varieties' here means species; it does not mean the same as Berlin's 'varietal' level, which refers to sub-species or 'local races'].
iFefe: "Roller, of which there are several varieties (*Coracias garrulus*, etc.)."
iDada: "Generic name for any bird of the 'duck' kind."
inCwincwi: "Sun-bird or honey-sucker of which there are several varieties (*Cinnyris Afra*, *C. chalybea,* etc.)."
iHlokohloko: "Yellow Weaver-bird, of which there are several varieties, the commonest being the Spotted-back Weaver-bird."
iHobe: "Generic name applied to certain birds of the dove kind."
iNqe: "Vulture, of two varieties [in South Africa] . . . "

The listing of generic names in Woodward and Woodward and in Bryant does not mean to say that all the bird names they recorded were generic. Indeed, the opposite is true: most of the Zulu names they give are for specific species. The listing here is simply to indicate that Bryant and the Woodward brothers were aware of the nature of generic names, and took pains to say so if they felt a name was generic.

Maclean, in his 'core quote', says "most bird species have no African name at all". A more accurate statement might have been "Most bird species have no <u>recorded</u> African name at all" given that there is a distinct difference between the names held in the 'group memory' of a primary oral society, and those names that have been noted down and later published in dictionaries and field guides.[14] Nonetheless, it is quite likely that many discrete species of birds have never been named. Berlin's 'First Principle' (1992: 21ff.) states:

---

14. During the five workshops held between 2013 and 2017, Professor Turner and I were several times given a Zulu bird name which all members of the particular group at that particular workshop said was in common usage, but which had not been previously recorded in the literature.

Traditional societies residing in a local habitat exhibit a system of ethnobiological classification for a smaller portion of the actual plant and animal species found in the same area. This subset is comprised of the most salient plant and animal species in that local habitat, where salience can be understood as a function of biological distinctiveness.

When applied to birds, this statement simply means what Maclean said, i.e. many bird species don't get named. But we should take Berlin's statement a little further than that, and explore the useful notion of 'salience', and why it is that certain bird species don't get named.

'Salience', as I understand it in this context, can be divided into the three areas of visibility, volubility and utility (of which a subset is 'cultural salience'). A bird species (or genus) is:

- Visibly salient if it is (a) comparatively large; or (b) spends much of its time out in the open; or (c) has distinctively-marked or brightly-coloured plumage; or (d) has some other feature that sets the bird visibly apart from otherwise similar birds.
- Volubly salient if it has (a) a very loud call that is heard from far off; (b) has a very persistent call that carries on without a break for long periods; or (c) has a very distinctive call, with features that make it unlike the call of any other bird.
- Usably salient if it is (a) commonly eaten; or (b) has desirable feathers which are used decoratively; or (c) is used medicinally or in protective charms; or (d) has strong cultural overtones, even if these are negative, as in birds of bad omen.

A bird such as the Common Ostrich will always get a distinctive name in languages spoken in the area where the bird occurs (and even in Zulu, although the bird is supposed not to have occurred naturally in KZN): it is huge, lives in the open, is distinctively marked, and its feathers and its eggs are highly desirable for different reasons. A considerably smaller bird, the Hamerkop, is not at all distinctively marked, and does not have a particularly distinctive call. But it builds a nest big enough to be called a house, which it decorates with cast-off human debris of a surprising variety; has a decidedly distinctive crest, which it keeps looking at sadly in the reflective surface of still pools, and it is associated with *abathakathi*, bad luck, death and destructive thunderstorms. Consequently, it has its own species-specific name in all African languages, frequently more than one in each language.

Tracing the salience of all bird species that have their own distinctive name in Zulu would take not just the entire chapter, but most of a book. Without doing an exhaustive analysis between the characteristics of bird species and their possession or lack of a discrete name, I would say that almost every bird that has its own unique species-specific name is salient in one way or another. Many of the bird species for which no discrete and unique name has been recorded belong to the comparatively large group of birds that in English-speaking birdwatching circles are known as 'LBJs' (Little Brown Jobs) – birds that do not have enough individual salience for the amateur to be able to tell one from the other.[15] A Zulu bird name that equates to the term 'LBJ' might be **uncede**, a well-known <u>name</u> because of its significance in both Zulu folk tales and in traditional proverbs, where invariably the theme of the folk tale or the proverb is the bird's tiny size and insignificance. The <u>identity</u> of **uncede**, however, is not so well known: At one of the 2013–2017 workshops, the Zulu-speaking bird experts were asked to pin down the identity of '**uncede**', and there was much hemming and hawing. Eventually, by mutual agreement, the name was assigned as a generic name to the cisticolas, to many people quintessentially 'little brown jobs'.

Hand in hand with the notion of salience is the notion of 'discontinuities'. Louwrens explains as follows:

> [N]ature is characterized by psychologically salient *discontinuities*, i.e. by groupings of similar biological entities between which clearly perceptible gaps exist, for example, mammals, reptiles, fish and birds. In the botanical world, perceptible gaps exist, for example, between trees, shrubs, herbs, grasses and mosses. This is also true of birds, where natural breaks or *discontinuities* exist, for example, between eagles, sunbirds, pigeons, swallows, guineafowl and ostriches. These naturally occurring breaks in nature serve as stimuli for humans' classificatory activities (2004: 99).

In other words, part of human ability to group birds into various categories is not only perceiving the commonalities that hold the groups together, but perceiving the differences that set the groups apart. Berlin (1992: 53) also uses the notion of discontinuities and links these to generic names in a quote that Louwrens uses at the head of his article on Northern Sotho bird names. In the

---

15. The somewhat dated English term for these is 'dicky bird' (see Plates 3 to 7).

"categorisation of plants and animals by people living in traditional societies", says Berlin, there exists a "partially predictable set of plant and animal taxa", which is marked by discontinuities, (i.e. by perceived differences). He goes on to say:

> This large but finite set of taxa is special in each system in that its members stand out as beacons on the landscape of biological reality, figuratively crying out to be named. These groupings are the generic taxa of all such systems of ethnobiological classification, and their names are precisely the names of common speech (1992: 53, cited in Louwrens 2004: 95).

While the sentence "[I]ts members stand out as beacons on the landscape of biological reality, figuratively crying out to be named" is particularly enjoyable, it is Berlin's assertion that the names of generic groups in folk taxonomy are "precisely the names of common speech" that is more important here.

We can interpret what Berlin is saying here in two ways: socially and linguistically. We can say that the following English words (and the corresponding Zulu words next to them) are (i) the names used by the 'common folk', namely those that are neither expert trained ornithologists, nor keen, experienced birdwatchers, and (ii) that they are ordinary common nouns, not the names of specific species of birds:

owl	–	**isikhova**
eagle	–	**ukhozi**
duck	–	**idada**
hawk	–	**uheshe**
lark	–	**umngcelu**

It is 'common speech', spoken by a 'common person', using 'common nouns', when someone says to a friend (in English, but the sentiment holds true for any language):[16]

'When I heard an <u>owl</u> hooting in the night, I thought it would be a bad omen for my <u>duck</u> hunting today, but no: I was up with the <u>lark</u>, and on the lake early and managed to bag many <u>duck</u>. At first I thought the <u>hawks</u> – they may have been <u>eagles</u> – circling above would take one or two of the <u>duck</u> I

---

16. I have underlined the 'common nouns' in this imaginary snippet of 'common speech'.

shot, but I was lucky and managed to bring the whole bag home. How about coming over for a <u>duck</u> supper tonight?'

Maclean said in his 'problem quote' (in the text box above) that "Bird names in the African languages present far more problems than in the European-derived languages". What he meant was that at the time he was putting together what would be the 1984 fifth edition of *Roberts' Birds*, separate and distinct names were available for each individual species of bird in southern Africa in English and Afrikaans, but this was not the case for Zulu, Xhosa, Sotho, Tswana and the other African languages. Louwrens makes this point clear in his article on Northern Sotho bird names:

> It will become evident in the course of the present discussion that there is a fairly simple explanation of the seemingly deep-seated differences in bird nomenclature in Northern Sotho, English and Afrikaans. It will be argued that, just like Northern Sotho speakers, the overwhelming majority of English and Afrikaans speakers also do not draw nomenclatural distinctions at the species level when conversing about birds in an informal and non-scientific way, and that the naming of species in English and Afrikaans field guides for birdwatchers was due to the deliberate intervention of professional ornithologists and linguists whose obvious aim it was to bring common English and Afrikaans nomenclature as closely in line as possible with the scientific nomenclature used in Linnean [*sic*] biosystematics (2004: 96).

The intervention of dedicated committees of professionals with the goal of establishing individual bird names in English for individual species of southern African birds is well documented. There are several reports from the 'List Committees' of the erstwhile South African Ornithological Society in the pages of their journal *Ostrich*. The decisions that lie behind the Afrikaans names are not so clear. However, it is not the function of this book to examine in any detail how species-specific names for birds in English and Afrikaans appeared in the first half of the twentieth century, but just to note that this was, as Louwrens puts it, a "deliberate intervention". Louwrens continues:

> Viewed in this way, what are presented in English and Afrikaans field guides for birdwatchers as 'common' bird names are not representative of what is regarded in ethnoscience as a *folk taxonomy*, owing to the strong influence exerted by professional scientists on these nomenclatural systems (2004: 96).

Louwrens here uses the term 'common names' (as used in the bird guides), to mean 'not scientific'. Berlin, above, has used the term 'common name' similarly, to mean 'non-professional' but he has also used the term to mean 'generic', rather than 'species-specific'. Louwrens later uses the term 'common' in its ordinary sense in English, to mean 'not rare' or 'frequently found', as in "This is a name commonly found for this bird". In addition, I have myself used the term 'common' in contrast with 'proper', to contrast a 'common noun' (a non-onomastic linguistic item) with a 'proper name' (an onomastic item). We can see then that the use of the word 'common' in a discussion of bird names is fraught with confusion.

There is, in fact, considerable confusion in the terminology used by various authors, writers and scholars when it comes to bird names, as we saw earlier when looking at terms such as 'vernacular' and 'book name'. It was for this reason that I proposed the typology of bird names in the previous chapter.

2.5   HOW DO WE HANDLE 'MACLEAN'S PROBLEM'?

Returning once again to 'Maclean's problem' as quoted in the text box in section 2.3.1, we note that the Zulu language suffers from the problem of what can be succinctly put as:

one name for many birds
one bird with many names
many birds with no names.

We are perhaps in a better position now to understand how these 'problems' arise:

The situation of 'one name for many birds' is very simply a case of a folk-generic name used for a number of related species. This is the "a duck is a duck is a duck" situation.

The situation of 'one bird with many names' has just been exemplified above, where intersecting continua of time, region and formality create two or more names for a single bird. Had Maclean had to deal with 36 names for *Troglodytes troglodytes*,[17] he would undoubtedly have found this highly problematic.

---

17. The Eurasian wren. Greenoak gives 36 names for this bird from the British Isles alone, ranging from 'Wrannock' to 'Stumpy Toad' to 'Kitty-me-Wren' (1997: 149–51).

The situation of many birds with no names arises from what Berlin has explained as a lack of salience, as quoted earlier (see page 34 above).

Again, this simply means that not all things that scientists see as separate species get separate names from non-scientists. People in traditional societies name those things that are important to them. Things that are very important may get many names. Things (birds) that are less important may get no names, or all be grouped together under a folk-generic term.

So that at least explains what Maclean sees a problem. It may be that solutions are not necessary if not everyone agrees that these issues are problematic. But let us assume that Zulu and other African languages are not matching a 'norm' or a standard' set by scientific names and English and Afrikaans names when it comes to birds in southern Africa. I wish to stress here that I do not agree with any notion that Zulu (and the other African languages) are somehow 'sub-standard' when it comes to naming birds, or that these languages are in any way 'faulty' in comparison to English and Afrikaans. Nonetheless, the two Germanic-origin languages in South Africa (Afrikaans and English) have names for all species of birds in the published field guides to southern African birds, and Zulu (and the other African languages) do not. Some people do find that problematic, so let us tackle the question here.

In 2005, P.A.R. Hockey, W.R.J. Dean and P.G. Ryan authored (as 'scientific editors') the monumental 1296-page seventh edition of *Roberts Birds of Southern Africa.* In one way they solved Maclean's 'problem' with the African names by simply leaving them out. As detailed in section 1.6.2 above, Maclean's unchanged African names reappeared in the 2007 edition of *Roberts Bird Guide,* only to disappear again in Chittenden, Davies and Weiersbye's 2016 edition. In other words, well into the twenty-first century, Maclean's 'problematic' African names remain just that – problematic.

In my 2015 *Zulu Plant Names*, I raised the same problems: many plants had no Zulu name; many names were used for one plant; and many plants had several names. The Zulu botanical nomenclature did not conform to the situation in English and Afrikaans, where there was one chosen book name in these two languages for each botanical species in South Africa. The questions I raised then are relevant here where the same situation is true for Zulu bird names: 'What do we do about the problem that many birds have no Zulu name; other names are generically used by many different species, and some birds have two, three or more names?'

In *Zulu Plant Names* (2015: 266–8), I offered three possible ways of dealing with the 'problem':

The first was simply to do nothing, to just leave the problem as it is. I suggested that the problem is only caused by attempts to shape an indigenous knowledge system to conform to Linnaean-type taxonomies and classification. I quoted Cockburn, Khumalo-Seegelken and Villet (2014: xx) as arguing that in South Africa it is important for African indigenous knowledge systems to remain African, and not be reshaped to suit Western concepts of naming and identification:

> Attempts to compare non-scientific naming systems (often termed 'folk taxonomies') such as those found in isiZulu or English with any terminological systems of biological nomenclature have ultimately been rejected as ineffective and undesirable. This situation is especially true when it comes to trying to promote and support interrelated networks of knowledge systems within a given context in post-colonial Africa. Speakers of an indigenous African language such as isiZulu perceive and experience such attempts as biased, presumptuous and exclusively Eurocentric.

The second possible solution to the problem was to do intensive fieldwork and research to accomplish as many as possible of the following goals:

- Acknowledge the indigenous nature of the existing Zulu ornithological nomenclature and attempt to record as much oral knowledge as possible;
- Correct all errors in existing publications;
- Do intensive fieldwork to establish which of many names for one bird are regional varieties, which are names used only by specialists such as traditional healers, hunters, herdboys, and so on;[18]
- Record this information with as much information as possible on bird interaction in Zulu society and publish this information in comprehensive dictionaries and other reference books;
- Recognise that the indigenous knowledge system is still a system and attempt to recognise the nature of the system, such as seeing the name **ukhozi** as a genus name rather than as a specific name which just happens to be used for several species of eagle;
- Construct 'onomastic profiles' for all groups of birds in the Zulu-speaking area of South Africa along the lines of the onomastic profile of doves and pigeons at the end of this chapter.

---

18. The people Jacobs calls 'vernacular birders' in her 2016 *Birders of Africa*.

Basically, these goals were the goals of the (unsuccessful) application that Turner and I submitted to the National Research Foundation for funding for the period 2017 to 2019. We had intended to send out twelve fieldworkers into different regions of the Zulu-speaking regions to investigate Zulu bird names in all their variations, as well as to collect whatever bird lore is still present today, but without funding this was not possible.

This approach was much the same approach that Cockburn adopted in her research into Zulu insect names (Cockburn, Khumalo-Seegelken and Villet 2014). She spent a year and more doing fieldwork in as many different places in KZN as possible, showing a collection of mounted insects to different members of the Zulu-speaking population and eliciting oral knowledge about insects.

The third possible solution was to hold workshops where Zulu-speaking bird experts would discuss in a controlled way the potential re-organisation of existing recorded bird names, extending them, adapting them, and even coining totally new ones, aiming ultimately for a situation where each species of bird in the Zulu-speaking area (effectively KZN) ended up with its own unique name.

This was in fact the solution to the 'Maclean problem' as envisaged by Turner, and the 2013–2017 Zulu bird name workshops that she organised have already been mentioned several times. In Table 2.2 above listing names for eagles in Latin, English and Zulu, the fourth column shows the result of this third possible solution. Eagles were tackled in the very first workshop held in 2013 and by the end of the second day we had seventeen distinct names for seventeen species of eagle found in KZN.

2.6   DOVE AND PIGEON NAMES AS AN EXAMPLE OF 'POLYTYPY'

2.6.1   Introducing the notion of polytypy

In sections 2.2 and 2.3 above, the notion 'eagle' was explored, and Table 2.2 listed Zulu eagle names as recorded at the time of Maclean's 1984 edition of *Roberts' Birds*. We noted that the generic name **ukhozi** was used for a number of species of eagle, that many others had no name at all, but also that at least three species had their own specific name: the African Fish Eagle (**inkwazi**), the Long-crested Eagle (**isiphungumangathi**) and the Bateleur (**ingqungqulu**). At the time, I suggested that these birds must have some salience over and above the other eagles, whether this was salience of appearance or voice, or of some cultural significance. We return to the idea of species-specific names within a folk genus here, and this time look at doves and pigeons.

Louwrens (2004: 103) talks of what he calls 'polytypy' – the splitting of a genus into separate species within a folk taxonomy and naming the species individually. In Northern Sotho, this occurs with pigeons. Within the folk genus *leeba* (pigeon, dove), there are separate names for such species as the Rock Pigeon, Cape Turtle Dove and others.[19] That there are separate names, says Louwrens, suggests that they must surely be culturally significant, and after probing more deeply with Northern Sotho-speaking bird experts he discovered that feathers of various dove species are used as head decorations by young girls, and so young men go out to hunt them to collect feathers to present to their favourites. It is therefore necessary for them to be able to distinguish between the various species. According to Louwrens:

> Apart from the single case discussed above, no other examples of polytypy were recorded during the course of this investigation. According to Berlin (1992: 131), this can be ascribed to the fact that there is no pressing need for Northern Sotho speakers to subject the avifauna in their natural environment to close, detailed inspection (2004: 103).

Louwrens gives three species-specific names for doves that are all extended variants of the generic *leeba*:
- *leebarupi*, used for both the Rock Pigeon (*Columba guinea*), and the Rameron Pigeon (*Columba arquatrix*);
- *leebamašu*, used for both the Cape Turtle Dove (*Streptopelia capicola*) and the Red-eyed Dove (*Streptopelia semitorquata*);
- *leebarui*, the Laughing Dove (*Streptopelia* [sic] *senegalensis*).

It is somewhat worrying that Louwrens gives these Northern Sotho names as examples of an instance of "generic[s] splitting into species" and receiving names accordingly, and then proceeds with three examples, two of which are used for more than one species!

The way in which Northern Sotho has extended its generic term *leeba* to produce these more specific names (even if they are not <u>entirely</u> species-specific) is similar to the way in which Zulu has extended the generic term **ijuba**, with Doke and Vilakazi (1958: 365) recognising **ijubantondo** (< **ijuba** +

---

19. Now the Speckled Pigeon and the Ring-necked Dove respectively.

intondo, i.e. dove found in large numbers) for the "Common green dove, *Vinago delalandî*" (now the African Green Pigeon, *Treron calvus*) and **ijubantendele** (< **ijuba** + *intendele* 'pheasant dove') for the "Rock-pigeon, *Columba guinea*".

The separate names for doves and pigeons, in the Zulu language, present long before the 2013–2017 Zulu bird name workshops started the process of revising, extending and coining Zulu bird names, shows a picture of polytypy far more extensive than Louwrens shows for Northern Sotho. The naming of doves and pigeons in Zulu is given in detail below, superimposed on, and juxtaposed with the different genera and species as seen by the scientific world.

2.6.2 Extracts from Doke and Vilakazi's 1958 *Zulu-English Dictionary* relating to doves and pigeons

The words below are all found in Doke and Vilakazi's dictionary and are without question the most extensive list of previously recorded species-specific names for any cluster of birds in KZN. Many of the birds here have more than one Zulu name and these names are still in regular use among the Zulu-speaking population who still know their birds.

Many of the Latin and English names for species shown below are no longer in current use. Given the high rate of change in both scientific nomenclature and English vernacular names, this is not surprising.

1. **isAgqukwe**: Cinnamon Dove, *Aplopelia larvata*
2. **isiBhelu**: White-breasted Dove, *Tympanistria bicolor* (cf. *ibhobobo, isikhombazana*)
3. **iBhobobo**: White-breasted dove (cf. *isibhelu*)
4. **imBumbazane**: Tambourine Dove
5. **iHobhe**: Dove (generic term including *ijuba, ivukuthu*, etc.). *Ukubamba isisila sehobhe*: to grasp a dove's tail = to hold on to non-essentials
6. **iJuba**: 1. Dove, pigeon of various kinds, e.g. Rock Pigeon, *Columba phoeonata*; Collared Turtle Dove, *Turtur semitorquatus*; Lesser Collared Turtle Dove, *Turtur capicola*. 2. Small, light-blue bead
7. **iJubantendele**: Rock-pigeon, *Columba guinea* (cf. *ivukuthu*)
8. **iJubantondo**: Common green dove, *Vinago delalandi* (cf. *ijubantonto*)
9. **iJubantonto**: as above
10. **inKombazane**: White-breasted dove
11. **isiKhombazana/e**: White-breasted dove (cf. *unkombose*)
12. **isiKhombazana sasenkangala**: Namaqua dove

13. **isiKhombazana sasehlanze**: Green-spotted wood dove, *Turtur chaleospilos* [*chalcospilos*][20]
14. **isiKhombazana sehlathi:** Tambourine dove, *Columba tympanistria*
15. **uKhonzane:** Cape laughing dove
16. **iNcumba:** Namaqua dove, *Pena* [*Oena*][21] *capensis*
17. **uNkombose:** Long-tailed Namaqua dove
18. **iNkwababakazana/inkwambakazana:** Brown tambourine dove
19. **uSamdokwe:** Cape turtle-dove, *Afropelia capicola*
20. **iVukuthu**: 1. Speckled rock pigeon, *Columba arquatrix*; 2. Common rock pigeon, *Dialiptila phaeonota;* 3. Olive pigeon, *Stictoenas arquatrix*

A cautionary note on the last entry: there is some definite confusion here. I do not believe that there are three different birds here all called **ivukuthu**, with two of them having the same specific epithet *arquatrix*. Doke and Vilakazi's gloss here for the word **ivukuthu** suggests strongly that they received bird information from three different sources and simply combined them in one entry without noticing scientific name anomalies.

2.6.3   Doves and pigeons of KwaZulu-Natal: a comparison of the scientific taxonomy with Zulu taxonomy

I start by sub-dividing the doves and pigeons according to scientific taxonomy and then go on to sub-dividing them according to Zulu 'taxonomy' – a categorisation determined by the lexical elements in the Zulu names. The result is two contrasting sets of categorisation based on scientific names on the one hand and Zulu names on the other hand.

(a) A scientific sub-division of doves and pigeons

Below, the doves and pigeons of KZN are sorted by scientific names and are boxed according to their genera. The details are taken from Chittenden, Davies and Weiersbye (2016). Two species in this edition are not found in KZN and

---

20. A misprint for *chalcospilos*. Doke and Vilakazi's 'Green-spotted wood dove' later became the Emerald-spotted Wood-dove (McLachlan and Liversidge 1957: 173), then the Greenspotted Dove (Maclean 1984: 311), and then back to the Emerald-spotted Wood Dove again (Chittenden, Davies and Weiersbye 2016: 262).
21. A misprint for *Oena*.

are left out of this categorisation: the Blue-spotted Wood Dove (*Turtur afer*) and the Mourning Collared Dove (*Streptopelia decipiens*). The scientific and English names of Chittenden, Davies and Weiersbye, and those of Doke and Vilakazi don't always match up. The Zulu names for the species are the names decided upon by Maclean and Koopman in the early 1980s when the drafts of the fifth edition of *Roberts* were being prepared. It may be worth noting that an earlier draft of the scientific classification below was based on Maclean's taxonomy and many changes had to be made when the latest classification (Chittenden, Davies and Weiersbye, 2016) was used instead. The separate entry for "Cinnamon Dove *Aplopelia larvata*" had to be moved into the box for the genus *Columba* as the "Lemon Dove *Columba larvata*", and the "Laughing Dove *Streptopelia senegalensis*" had to be moved into its own genus box when it became *Spilopelia senegalensis*. Many other changes had to be made to the English names.[22]

The doves and pigeons are in the order *COLUMBIFORMES*, of which there is only one family, the family *Columbidae*. The six genera are shown below:

GENUS: Columba (5 species)
Columba livia, Rock Dove, **no zulu name**
Columba guinea, Speckled Pigeon, **ijuba, ivukuthu**
Columba arquatrix, African Olive Pigeon, **ivukuthu lehlathi**
Columba delegorguei, Eastern Bronze-naped Pigeon, **no Zulu name** (highly restricted occurrence)

GENUS: Streptopelia (2 species)
Streptopelia semitorquata, Red-eyed Dove, **ihobhe**
Streptopelia capicola, Ring-necked Dove, **ihobhe, usamdokwe**

GENUS: Spilopelia (1 species)
Spilopelia senegalensis, Laughing Dove, **ukhonzane**

GENUS: Turtur (2 species)
Turtur chalcospilos, Emerald-spotted Wood Dove, **isikhombazane sehlathi**
Turtur tympanistria, Tambourine Dove, **isikhombazane sehlathi, isibhelu**

GENUS: Oena (1 species)
Oena capensis, Namaqua Dove, **unkombose, isikhombazane sasenkangala**

GENUS: Treron (1 species)
Treron calvus, African Green Pigeon, **ijubantondo**

22. For example: The Feral Pigeon is now the Rock Dove and the Rock Pigeon is now the Speckled Pigeon; the Rameron Pigeon has become the African Olive Pigeon, and Delegorgue's Pigeon is now the Eastern Bronze-naped Pigeon. There are numerous other changes to the English names as well.

(b) Dividing the KZN dove and pigeon species according to Zulu-based onomastically organised genera

The categorisation of KZN doves and pigeons is based on their Zulu names, more specifically on the lexical items on which the names are based. Basically there are three groups: those based on the noun **ijuba**, those based on the verb *khomba* and its derivatives, and those from other lexical items, unrelated to each other. I have placed all names under a 'family level' name (**ihobhe**) following Doke and Vilakazi gloss of this word as "Dove (generic term including *ijuba*, *ivukuthu*, etc.)" (1958: 342). Bryant (1905: 264) says of "**i(li)-Hōbe**" that it is a "Generic name applied to certain birds of the dove type". Although both Bryant and Doke and Vilakazi refer to **ihobhe** as a generic name, I have elevated it to a family name on the basis that it covers the three separate genera shown below:

The Latin and English names are given as they appear in Doke and Vilakazi's dictionary.

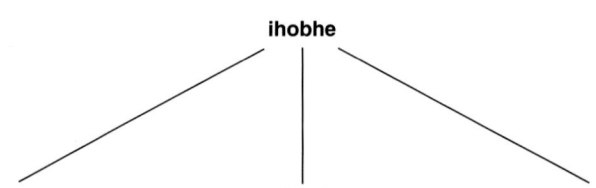

ihobhe

Names based on noun **ijuba**:	Names based on other lexical items:	Names based on verb **khomba** 'point':
ijuba; dove, pigeon of various kinds, e.g. Rock Pigeon, *Columba phoeonata*, Collared Turtle Dove, *Turtur semitorquatus*. Lesser-collared Turtle Dove, *Turtur capicola* **ijubantendele**: Rock Pigeon, *Columba guinea* **ijubantondo/ijubantonto**: Common green dove, *Vinago delalandi*	**isagqukwe**: Cinnamon dove, *Aplopelia larvata* **isibhelu**: White-breasted Dove, *Tympanistria bicolor* **ibhobobo**: White-breasted Dove **imbumbazane**: Tambourine Dove **ukhonzane**: Cape Laughing Dove **incumba**: Namaqua Dove **usamdokwe**: Cape turtle dove, *Afropelia capicola* **ivukuthu**: 1. Speckled rock pigeon, *Columba arquatrix*; 2. Common rock pigeon, *Dialiptila phaeonota*; Olive pigeon, *Stictoenas arquatrix*	**inkombazane** and **isi-khombazane**: White-breasted Dove **isikhombazane sasenkangala**: Namaqua dove **isikhombazane sasehlanze**: Green Spotted Wood Dove, *Turtur chaleospilos* **isikhombazana sehlathi**: Tambourine dove, *Columba tympanistria* **unkombose**: Long-tailed Namaqua dove, *Pena* [*sic*] *capensis* **inkwababakazana// inkwambakazana**: Brown Tambourine Dove

A number of points can be made about the information shown in these boxes:

(1) Species are not exclusive to a box:
In the 'boxing' of the KZN dove and pigeon species according to their scientific name, each species of birds is exclusive to a box, that is to say if it occurs in box 'A' it will not occur in boxes 'B', 'C' or any other box. In 'boxing' them according to a Zulu name, we find species occurring in more than one box. This is understandable when a single species of bird may have more than one name. For example, the Tambourine Dove is in the '*khomba* box' under the name **isikhombazana sehlathi** (forest *khombazana*) but is also under the 'other lexical items' box under the name **imbumbazane**.

(2) The information in Doke and Vilakazi comes from several different sources:
This is a point made earlier at the end of the list of names in section 2.6.2. This is clear from such occurrences as 'White-breasted dove' for several entries, but 'Tambourine dove' for others, when this refers to the same species of bird. It is also noticeable in the different scientific names given for the same species. For example, four different sources are apparent in the names given (English, scientific and Zulu) for the same bird:

Rock Pigeon	*Columba guinea*	**ijubantendele**
Speckled Rock Pigeon	*Columba arquatrix*	**ivukuthu**
Common Rock Pigeon	*Dialiptila phaeonota*	**ivukuthu**
Olive Pigeon	*Stictoenas arquatrix*	**ivukuthu**

Maclean has 'Rock Pigeon (Speckled Pigeon) *Columba guinea*' for this bird, with the Zulu names **ijuba** and **ivukuthu**. Both Zulu names could be considered correct here: '**ijuba**' as a folk generic for all doves and pigeons (see below) and '**ivukuthu**' as a species-specific name for the Rock Pigeon. It is an onomatopoeic name, based on the call *vukuthu-vuuu, vukuthu-vuuu*. But the name **ijubantendele** is also accurate as a species-specific name, as it is a compound of **ijuba** 'dove, pigeon' and **intendele** 'Red-winged Francolin'. Given that the Red-winged Francolin is highly speckled and this is the only dove/pigeon that is speckled, the Zulu name **ijubantendele** and the English name 'Rock Pigeon' refer to the same bird.

(3) Every species of dove/pigeon has its own unique name:
None 'requires' either **ijuba** or **ihobhe** added to their unique name (unlike 'Dove' or 'Pigeon' in English). These two names then function purely as

generic names. Whether or not **ihobhe** is higher in a taxonomic hierarchy as my diagram above suggests is debatable. Equally debatable is whether **ijuba** can be used for all doves and pigeons. Both **ihobhe** and **ijuba** would have to be tested in the field, both with 'bird-savvy' Zulu speakers, and with those who have no special knowledge of birds.

(4) Doves and salience:

Elsewhere we have talked about species-specific names for birds within a genus, or for individual birds as being a result of 'salience', where salience could be considered to be any combination of visibility, vocality and utility. Although not everyone will agree, I do not see doves as particularly salient visually. The calls of some of the species are instantly recognisable, which gives them some salience in terms of vocality. I contend, though, that it is in terms of utility that doves are especially salient, but not in terms of the use of their feathers, as perceived by Louwrens in connection with the naming of doves in Northern Sotho. The literature available on the links between doves and pigeons and Zulu traditional culture suggests that food is the link, and it is a double link: doves and pigeons are among the birds most frequently eaten, and doves and pigeons are among the more notorious of grain thieves.[23]

* * *

The theme in this chapter has been 'names and identity', with special emphasis being laid on whether names denote genera or species, and the differences between genera and species in Linnaean-type categorisation and the categories found in traditional taxonomies. We now move on to a different type of name relationship: the relationship between name and meaning.

---

23. The use of doves and pigeons as food, and their theft of grain, are discussed in several places in this book: in the verbalisation of their calls, for example, and as common themes in proverbs.

3

The meaning of African bird names

3.1 INTRODUCTION: SEMANTICS – THE THEORY OF MEANING

3.1.1 The notion of 'meaning'

In this chapter we look at the underlying meanings of selected African bird names. Linked to the question of meaning are the questions about identity and the function of names. At first glance, this is simple: the Zulu name **uphezukomkhono** refers to a species of bird known to scientists as *Cuculus solitarius* and to English-speaking birdwatchers as the Red-chested Cuckoo. But there is more to the name than that: some writers have said that the name **uphezukomkhono** is onomatopoeic, i.e. an imitation of the call of the bird, in the same way that its Afrikaans name *Piet-my-vrou* is. Another well-known interpretation is based on the meanings of the two elements in this name: *phezu* (on top) and *komkhono* (on the shoulder).[1] This is understood to refer to women placing their hoes on their shoulders to prepare the land for planting at the first rains in the spring. The name thus becomes symbolic of both springtime and of planting and fertility. All of these notions are part of the meaning of **uphezukomkhono** – the name does far more than simply identify a species of bird. It is these other meanings, as well as the identificatory meaning, that I wish to explore in this chapter.

---

1. *Komkhono* is a locative form of the noun *umkhono*, which strictly speaking means an arm, especially the forearm. The meaning 'on the shoulder', however, is so deeply entrenched that I have not wanted to disturb it.

Names, like all nouns in all languages, have a primary function, which is to refer to entities in the real world. These may be abstract notions, like peace and democracy, or they may be 'real-life things': objects such as trains, books and washing machines, or living entities like butterflies, fish, and Red-chested Cuckoos. This primary function of names is to denote, in other words to identify which particular notion or thing we are talking about. When bird names are printed in a field guide for birders, it is their primary function to identify the particular species of bird which is to the fore. In such a publication, the three names *Cuculus solitarius*, Red-chested Cuckoo, and **uphezukomkhono** have the same primary function. In a different publication, which examines and discusses the roles of birds in traditional Zulu society, other meanings might come to the fore, such as the link between the cuckoo and the planting season.

Notions of identification are very important for 'indigenous' hunter-gatherer societies, but also for modern birders, including international 'avitourists'. For this purpose, an identification system based on 'one bird, one name' is important, which is what (a) scientific naming and (b) vernacular nomenclature in modern field guides is aimed at. At any single point in time (the publication of a particular book, especially bird guides, for example), this is by and large achieved. A diachronic analysis (historical context) shows this not always to be the case.

An indigenous or 'folk taxonomy', as we have seen in Chapter 2, does not always see genus and species in the same way as Western ornithology, with the result that when Zulu names are applied to western names of species, a situation arises where instead of 'one bird, one name', we have 'one bird, many names' and 'one name, many birds'.

This identificative function of bird names is not a major feature in this book, as this is not a field guide for the purpose of identifying birds. However, if the primary meaning (and the related primary function) of bird names is to denote or identify separate species of bird, then all the other meanings and their related functions can be considered to be secondary meanings and functions, and these are very much a feature of this book, especially in this and the following chapter. Secondary meanings of names include lexical meanings, connotative meanings, associative meanings and symbolic meanings. I will discuss each of these very briefly below.

3.1.2 Secondary meanings and secondary functions

Lexical meanings

Lexical meanings are those meanings found in a dictionary, which is a lexicon of words. A lexicon is a list of words and their meanings and anyone who speaks a particular language carries a lexicon of the words of that language in her or his head. It is only when a lexicon is published that it becomes a dictionary.

Compound nouns not only have lexical meaning like any other nouns, but also the lexical meanings of their constituent parts. All English speakers know the meaning of the three words 'milk', 'man' and 'maid', but these meanings take on subtle differences when they are combined into 'milkman' and 'milkmaid.'[2] In the same way, Zulu speakers know the meaning of the adverb *phezu(lu)* and the noun *umkhono*, but also understand that when they are combined they produce a different lexical meaning. We can illustrate this principle with the following three Zulu bird names: **isikhovamphondo** (Cape Eagle-Owl and Spotted Eagle-Owl), **ihlolamvula** (a generic name for swifts) and **unozalizingwenya** (Goliath Heron).

In the first name, the underlying lexical elements are **isikhova** (generic for owls) and *i(zi)mpondo* (horns). Strictly speaking, the combined meaning of these elements should be 'owl [with] horns', but as birds do not have horns, it is understood that 'horns' here is a metaphor for projecting tufts of feathers on the head, sometimes (erroneously) called ear-tufts.

In the second name, the underlying elements are the verb *hlola* (with a number of meanings, but used here as 'predict') and *imvula* (rain). Taking these elements literally we have a bird that predicts rain. Again, it is understood that what these two elements mean when they are combined like this, is that it is the <u>appearance</u> of the birds that allows humans to predict rain.

In the third name, the two lexical elements are *zala* (give birth to) and *izingwenya* (crocodiles). Literally then, this is a bird that gives birth to crocodiles, a biological impossibility. But, once again, it is understood that when these two lexical elements are combined in a compound bird name they mean the bird that appears at first glance to have given birth to baby crocodiles because these small crocodiles share the shallow waters of pans where the bird is fishing.

---

2. For example, a milkman does not actually milk the cows; whereas the milkmaid does.

It is not only compound names like the three above where the lexical meaning of the name is not the same as the primary denotative or referential meaning. Take the name **umamhlangeni** for the African Marsh Harrier, for example. This name is derived from the word *umhlanga* (reed bed) in its locative form *emhlangeni* (in the reed bed). This has been prefixed with –*ma*– which indicates characteristic features, to produce the 'underlying' meaning of '[bird which is] characteristically [found] in a reed bed'.

These underlying lexical elements, whether on their own, or when combined, connote certain features of the entity so-named in addition to identifying it, and meanings like 'characteristically found in a reed bed' are often known as connotative meanings.

Connotative meanings

Connotative meanings can best be illustrated with a few personal names, a few place names and a few plant names. Let us look at the girl's name uZibuyile, the boy's name uMfanufikile, the place name Table Mountain and its close linguistic kin uMkhambathini, and the plant names *umakuphole* and *icishamlilo*.

The girl's name uZibuyile is based on the verb *buya* (return) and literally means 'they have returned'. Its connotative meaning is much more complex: the subject concord –*zi*– here refers to *izinkomo* (cattle). When boys in traditional Zulu society marry, a father is obligated to contribute cattle from his herd to help his son pay *ilobolo*. When a daughter marries, on the other hand, he receives cattle from the family of the groom. A man with many sons and no daughters sees his herd diminishing without any prospect of return. When a daughter is finally born, even though she is still many years from marriage, the father sees the prospect of his cattle returning. So although the lexical meaning of this name is 'they have returned', the connotative meaning is 'daughter born after many sons.'

The literal meaning of the boy's name uMfanufikile is 'a boy has arrived'. Missing from this lexical meaning is the connotation of 'at last!' as this is a name that refers to a boy being born after a number of girls.

In the place name Table Mountain, the connotation 'flat-topped' is obvious. The Zulu name uMkhambathini for the flat-topped Table Mountain east of Pietermaritzburg is not so obvious. It is derived from *emkhambathini*, the locative form of the noun *umkhambathi* meaning 'Paperbark Acacia' (*Vachellia* [*Acacia*] *sieberiana*). The lexical meaning of the name is thus 'at the Paperbark Acacia' or 'among the Paperbark Acacias'. As this mountain is indeed found in Paperbark country, this connotation of the dominant vegetation near and on the mountain seems reasonable. But Paperbarks are also relatively flat-topped trees, and a metaphorical connotative meaning of 'flat-topped like an acacia tree' is also a possibility. Here we have a name with one lexical meaning and one referential meaning, but two possible connotative meanings.

The Zulu plant names *umakuphole* and *icishamlilo* both refer to the Broad-leaved Pentanisia (*Pentanisia prunelloides*). The first name comes from the verb *phola* (cool down, become cool) and the second is a compound of the verb *cisha* (extinguish) and *umlilo* (fire). The lexical meanings of these two names are 'let it cool down' and 'put out the fire' respectively. The connotative meanings of these names becomes clear when we note the medicinal use of the Pentanisia plant: to apply pounded roots to burns and as a poultice for swellings, and hot concoctions of the plant used to reduce fevers. Here we see two names for one entity, with two different lexical meanings, but the same connotative meaning of reducing the heat of a burn or a fever. The subtle differences in the relationships between three different types of meaning can be seen if we put uMkhambathini, *umakuphole* and *icishamlilo* into a table:

Table 3.1   Relationships between different types of meaning.

Word	Denotative or referential meaning	Lexical meaning	Connotative meaning
uMkhambathini	Natal Table Mountain	'among the Paperbark Acacias'	'tree situated among Paperbark Acacias' OR 'tree with flat top like a Paperbark Acacia'
umakuphole	Broad-leaved Pentanisia	'let it cool down'	'reduce heat of burn or fever'
icishamlilo	Broad-leaved Pentanisia	'put out the fire'	'reduce heat of burn or fever'

Returning to the birds, we have seen above in four Zulu bird names the connotative meanings of 'bird with large ear-tufts', 'bird that is an omen of rain', 'bird associated with shallow waters of pans' and 'bird associated with reed beds'. I do not need to give any further examples of connotative meanings, because the bulk of this chapter and the next is devoted to examples of this type of underlying meaning.

Associative and symbolic meaning

The name 'Table Mountain' may denote a particular mountain (not just in South Africa – there are Table Mountains all over the English-speaking world), and it may have the connotation of being flat-topped. It is, however, indelibly associated with the City of Cape Town and the Western Cape Peninsula area. Any visual of Table Mountain in a newspaper or magazine or on television, immediately says that the accompanying words will be about Cape Town, just as visuals of the Eiffel Tower signal Paris, and the Statue of Liberty signals America. These examples show why associative and symbolic meanings are discussed together.

Associative meanings of bird names may be personal, depending perhaps on a particular incident or episode in a person's life, or they may be cultural associations made by a people as a whole. Vultures are associated with death. Owls are associated with night, and therefore with witchcraft, which takes place at night. Red-chested Cuckoos are associated with spring and swifts with rain. The Bateleur Eagle is associated among both the Zulu and the Xhosa people with war. It follows then that death, witchcraft, springtime, rain and war are associative meanings of the names of these birds.

With some birds, this association is achieved through the use of feathers, so the name **indwa** (Blue Crane) (see Plates 57 and 58) becomes associated with Zulu kings who customarily wore a single feather of this bird on their heads, and the name **igwalagwala** (Purple-crested and Knysna Turacos – formerly called louries) is associated with the Zulu and Swazi royalty who still wear these feathers in their hair (see Plate 56).

Associative and symbolic meanings like these provide poets with much rich material, and will be discussed in chapters 8 and 9. They also provide entrepreneurs with names for products, usually illustrated with logos based on visual images of birds, and these associations are discussed in Chapter 5.

3.2   AN OVERVIEW OF MEANING IN AFRICAN BIRD NAMES

In order to situate the Zulu bird names in this book, both the old and the new, within a wider African context, this chapter looks at underlying meanings from a number of different languages or language clusters from four different regions of Africa. The regions and the languages or language clusters, together with the major sources of data of bird names from each region, are as follows:

The language Nyabongo is spoken in the Lake Kivu region in the north-eastern corner of the Democratic Republic of the Congo. A considerable number of bird names in this language, with their underlying meanings, were recorded by F.L. Hendrickx and published in his 1944 article in *The Ostrich*.

An equally useful collection of bird names and their underlying meanings in a number of Tanzanian dialects was published by Reginald Moreau in two articles in 1940 and 1942. A few bird names from Father J. Torrend's 1931 dictionary of Bantu-Botatwe and D.C. Scott's 1929 dictionary of Nyanja add to the profile of bird names from the east-central region of Zambia–Malawi.

I have counted the Malagasy language of Madagascar as 'African' even though it is offshore and Malagasy is not a Bantu language. The Rev. James Sibree's 1891–2 articles in *Ibis* show that the underlying meanings of the bird names there are decidedly 'African' in nature.

And then from southern Africa as the fourth region, I have added Zulu names to a number of examples from the Sotho-Tswana cluster of languages. Most of these are Southern Sotho examples, taken from *Roberts Birds* and from Rodney Moffett's 2010 book on Sotho plant and animal names. I have worked out some of the underlying meanings myself with the aid of a Sotho-English dictionary and have also had much help from Dr Johan Meyer of Pretoria who was working on bird names in Sotho, Tswana, Venda and Tsonga during 2017.[3]

3.2.1 Typologies of bird names based on semantics

I am only aware of three authors who have ventured to construct a typology of bird names on semantic grounds. One is my own, a typology of Zulu bird names, found in the chapter on bird names in my book *Zulu Names* (Koopman, 2002: 240–8). A second is that of Moreau, whose 1942 article is all about his classification of Tanganyikan bird names on the basis of their underlying meanings. The third is that of Jeremy Mynott, in his chapter 'Naming Matters' (2009: 237–43).

Table 3.2 shows how Koopman's broader categories compare to the narrower categories of Moreau and Mynott.

---

3. I am very grateful to Dr Meyer for lists of Sotho bird names and their underlying meanings that he sent me by email in early 2017.

Table 3.2   Three different typologies of bird names on semantic grounds.

Koopman (2002)	Moreau (1942)	Mynott (2009)
appearance	colour and patterns shape and size	colour pattern and shapes appearance of bill and legs size of bird
song	onomatopoeia verbalisation of call non-vocal sounds	voice
habits	motion food habits nesting habits personal attributes	activity food
habitat	habitat	place
other	season and weather superstition personal names (one example)	people [honorific names] other

After considering the data from all the authors mentioned above as providing examples of African bird names with their meaning, a tentative categorisation is postulated, made up of a combination of the three typologies in Table 3.2.

A re-worked typology of semantic categories

After recording a total of 280 bird names from Hendrickx, Moreau and Sibree, an initial sub-division was made as follows:

Table 3.3   Initial sub-division of semantic categories.

Song (song, call, voice, vocalisation)	78 names (27.7%)
Habits (diet, nesting, behaviour, etc.)	63 names (22.4%)
Appearance (colour, patterns, plumage, size)	52 names (18.5%)
Habitat	35 names (12.5%)
Motion (flight, gait, other motion)	26 names (9.2%)
Harbinger (weather, omens, luck, seasons)	15 names (5.3%)
Other (miscellaneous, opaque, unclear)	12 names (4.2%)

We will expand on each of these categories, but first it may be useful to look at what sub-divisions occurred in the data analysis, and what might be a useful re-organising of the semantic categories.

The biggest category is 'song'. This can be further sub-divided into onomatopoeia, descriptive names, metaphors and the like, but it does not seem necessary to make these major categories on the first level.

The category 'habits' (63 names) can be broken down into (a) names referring to diet (29 names), (b) names referring to nesting habits (seven names) and (c) other behaviours and attributes. As 'diet' itself (feeding/food types) and 'nesting' are separate distinct features of birds, these are made into major level categories, with 'other' here becoming the category 'behaviour and attributes'.

The category 'appearance' (52 names) could be broken down into 'colour', 'shapes and sizes', 'patterns', and so on, but this seems pointless. We will keep it as a single category at major level.

The category 'habitat' represents a continuum from the wider reference to biomes such as marshes or forests to the narrower references of favourite plants where a particular bird is found. It is not practicable to divide this category into separate categories at any level.

The category 'motion' can be sub-divided into 'flight', 'gait' and 'other motions/movements', but it does not seem necessary to make these major level categories.

The category 'harbinger' can be sub-divided into those names that refer to weather predications, those that refer to seasons or times of the day, and those that refer to evil omens, and to good or bad luck. There seems no point in making these into major categories, but the category should be renamed 'belief system names' or just 'belief names'.

The category 'other' refers to such a miscellany of 'otherness' that it defies any further sub-division.

The following, in order of number of names, are the proposed categories of meaning in the bird names of four African regions:

Table 3.4   Percentages of semantic categories: African bird names and Zulu bird names.

	African names/percentages		Zulu names/percentages	
1. Song	89	28.6%	23	25.0%
2. Appearance	59	18.9%	14	15.5%
3. Habitat	39	12.5%	9	10.0%
4. Diet	32	10.2%	14	15.5%
5. General behaviour	31	10.0%	9	10.0%
6. Motion	27	8.6%	7	7.7%
7. Belief names	15	4.8%	4	4.4%
8. Nesting	7	2.2%	1	1.1%
9. Other	12	3.8%	9	10.0%
Total	311		90	

The graph below shows the percentages from Table 3.4 in a 'Manhattan' form, with the semantic profile of the combined Nyabongo, Tanzanian and Malagasy names on the left, and the semantic profile of the Zulu names on the right. It is striking how the semantic profile of Zulu fits in with the wider African profile. The only real difference is in the category 'Diet' where Zulu has a higher percentage of names than the African average.

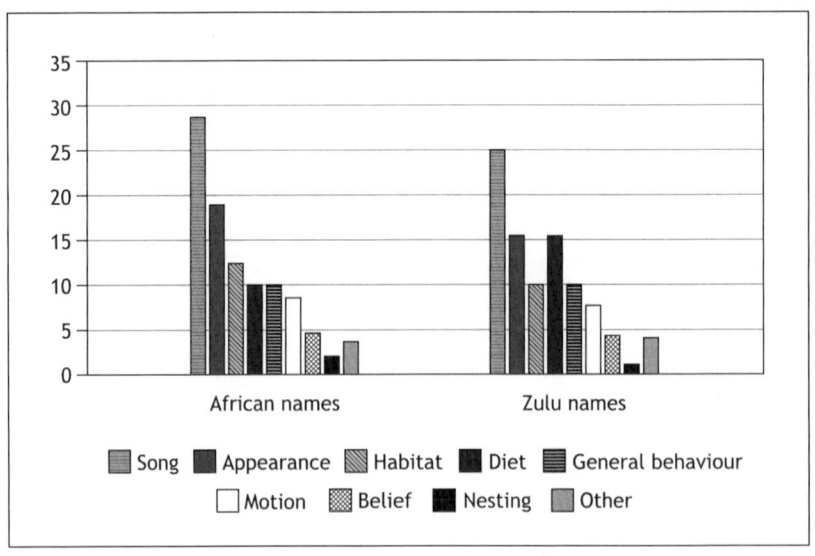

Comparative semantic frequencies.

Let us look at each of these categories in turn.

3.2.2 Song

Mark Cocker, in *Birds and People*, gives a considerable number of names for the Northern Lapwing, ranging from 22 surviving vernacular names in Britain alone through Europe to Russia. Almost every name "echo[es] the high, pleading double note" (2013: 200) of the bird, as in the British regional names 'peewit' and 'teewhup' and names like *dix-huit* (French), *kievit* (Dutch), *kiebitz* (German), *ijip* (Friesian) and *vivak* (Serbian). Cocker comments:

> The extraordinary similarities revealed in this transcontinental naming pattern is one of the best examples I know of the psychological processes underpinning much avian nomenclature. The birds really do call out to us their names (2013: 200).

Cocker's examples are all of the one species, and the name from across Europe. Under this sub-heading we look at African birds that also "really do call out to us their names".

Bird names that refer to song (see also Chapter 7) can be further sub-divided into three basic types: (a) onomatopoeia (with a further sub-category 'verbalisation'), (b) descriptive names and (c) metaphorical names.

Onomatopoeia

Onomatopoeic names attempt to recreate the sound of the bird call in the phonemes of the language(s) of a particular society. Various names for the Hadada Ibis[4] illustrate this: *lengangane* (S. Sotho), *man'an'ani* (Tswana), *ing'ang'ane* (Xhosa), **ihahane** and **inkankane** (Zulu). Moreau (1940: 61) gives the following onomatopoeic names for the hadeda in various Tanzanian dialects: *kwarara, nyawawa, khako, haha*.

Onomatopoeic names for a single bird should be roughly the same, not only within one language, but across several languages, as the songs of birds do not

---

4. This is the latest name given in Chittenden, Davies and Weiersbye (2016), but it is better known by Hadeda Ibis or simply hadeda or hadedah.

vary significantly from region to region. Curiously, however, names marked as onomatopoeic by their recorders seldom show this degree of conformity. Take for example, the following supposedly onomatopoeic names recorded by Moreau (1940: 57) for the Black-and-white Flycatcher (now known as the Fiscal Flycatcher): *chapuluka*, *chekiuwa*, *chipiyo*, and *chokiro*. Or, with markedly different names, the onomatopoeic names recorded by Moreau (1940: 66) for the Boubou Shrike:[5] *kumbo*, *mungo*, *a-nini-a*, and *nguongo*. And then compare these to the Zulu name **ubhoboni**. Moreau (1940: 66) gives the two names *mkubwamkubwa* and *kipulipuli* as onomatopoeic names for the Black Boubou Shrike.

The twittering of the Large-billed Lark can clearly be heard in the S. Sotho name *tsoroane-lipholile* and the cooing of the Ring-necked Dove is just as clear in the S. Sotho *lenkunkuroane*. Cocker (2013: 239) has an amusing comment to make about the calls of the Ring-necked Dove, a comment that links directly to African place names:

> To many humans, the sounds of Ring-necked Doves – rather like a wide range of calls produced by African pigeons – command themselves to our ears as words. In a number of places the bird appears to proclaim its nationality, such as *zim-Bab-we*, *u-Gand-a*, *ma-Law-i*, and perhaps most perfectly, *bot-Swan-a*.

English, Afrikaans and Zulu names all reflect the call of the Grey Go-away-bird (previously the Grey Lourie), the Afrikaans name is *kwêvoël* (bird that goes *kwê*) and the Zulu name is **iklewu**, with regional variations like **umkluwe**, and **umkliwu**. The distinctive 'kl' sound in the Zulu onomatopoeic names for this bird is also seen in the Zulu name **ubuklekle** for the Ethiopian Snipe, marking this bird as 'the one that goes *kle! kle!*'

Motintinyane is a S. Sotho generic name for cisticolas and we note that Afrikaans *tinktinkie* is also a generic name for the cisticola group. These two names may look somewhat dissimilar, but remove the class prefix *mo–* from the Sotho name, and the diminutive suffixes *–nyane* and *–ie* from the Sotho and Afrikaans names respectively, and you are left with 'tinti' and 'tinktink' – decidedly more similar and more apparently onomatopoeic.

---

5. Itself an onomatopoeic name.

Sometimes there is similarity across wider regions: Moreau's name *kitwitwi* for a small species of plover has some similarities with the Zulu name **ititihoye**. And the Zulu name also comes to mind in the Malagasy name *toitòy* for the Common Sandpiper (Sibree 1892: 107).

Perhaps the best examples of onomatopoeic similarities across the southern African sub-region are the concordance of names for the Crowned Crane and the African Hoopoe: recorded names for the crane include *lehehemu* (S.Sotho), *ihem* (Xhosa), **unohemu [unohhemu]** (Zulu) and *mahem* (Afrikaans). Maclean (1984: 188) notes of this bird's call "Highly characteristic 2-syllabled trumpeting *maHEM*". Maclean (1984: 391) gives the following names for the Hoopoe (itself of course also onomatopoeic): *hoephoep* (Afrikaans), *kukuku* (N.Sotho), *popopo* (S.Sotho), *pupupu* (Tswana) and *mhupupu* (Shona). Cocker (2013) also remarks on the widespread use of onomatopoeia for the Hoopoe:

> Many names for the species, whether in the Pashto-speaking area near Kabul (*poppoo*) or across the Arab world (*bud-bud*), or among the Xhosa of South Africa (*ubhobhoyi*) or in Europe (*poup* in Portuguese and *hoopoe* in English) are onomatopoeic. Even the often desiccated traditions of scientific Latin cannot avoid reproducing a version of this wonderfully resonant sound: both parts of the name *Upupa epops*, and even the family name *Upupidae*, carry an echo of the real call (2013: 323).

Then again, it is clear that different communities hear different things when they give onomatopoeic names to the same bird. Take for example, the Afrikaans name *bokmakierie*[6] and the S. Sotho name *pjemtjete* for the bird *Telophorus zeylonus*, and then again the Red-chested Cuckoo, with the Afrikaans onomatopoeic name *Piet-my-vrou* and the equally onomatopoeic S. Sotho name *tlo-nke-lesoho*. The Zulu name for the Red-chested Cuckoo – **uphezukomkhono** – is also said by some sources to be onomatopoeic, but others, as we have already seen, perceive the lexical elements *phezu* (on top) and *komkhono* (of the shoulder) and say that the name of this bird is a reminder to women to put their hoes and their shoulders and start tilling the fields. It is often difficult to assign onomatopoeia as a semantic base when perceived lexical elements are rationalised by folk etymology.

---

6. This Afrikaans name is used as the English vernacular book name as well.

There is little further to say about onomatopoeic names. Different languages have different phonemic systems, but there also appears to be a wide perceptive distance between the call of a bird and the lexeme that represents that bird. The question raised is, what aspect of the call is actually being heard? Is it the whole call, initial notes, leitmotif, characteristics such as nasality or harshness?

Verbalisation

I use the term 'verbalisation' for that type of onomatopoeia that recasts the apparent 'voice' of the bird into understandable human speech segments. It is verbalisation when a bird guide says that the Black-collared Barbet is saying "Two-puddly Two-puddly" or "clean-collar clean-collar" (Maclean 1984: 404). It is verbalisation when a Zulu dictionary adds to the entry **uzavolo** for the Fiery-necked Nightjar that its call is 'interpreted' as "*Zavolo, Zavolo: sengel'abantabakho*" (Zavolo, zavolo: go and milk [a cow] for your children). Verbalisation is a sub-category of onomatopoeia in that both attempt to reproduce in human phonemes the quality and essence of a bird call. It differs in that 'pure' onomatopoeia is normally unintelligible – it does not create recognisable lexical items.

The following are examples of verbalisation found in the data (Moreau (1940):

ngechechangu (MOR55): 'bring me my cooking pot' = verbalisation of call of Black Cuckoo

mhokeuta (MOR51): 'draw the bow' = verbalisation of call of D'Arnaud's Barbet

sayeikia (MOR55): 'famine is coming' = verbalisation of call of Klaas's Cuckoo[7]

ngilikilajako (MOR59): 'warthog your tail!' = verbalisation = Geelgat (*Pycnonotus* bulbul)

lolenzoka (MOR51): 'look at the snake' = verbalisation of call of D'Arnaud's Barbet

songororokanturi, sungurire katuri, kasongolela katuri (MOR56) 'I-have-a-*kinu*- carved'[8] = verbalisation of call of the Laughing Dove

---

7. Compare this to the Afrikaans name *meidjie* for this bird, also a verbalisation.
8. *Kinu* = the wooden mortar used for pounding corn.

ngolankuchange (MOR55): 'knife for gashing' = verbalisation of call of Didric [Diederik] Cuckoo. [A Zulu workshop-coined name for this bird is **unonengekhanda** which translates as 'it got fat in the head'.]

Descriptive names

Descriptive names are names, usually based on verbs or adjectives, which describe the nature of a particular call: whether it is shrill, harsh, croaking, grunting, staccato, melodious, high-pitched and so on. Sibree (1891a: 425) gives the name *jìjỳ* for several related warblers explaining that this is from a verb meaning 'well-delivered' or 'well-recited'. Moreau (1940: 53) gives *mguna*, meaning 'grunter' as a name for a large species of bustard. In another example (1940: 68), he gives *mshongogolo* (little banger) as a name for a tinkerbird of the genus *Pogoniulus*. Hendrickx (1944: 208) says that one of the names of the Hamerkop is *nyakanana,* from the verb *kunana* 'to cry like a little child'. He also says (1944: 204) that the name of the Hadeda Ibis is *mwanana*, from the verb *kuyanana* (cry in a loud and harsh voice). A descriptive name is linked to a habitat in the name *mlangilambago* (chatterer of the forest), given by Moreau (1940: 54) for the Green Coucal.

The well-known duetting of the Southern Boubou is captured in the S. Sotho name *sehoeletsane* (what shouts at another).

Doke and Vilakazi's dictionary gives four examples of Zulu descriptive names, but only one can be identified as a particular species. The name **igedezi** for an unidentified small bird is derived from the verb *gedeza* meaning 'talk incessantly' or 'chatter loudly'; the name **isiwelewele** for an unidentified marsh bird that utters a shrill, screeching cry is derived from the noun *isiwelewele* (confused noise) and the name **umantuluza** for an unidentified bird is derived from the verb *ntuluza* 'pour forth, as in a stream of words'. The bird that Doke and Vilakazi identify is the Red-fronted Tinker Barbet, with the name **unogandilanga** (that which pounds all day long) is in reference to its continual calling.

Lastly, a name that combines call with behaviour is the Zulu name for the Greater Honeyguide – **inhlavebizelayo**. This compound noun is derived from *inhlava* 'honeyguide' and *ebizelayo* (which is calling to someone) and refers to the behaviour of this bird in calling to humans to guide them to bees' nests.

These names are the equivalent of entries in bird guides such as "variety of ringing and harsh vibrating calls" (Northern Grey Tit), "varied phrases of a few quickly whistled and harsh notes" (Groundscraper Thrush) and "song monotonous high-pitched tinkling" (Desert Cisticola).[9]

Metaphorical names

Under this heading we consider names that compare the bird to another entity that makes a sound. For example, Hendrickx (1944: 198) gives the name *kahene* (< *hene* 'goat') for an unidentified bird that cries like a goat.[10] The metaphor here is the same as that in the two Zulu names **imbuzana** (little goat) and **imbuzi yehlathi** (goat of the forest) for the Bleating Warbler. The English vernacular book name obviously uses the same metaphor as the Nyabongo and Zulu names. It is a pity that at some time between 1984 and 2016, this English name changed to Green-backed Camaroptera although the 2016 *Roberts Bird Guide* that records the new name still says of the call "Alarm call a nasal bleating *maa*, vaguely reminiscent of a goat kid" (Chittenden, Davies and Weiersbye 2016: 428).

The following are some of the more interesting metaphor-based names in the database.

Moreau gives the name *kigoma msindo* (little drum roll)[11] as a name for a plurality of various weaver-like birds, especially the Red-billed Quelea, from "the sound of the multitudinous wings" (1940: 70). The same drum (*n-goma*) is found in another name, from Hendrickx (1944: 204), in the name *mushumbiza-ngoma* (strike a drum) for the Red-eyed Turtle Dove. Still with musical instruments, Moreau's example (1940: 65) of a name for the Button Quail – *kimpululu zeze* – is a composite of a generic and a metaphor, with *kimpululu* being a generic name for quails and *zeze* referring to a particular stringed instrument with a note that resembles that of this species.

---

9. All from Maclean (1984).
10. The *ka–* here is the prefix of noun class 13. All Bantu languages north of the Limpopo River use class 13 to form diminutives, whereas Bantu languages south of the Limpopo use the suffix *–ana*. Thus *kahene* (little goat) is the exact equivalent of Zulu **imbuzana** (little goat) and it is very probable that Hendrickx's Nyabongo name refers to the same species. Maclean (1984: 571) says that the range of this bird is "Africa S[outh] of the Sahara". See also section 6.5.5 in Chapter 6 (page 142), for more detail on diminutives in Bantu languages.
11. Zulu speakers may recognise *ingoma* 'drum' and *umsindo* 'sound' here.

In another Moreau example, the White-browed Coucal has as one of its names *kijogoo* (cock, rooster) "because it often calls at night especially if a cock has crowed" (1940: 55). He also gives the name *kijogooshamba* (country cock) for the Broadbill (*Smithornis*) saying that "it sounds like a cock crowing in the distance (1940: 52)". Moreau's name *shemsana* (blacksmith) for the Green Coucal is reminiscent of the English vernacular name Blacksmith Lapwing (formerly Blacksmith Plover).[12] Maclean (1984: 231) says of the call of the Blacksmith Plover [Lapwing] "characteristic *klink, klink, klink*, like a hammer on anvil (hence the name)", showing that he too is aware of the role of metaphor in the naming of birds.[13] The connection between a blacksmith and the call of this bird is clearly an enduring one, and the S. Sotho name for this bird is *mo-otla-tshepe*, based on the verb *otla* 'strike' and the noun *tshepe* 'iron', i.e. 'the one that strikes iron'.

Maclean says the call of the Crested Barbet "sounds like alarm clock with the bell removed" (1984: 410), and clearly this notion also occurred to the earliest S. Sotho bird namers, for the name of this bird is *'malioache/mmaliwatjhe*.[14] The S. Sotho name is based on *mma* '[honorary] mother of' + *oache/watjhe*, derived from English *watch*, and used for both watches and clocks.

The Zulu name of the Crested Barbet (**usiqhovana**) refers to its little crest rather than its call.
© Adrian Koopman

---

12. See Moreau (1940: 54).
13. The Zulu name for this bird is **indudumela**, a descriptive name that refers to continual thudding.
14. The two different spellings reflect the orthography used in Lesotho (*'malioache*) and in South Africa (*mmaliwatjhe*).

Descriptions of the call of the Green Wood Hoopoe, (formerly, Red-billed Woodhoopoe), in English-language bird guides often refer to the call being interpreted as 'the laughter of women'. Strictly speaking, the name **inhlekabafazi** should be translated as '[the bird] that laughs <u>at</u> women'.

Perhaps the most interesting metaphor in the database is the name *kishosoungula* (cool-porridge), given by Moreau for the Broadbill (1940: 52). He says the bird's call sounds like someone blowing on hot porridge to cool it. There must be different ways of blowing on hot porridge in Tanzania, however, for on another page (1940: 62), Moreau gives a quite different name – *mposanji* – to a quite different bird – the Flappet Lark – but translates this name as 'cool-porridge' as well, saying that it is "because the song is like someone trying to blow his hot porridge".

3.2.3   Appearance

Mynott found it necessary to create four separate categories here, dividing appearance-names into (a) colour, (b) pattern and shapes, (c) appearance of bill and legs and (d) size of bird. One could, of course, add even more categories like reference to crest on head, reference to length of tail or references to other parts of the body (eyes, neck, etc.) that are peculiar to a particular species. I prefer to treat 'appearance' as one overarching category of meanings and have loosely grouped the examples within this single category.

Colour

Altogether 30 per cent of the appearance-names in the database make mention of colour, either as an overall colour of the bird, or a reference to the particular colour of wings, chest, head, tail, etc. Metaphor is used as well here in indicating colour. The following are some examples of colour-names.

The Zulu name **uhlazazana** (small intensely green thing) refers to the Malachite Sunbird, a bird that is indeed small and intensely green. Not nearly as intense in its hues is the Wattled Starling, whose Zulu name **impofazana** means 'slightly dunnish in colour'.

Sibree notes two specific groups of herons among the herons of Madagascar (1892: 112). There are those with the names *vànofòtsy* and *langórobè*, both of which mean 'white heron' and there are those with the names *vànomainty* and *dangòromainty*, both of which mean 'black heron'.

Hendrickx (1944: 199) and Moreau (1940: 52) between them look at how 'red' is expressed in names for the Southern Red Bishop. Hendrickx gives the name *karwhekeru*, explaining that this is a compound of *karwhe* (a diminutive of *irwhe* 'head') and *keru* 'bright red', (i.e. 'little red-headed bird').[15] Moreau gives two names: *kitambi*, derived from *tambi*, the red garment traditionally worn by waSaamba people, and *nyamoto*, which means 'fiery'. Two Zulu names for this bird, we note, are **ibomvana** (little red thing) and **intakansinsi** (finch [which is like the] red-and-black seed of the Erythrina tree). The S. Sotho name for this bird is *thahakhube*, a compound of *thaha* (cognate of Zulu **intaka** 'finch, ♀ weaver', and with the same meaning) and *khube* (red).

An interesting name for a bird with red markings, is the S. Sotho name *jeremane* for the Red-headed Finch. Ambrose and Maphisa (1999: 60) explain:

> The common Sesotho name, derived from 'German', possibly derives from the bird's resemblance to scarlet uniforms used by German troops at the outbreak of the First World War. Basotho participated in this war as part of the Labour Corps.

The name coined for the Red-headed Finch at one of the 2013–2017 Zulu bird name workshops was **ugazini** (< *egazini* '[covered] in blood').

Turacos (previously louries) also sport a flash of red when they are flying, and Hendrickx gives the name *nduku* (red) as a generic name for turacos (1944: 205).

A straightforward descriptive is seen in *fòtsièlatra* (white-wings), given by Sibree (1892: 112) for a species of heron, but metaphor is involved in the Tanzanian name for the Tambourine Dove, known as *pugi kombe* (cup dove), according to Moreau – a reference to the pure white breast of this dove that resembles white porcelain (1940: 56). (Note here that the Tambourine Dove has often been known as the White-breasted Dove, without needing this metaphoric reference to white porcelain.) The Zulu name **ijubantendele** for the Rock Pigeon (now Speckled Pigeon) also involves metaphor: **ijuba** is a generic name for doves and pigeons and this particular pigeon is speckled like the **intendele** partridge (see Plate 11). Describing one bird in terms of another is also seen in the Zulu name **usangqwashi** for an unidentified species of lark. The name is derived from the prefix *sa–* (something 'like') and the bird name **ungqwashi** (Rufous-naped Lark) (see Plate 38).

---

15. See footnote 10 above.

Metaphor is also the basis of the Tanzanian name *mwalabu* (from the word Arab), given to the beautifully coloured Narina Trogon. Moreau notes that in Tanzania it is the Arabs who wear the most colourful robes (1940: 69). Another metaphor is the name *lunyuwa mazi*, which Moreau gives for the colourful "reed-hens" of the genera *Porphyrio* and *Porphyrula* (1940: 65). It translates as 'rainbow of the water'. A rather curious metaphor is seen in the Zulu generic name **inqelendlovu** for the flamingo. It means 'vulture of the elephant' and clearly refers to the long, hooked beak of the flamingo (= 'vulture'), a bird that is much bigger than a vulture, (i.e. an 'elephantine' version of a vulture').[16]

Personification plays a role in the S. Sotho name *mmamokete* (party-goer) for the African Hoopoe. The name is derived from the name-forming prefix *mma* (lit. 'mother of') and the noun *mokote* (feast, festival) and refers to the 'festive raiment' of the bird. One is reminded of the Zulu name for this bird given by Chadwick (1947): **umambathingubo** (the one who wears a [festive'] cloth/ blanket) (see Plate 20).[17]

Wilson gives the Chewa name *komakacoka* (beautiful when it goes) for the Little Bittern, saying "the bird appears strikingly coloured in flight, in contrast to when it is on the ground" (2011: 37).

Long tails

Moreau points out that certain bird names not only refer to distinctively different species, but sometimes cut across genera and even families if there is a feature common to two or more birds (1940: 49). He gives as example the name *chemilunda* (long tail) as a name recorded for both the Paradise Flycatcher (see Plate 19) and the Pin-tailed Whydah. He notes that in one region the Pin-tailed Whydah is called *chemilunda ntaa* (long-tail finch) to distinguish it from the Paradise Flycatcher (1942: 72).[18] These two birds have other names that refer to the very long tail: the Paradise Flycatcher is also *nyalumbwizi* (chap

---

16. As there are already a number of attested Zulu names for flamingos, the 2013–2017 Zulu bird name workshops transferred this name to the Marabou Stork, which had no previously recorded name. The Marabou is even more 'elephantine' in comparison to other storks (and flamingos and vultures), and is distinctly vulture-like around the neck and head.

17. Chadwick's brief 1947 list is the only published reference to this name, but the participants in the 2103–2017 Zulu bird name workshops were quick to recognise and accept the aptness of the name.

18. Clearly cognate with Zulu **intaka** 'finch' and Xhosa *intaka* 'bird'.

with long tail) in Tanzania, according to Moreau (1940: 58) and the Pin-tailed Whydah is *nyalundshirandshira* in the Congo, with Hendrickx explaining that the length of the tail is indicated in the repetition of the element *ndshira* 'tail' (1944: 209).

The Namaqua Dove (often known as the Long-tailed Dove) has the name *kahuji mirunda* in Tanzania, with Moreau explaining this as the generic *kahuji* 'dove' with the qualificative *mirunda* 'with tail' (1940: 56).

Metaphor comes into play in the name *làvasalàka* (long loin-cloth), with Sibree (1891a: 425) explaining this name for a species of warbler with a long tail as referring to a particular kind of loin cloth worn in Madagascar which hangs down low behind.

Crests, plumes and distinctive feathers

Compared to birds like the Long-crested Eagle, several egrets, the African Hoopoe and the Grey Go-away-bird, the Black-eyed Bulbul[19] is not notably crested. But clearly this is a matter of perception, for in Tanzania this bird has two names referring to its 'crest'. Moreau gives *shishungi* (cresty) and *kibwenzi* (little tuft) as names for the bird he identifies as the "*Pycnonotus* Geelgat".[20] The crest is certainly more obvious in the Long-crested Eagle, and Hendrickx gives the name *shamushule* (< *mushulu* 'crest') for this bird (1944: 210). Moreau has two names for this eagle, *mbilili* (long-haired) and *selunchungi* (him with the crest) (1940: 60). Wilson gives the Chewa name *tsitsimungu* (hair of God) for this bird, in reference to its crest (2011: 44).

The Speckled Mousebird, with a definite crest, is named *nshule* (hair, scalp) in the Congo, according to Hendrickx (1944: 207).

Two other references to feathers or plumes involve metaphors. Hendrickx gives the name *nymirhere* (from *mwirhero* 'ornament consisting of fringes

---

19. Now Dark-capped Bulbul.
20. It is surprising to see this Afrikaans name used so far north. This name for the various *Pycnonotus* bulbuls translates as 'yellow arse' and has long disappeared from the pages of *Roberts Birds*. Muir, however, gives it (1940: 6), and offers the alternative name *kuifkop*, another name which refers to the crest on the head. This bird must have had great salience for the earlier Dutch-speaking farmers at the Cape, for Odendal et al. (1979) give the following names for this bulbul in addition to Muir's *geelgat* and *kuifkop*: *swartkopgeelgat* (black-headed yellow-arse'), *kluitjiekorrels* (a loose translation: 'lumpy tufts of hair') and the highly accurate onomatopoeic name *pietmajol*. The latest 'official' book name, by contrast, is the rather more bland *Kaapse tiptol* (Chittenden, Davies and Weiersbye 2016: 382).

cut out of otter skins') for the Standard-winged Nightjar, a "bird with long fringes" (1944: 209). Moreau gives as one of several names for the Puffback Shrike (now the Black-backed Puffback) the name *kichajatui* (little strainer of coconut juice), saying that the name refers "to the finely cut feathers on the back" (1944: 66).

The feather-tufts on the head of the Cape Eagle-Owl give rise to the Zulu name **isikhovampondo** for this bird, with its literal meaning of 'owl [with] horns', as we saw in the introduction to this chapter.

Spots, stripes and patterns

A very straightforward name here is the name Moreau gives for the Olive Pigeon – *hua matembwe* (pigeon with spots) (1940: 64). Also using a generic name with a qualificative is *kitolondo kanga* (*kitolondo* bird with spots),[21] a name given by Moreau for a Twinspot (1940: 71). Straightforward again is the name *msilimba* (stripes), which Moreau gives for the Puffback Flycatcher (now called the Pririt Batis) (1940: 58), but metaphor comes into play in another name for this bird – the Swahili name *ndege mpunda* (bird [which is] zebra) (Moreau 1940: 58). The zebra is the quintessential 'striped thing' and its name in various languages has frequently been used to coin a name for something striped. In the 2013–2017 Zulu bird name workshops, when a name was sought for the previously unnamed Stierling's Wren-Warbler, (formerly Stierling's Barred Warbler), with its undeniable stripes, the first name that came to mind among the workshop participants was **isadube** (something like a zebra) (see Plates 22 to 25).

The S. Sotho names for the Red-knobbed Coot *mmabolesana* (< Afrik. *bles* 'blaze' + –*ana*) and *boleseboku* (< Afrik. *blesbok*) are both a metaphorical reference to the white 'blaze' on the nose of both the bird and the antelope.

Moreau (1940: 52) gives the name *kibasu* (joined together) for the Yellow Bishop "because it has one sharply defined block of yellow inserted into otherwise black plumage". Metaphor comes into the picture when the Red Bishop is named *kikoti* in Tanzania. Moreau (1940: 52) explains this name as an adoptive (loanword) from English 'coat', saying that the black part of the plumage of this bird makes it seem as it is wearing a jacket or coat.

---

21. The name *kitolondo* is applied to any small birds with much red colouring, (such as firefinches, crimson-wings, etc.).

Other birds may have a name that refers to a plumage pattern that looks like another species. Moureau (1940: 56) records for the Tambourine Dove the name *pugi kubo*, which means 'boubou dove'.[22] And he gives the name *kasuku ndogo* (little parrot) for one of the lovebird species (1940: 62).

Various other names referring to appearance

Moreau gives two names for the Puffback Shrike (Black-backed Puffback): *gongofutu* (back-swelling) and *kivuyu* (little ruffled fellow) (1940: 66). In three other names, long necks are the focus: Hendrickx (1944: 200, 202, 206) has *lubondo* (bend the neck) for a species of heron, *muhangali* (< *ku-hangalala* 'raise the neck') for the Crowned Crane and *nkongorho* (< *ku-kongoroka* 'have a long neck') for the White Stork. The S. Sotho name *mokoroane* for a stork in the genus *Ciconia* refers to the shape of its long neck. The word is a diminutive of *mokoro* 'boat' (especially the long hollowed out canoes famous in the Okavango).

Wilson (2011: 47) gives the Chewa name *chidazi* (baldness) for the Lesser Gallinule in reference to the bare skin on the forehead.

The S. Sotho name for the Spoonbill – *molomo-khaba* (< *molomo* 'mouth', 'bill' + *khaba* 'spoon') may be a calque or loan translation from the English or it may be a traditional name based on the same observation that led to the English 'spoonbill'. Related to this Sotho name for the Spoonbill is the Chewa name *namasupuni* for the same bird, given by Wilson (2011: 40). The root of this word is *supuni*, adopted from the English 'spoon'. While still on bills, we can note the Chewa name *mukokafodya* (tobacco smoker), given by Wilson (2012a: 53) as a generic name for bee-eaters. He says that this name is a reference to their "long bill like a tobacco pipe".

Another name referring to a characteristic bill is the Zulu generic name **ingcungcu** (often as **incuncu** or **incwincwi**) for sunbirds. The name is linked to the noun *ingcungcu,* which in addition to meaning 'sunbird' means any pot or basket with a long, curved neck, and any head of cattle with long curving horns in a sickle shape.

Most of the names recorded for the white-eyes (genus *Zosterops*) focus on the eyes. Malagasy is no different and Sibree (1891a: 426) records *tsàramàso*

---

22. The one with the breast as white as a porcelain cup.

(beautiful eyes) for a species of white-eye. The S. Sotho name for this group of birds is *setoma-mahloane* (little one that opens the eyes wide). The Zulu name **umehlwane** for this bird simply uses the noun *amehlo* (eyes) with a diminutive *–ane*.[23] Another Zulu name for this bird uses a form of metaphor, the name **umbicini** is also a word for a purulent discharge forming a ring around the eye of a victim, a clear reference to the distinctive ring around the eyes of white-eyes.

Thin legs are the basis of the S. Sotho name *robe-re-bese* for the White Stork.[24] The literal meaning of the name is 'break-and-burn' and refers to thin sticks used for kindling a fire.

Three names, all recorded by Moreau have particularly interesting name-narratives (1940: 64, 55).

The first is the name *mbadule*, given as a name for a small species of owl with a white 'clerical' collar around the neck. The name is derived from the word *padre*, which at first glance seems impossible, until one pronounces 'padre' with three syllables and changes the /r/ to /l/: [padre → pad[u]le → [m]badule].

Second is another name for a small owl – *shematigili* (bundle), with Moreau explaining that the owl looks very like the small bundle made out of banana fibre (*tigili*) in which *kweme* nuts are packed.

The third name is the name *ndege chai*, an urban Swahili name for the Crowned Crane. Moreau explains this as a name used by the "local boys" who have never seen a Crowned Crane (see Plate 26) in the wild, only in a compound open to the street. They named it after the picture of a Crowned Crane on a local brand of tea, hence *ndege chai* – 'bird [which is] tea'. Mlingwa (1997: 85) has *taji* (or *korongo taji*) for the Crowned Crane, the standard Swahili name for this bird.

3.2.4   Habitat

As indicated above, habitat names fill a continuum from the broader biomes, such as marsh or forest at one end, and narrower habitats, such as a particular plant, at the other. Let us start with a few examples of the wider habitats.

Sibree (1892: 118, 105) gives two examples of the wider habitat: *akóhalàhinála* (forest-cock) for the Crested Ibis and *akòholàhindràno* (water-

---

23. The suffixes *–ana* and *–ane* are interchangeable.
24. And/or other *Ciconia* storks.

cock) for a species of rail. Other forest birds include *kijogoo mburo* (cock of the woods) for the African Hoopoe (Moreau 1940: 60), *mpuji mbago* (dove of the forest) for the Tambourine Dove (Moreau 1940: 56) and *kitolondo mzitu* (forest *kitolondo*) for a particular small forest-dwelling bird (Moreau 1940: 71).

Other birds of watery habitats include *tolòkoràno* (water-cuckoo) for the White-necked Jacana (Sibree 1892: 104), *vóronòsy* (marsh-bird) for a species of heron (Sibree 1891: 245) and *vòrondríaka* (ocean-bird) for both Geoffrey's Plover and the Turnstone (Sibree 1892: 108). Staying with reference to watery habits is the S. Sotho name *mmamasionoke* for the Hamerkop. The name, which is shared by N. Sotho and Tswana, means 'the one coming from the river'. Water generally is the basis of the Zulu name **inkwalimanzi** (water-quail) for the Purple Heron, and here, as in other African languages, the quail has taken on a meaning of 'any bird' or 'bird' generally. Compare this to the Malagasy name *tolòkoràno* (water-cuckoo) for the White-necked Jacana mentioned above. Also referring to water are the Zulu names **isithandamanzi** (that which likes water) for the Woolly-necked Stork, and **umphishamanzi** (what breaks through water), a generic name for cormorants.

Still on 'onomastic waterbirds', we note the Chewa names *nkukumadzi* and *nkungamadzi*, both meaning 'chicken of the water', which Wilson gives for the Black Crake (2011: 47). Northern Sotho has the names *kgogomeetse* (chicken of the water) and *kgogonoka* (chicken of the river), both names for the Common Moorhen. Following these examples, the 2013–2017 workshops proposed **inkukhuyamanzi** (chicken of the water) for the same bird, which previously had no recorded name.

From birds of the river to birds of the river bank. Wilson gives the Chewa name *fulangombe* (bank opener) for both the Little Bee-eater and the Carmine Bee-eater (2012a: 54). The reference is to their excavating holes on river banks for their nests. Both Woodward and Woodward (1899: 88) and Austin Roberts (1940: 410) give the Zulu name **i-guondwana** for the Little Bee-eater, a name I puzzled over for years until I realised it was a simple misspelling of *igundwana* (little mouse), and that the name had been given to the bird for the mouse-like tunnels this species builds into river banks.

Birds associated with particular plants, i.e. at the other end of the habitat continuum, include *vòrombéndrana* (papyrus-bird) and *vòrombàraràta* (bamboo-bird), both names for Newton's Warbler (Sibree 1891a: 425), *kambalazi* (pigeon-pea chap) for the Brown Fly-catcher (Moreau [1940: 57] explains "because the plant is a favourite perch") and *luzila* (*Cyperus* rush) for any golden yellow weavers who favour nesting in such rushes (Moreau 1940: 71). Rushes or reeds are also seen in the S. Sotho name *khoiti-mohlaka* for the Cape Bittern. The name is derived from *khoiti* 'mole' and *mohlaka* 'reed' or 'reed-bed', the metaphor 'mole' presumably referring to something usually hidden away. Reed beds also feature in the Zulu name **ujamelumhlanga**, (what stares at reeds), which Doke and Vilakazi simply gloss as 'species of bird', but which the 2013–2017 workshop participants have assigned to the Bittern.

In between are a whole host of different habitats and favourite perching or resting places: *fandiafásika* (sand-stepper) for both the Common Sandpiper and the Yellow-bellied Wagtail (Sibree 1892: 107); *kanyamarhaza* (< *marhaza* 'stones at edge of lake or river') as a generic name for sandpipers (Hendrickx 1944: 198); *kibarabara* (roadling) for Cabanis's Bunting, often found in the road (Moreau 1940: 53); *nna-lugurhu* (hedge-dweller' < *nna* 'owner' + *lugurhu* 'hedge') for *Prinia mistacea graneri*; and *langòrovalàfa* (palm-heron) for a species of heron (Sibree 1892: 112).

Open lands are the habitats of the unidentified bird with the Zulu name **isihlalamahlangeni** (what lives in the harvested lands) and the unidentified species of a small grass-seed eating bird with the Zulu name **umafusini** (the one in the open pastures).

Undefined bushes are the basis of the S. Sotho name *palalithupa* for the Fairy Flycatcher. The name is derived from *pala* 'small thing' and *lithupa* 'bushes'. Stones, on the other hand, are where we might find the Mountain Wheatear, with the S. Sotho name *letšoanafike*, derived from *tšo* 'black + the diminutive *–ana–* + *lefike* 'stone', i.e. 'small black thing [among the] stone[s]'. The name is similar in meaning to the Zulu names **isihlalamatsheni** (what dwells among rocks) for the Cape Rock Thrush and **ikhwelematsheni** (what climbs on rocks) for the Sentinel Rock Thrush. Cliffs rather than rocks determine the habitat of the Bokmakierie in its Zulu name **uhlazalwesiwa** (green bird of the cliffs).

And finally, both metaphor and whimsy are incorporated in the Malagasy name for the Squacco Heron, given by Sibree (1892: 112) as *fiàndrivòditàtatra* (waiter-at-the-end-of-the-furrows).

3.2.5 Diet

As with habitat names, names linked to the preferred food of birds are straightforward, and there are no obvious sub-categories. Little more can be done than illustrate this category with several examples, grouping them loosely where possible.

Let us begin with a few 'ant-eaters': Moreau (1940: 68) gives the name for the Ant-Thrush as *sheisafu* and translates it as 'the lad for driver ants', and for the Rufous Chatterer he gives the name *dondolamswa* (pick up termites) (1940: 54). The name for Heuglin's Robin is given by Hendrickx as *mulyambasi*, derived from the verb *lya* 'eat' and the noun *mbasi* 'red ant' (1944: 203).[25] Also based on the verb *lya* is the word *kalyabuhuka* (what eats small insects), a name Hendrickx gives for the bird '*Spermestes cucullatus scutatus*', the Bronze Mannikin (1944: 198). Insects again are the food source of the bird *dangòrovana*, explained by Sibree as 'insect-heron' (1892: 112). Still on insects as diet, Hendrickx (1944: 203) gives the name *mnanandjutshi* (from *kunana* 'to like' + *ndjutshi* 'bee') as a generic name for bee-eaters.

The S. Sotho name for a stork in the genus *Ciconia*, presumably the White Stork, is *mokotatsie*, derived from *kota* 'peck at' + *tsie* 'locust'. Doke and Vilakazi give **indlankumbi** (what eats locusts) as the name of an unidentified stork, and again, this is presumably the White Stork. Another Zulu name, meaning 'locust-catcher', is **ugolantethe**, with Doke and Vilakazi assigning this name to a "Species of large swallow".

Both the White Stork and the Black-winged Pratincole were known as 'the locustbird' to English-speakers in colonial Natal. In fact, Alwin Haagner and Robert Ivy, in their 1923 *Sketches of South African Bird-Life*, claim that there are "five species of Locust Birds belonging to three widely divergent ornithological groups" (1923: 10). The 'true' locust bird, say Haagner and Ivy, is the Wattled Starling (*Creatophora carunculata*, with the new name *C. cinerea*), also known as the "Klein Springhaan [*sic*] Vogel" (1923: 10).[26] Then there are the two pratincoles (*Glareola pratincola* and *G. melanoptera*), both known as the "Small Locust Bird" (1923: 12). The fourth is the White Stork, and the "last of the locust birds" is the White Bellied Stork (*Abdimia abdimii*).[27] In contrast, Muir (1940: 15) only identifies two 'locust birds' in the Riversdale area of the 'Cape Province': the *Groot Sprinkaanvoël* (*Ciconia ciconia*) and the *Klein Sprinkaanvoël* (*Creatophora carunculata*), the latter being, as we have seen, Haagner and Ivy's 'true locust bird'.

---

25. Heuglin's Robin is now the White-browed Robin-Chat.
26. This should be *sprinkaan*, Dutch and Afrikaans for 'locust'.
27. Today Abdim's Stork *Ciconia abdimii*.

There used to be two species of oxpecker in KwaZulu-Natal (KZN): the Yellow-billed Oxpecker (*Buphagus africanus*) and the Red-billed Oxpecker (*Buphagus erythrorhynchus*), birds that sit on the backs of cattle and buffalo eating ectoparasites. Between them, they shared the names **ihlalankomo** (what sits on a beast) and **ihlalanyathi** (what sits on a buffalo). According to Roger Porter,[28] however, the Yellow-billed Oxpecker is extinct in KZN, and so the Red-billed Oxpecker has these two names all to itself.

Shifting now to fish, we note the simple metaphor in *muvuzi* or *luvuzi* (fisherman), given by Moreau (1940: 54) as a name for the cormorant and (1940: 60) as a generic name for herons. In similar fashion, the S. Sotho *seinoli* (what takes things out of the water) is a generic name for kingfishers, but it is also used for cormorants. Hendrickx gives the base of the Nyabongo name *nyamuloba* as the verb *kuloba* 'to fish' and assigns it specifically to the Reed Cormorant (1944: 209). Herons come back into the picture again with the Malagasy name *vorompatsa* (shrimp-bird) assigned by Sibree to one of several species of Madagascan herons (1892: 112). Also taking things out of the water is the Spoonbill, with the Zulu name **isixulamasele** meaning 'what catches frogs'. Wilson gives the following Chewa generic names for the pelican: *bvuo, bvuobvuo, bvuwe* and *bvuwo*, saying that all are derived from the verb *ku-bvuula* (scoop fish from the water) (2011: 37).

In between the insects, the fish and the shrimps are a variety of other foodstuffs. Some are perhaps unexpected, as in the names *bata mchikichi* (palm-oil duck) and *shempule* (chap for pounded maize), both names Moreau (1940: 57) assigns to the Vulturine Fish-Eagle [Palm-nut Vulture]. More expected, perhaps, is the monkey-based diet of the Crowned Hawk-Eagle, expressed in the two names *kumbakima* (grip monkey) and *kimakima* (collect monkey), which Moreau (1940: 60) gives for this bird. The S. Sotho name for Verreaux's Eagle is *mojalipela* (< *ja* 'eat' + *lipela* 'dassies') and this is exactly reflected in the Zulu name **umdlambila** (what eats dassies) given for an unidentified 'species of hawk-eagle', but just as likely to be Verreaux's Eagle, the more so as Muir (1940: 4) identifies the Afrikaans folk name *dassievanger* (dassie catcher) as referring to *Aquila verreauxi*. Also expected are names that identify the smaller raptors as eaters of rodents. Godfrey (1941: 31) gives the Xhosa names *umdlampuku*

---

28. Roger Porter is a retired ecological scientist previously with the erstwhile Natal Parks Board and KZN Wildlife. He is an avid and highly knowledgeable birder.

The Malachite Sunbird (**uhlazazana**) is typical of the sunbirds or 'honeysuckers'. © Adrian Koopman

(mouse-eater) and *unoxwil'impuku* (mouse-snatcher) for the Black-shouldered Kite, and Wilson (2012b: 43) gives the Tumbuka and Khonde name *katumbulambewa* (disemboweller of mice) for the Southern Fiscal (formerly the Fiscal Shrike).

The specialised diet of honeysuckers is expressed in three names: *mununi* (< *kununa* 'to suck') given by Hendrickx (1944: 203) as a generic name for honeysuckers; *kihunguluwa* (little seek-flower) given by Moreau (1940: 67), also as a generic name for honeysuckers; and *kinunerako* (little kiss-banana-flower), a name Moreau (1940: 67) assigns specifically to the Red-chested Sunbird. Wilson (2012b: 41) gives the Chewa and Yao name *kadyanauwo* (flower eater) as a generic name for sunbirds.

Plant material is a food source for the following six birds:
- The Button Quail, whose name Moreau (1940: 65) records as *matumandago* (peck *ndago)*, explaining that *ndago* is a *Cyperus* rush with a fleshy root.
- The African Yellow White-eye, named *ndyabusogya* and explained by Hendrickx (1944: 205) as derived from *kulya* 'to eat' + *usogya* 'pistil of the *Pentas coccinea* plant'.
- A species of dove or pigeon with the name *vòronodábo*, explained by Sibree (1891b: 559) as 'what eats the fruit of the *adaba* tree'.

- The Coqui Francolin, with its Zulu name **unongcangiyana** based on the noun *ingcangiyana* (species of edible tuber of sorrel).
- A species of Madagascan rail, named very simply by Sibree (1892: 105) as *fangàlatróvy* (yam-thief).
- The Brown-headed (Cape) Parrot, for which Wilson (2012a: 47) gives the name *chinkwemaula* (fond of *maula* fruits' – the fruits of the *maula* tree (*Parinari* sp.).

Matjila gives the name *bibing* for the Lappet-faced Vulture, saying that "the word *bibing* is derived from the word *sebibi* (2015: 105). *Sebibi* is a place where animals have died". This name thus refers to both diet and habitat.

The S. Sotho generic name for mousebirds does not indicate what they eat but rather how they eat. The name *fariki* is derived from the Afrikaans *vark* (pig) and it is said to refer to their greedy feeding habits. Another name that suggests how a bird eats rather than what it eats is the Zulu name **inkotha**, a generic name for bee-eaters: the verb *khotha* can mean to lick and also to scoop up food. Wilson gives the Chewa name *chithathale* (plucking) for the Red-billed Teal, saying that the name refers to the way this bird eats (2011: 41). Presumably the bird is 'plucking' waterweeds and similar vegetation. More specific is the Chewa name *chitotola* (plucker of feathers), which Wilson (2011: 43) gives for goshawks and sparrowhawks generally, but specifically to the Lizard Buzzard. This suggests that these birds prey on other, smaller birds. More specific in its avian diet is the Martial Eagle, for which Wilson gives the Chewa name *chiombankhanga* (firer at guinea fowls) (2011: 44).

Finally, although there is a plant in the name, the following bird is really trying to catch and eat bats: Moreau explains the name *doholantalambo* (open the banana screws) as something the African Harrier-Hawk (Gymnogene) has to do to get at the roosting bats inside (1940: 60). This bird, with a tarsal joint that bends forwards, backwards and sideways, is able to reach into holes in tree trunks (and other cavities) to get at young chicks and other foods. Wilson gives the following Chewa names for this bird: *dzanjamphako* and *manjamphako*, both meaning 'hand in the hole', and *kamwendomphalo* and *mwendophikalo*, both meaning 'leg in the hole' (2011: 42). When the 2013–2017 Zulu bird name workshop participants had to find a name for this previously unnamed species, the focus was on the same habit and same attribute of the bird, giving rise to the Zulu name **ujikanyawo** (what turns the foot).

3.2.6 General behaviour

Some birds have names that suggest that they are tricky in one way or another. Moreau gives the name (*ndege*) *lubozi* for the [Eastern] Nicator,[29] translating it as 'the loony' and explaining "because when the nest is approached it puts up an 'injury-feigning' performance like a mad thing" (1940: 63). Presumably Moreau's other name for the nicator – *babasi* (simpleton) – is given on the same basis. A similarly tricky bird is the [African] Rail (*Rallus caerulescens*, formerly *R. gularis*), with the name *angòly*, which according to Sibree means 'artifice' or 'deceit' and refers to the tricks the bird plays to avoid capture (1892: 105). In a similar vein, Hendrickx records the name *mugeke* (cunning, clever) for a sandpiper species said to be very difficult to be caught in a trap (1944: 202). On the other hand, the name *sokelo* (tired fellow) for the Black-headed [Black-crowned] Tchagra suggests the opposite to these tricky, deceitful trap-avoiding birds, as Moreau explains the name as meaning that the tchagra is supposed to be caught easily if you chase him (1940: 68).

Not quite referring to a behavioural trait, but certainly what Moreau would have regarded as referring to a 'personal attribute' in his semantic classification are those names that refer to the odour of the bird. Zulu has the word **unukani** (what do you smell of?) for the supposedly odiferous woodhoopoe;[30] Moreau (1940: 61) records a similar name for this bird, translating *kinuka* as 'little stinker'. Another name recorded for the same bird by Moreau is *machema*, translated simply as 'stench'. Hendrickx (1944: 208) gives the name of the Blue-headed Coucal as *nyabanya*, explaining that it is derived from the verb *kunya* 'to expel excrement', and adding "a smelly bird" in case the reference is not clear.

The name *nyunda*, derived from *kuyunda* 'to look out for prey', which is recorded by Hendrickx (1944: 210) for the African Fish Eagle is clear enough, but what does one make of the name *ndiyembili* (picker up of two things at once)? Moreau (1940: 53) gives this name for a *Serinus*-type canary.

Wilson gives the Chewa name *kamlenje* (covering) and the Yao name *chivundikile* (coverer) for the Black Egret, saying that the former name refers

---

29. *Ndege* is the Swahili for 'bird', so this is literally the '*lubozi* bird', sometimes just called *lubozi*.
30. Roger Porter (personal communication) says this is a misconception, it is the African Hoopoe which has a notoriously smelly nest.

to the bird spreading its wings and covering its head, and the latter name refers to the wings being partly spread, shading the water, when the bird is feeding (2011: 38).

Muir records the names *klaasperdewagte*r (Nicholas the horse watcher) for the [African] Stonechat, and *skaapwagter* (sheep watcher) for the Capped Wheatear (1940: 8).[31] The database on African bird names also contributes names for those birds regarded if not actually as herders of domestic stock, then at least strongly associated with them. For example, from Moreau comes the name *kadima mbuzi* (goat-herd) for a species of wagtail (1940: 69). Sibree gives the name *kitàndry* (the watchman) and the name *langòroaomby* (ox-heron) for a species of cattle egret (1892: 111). Similar to this is *vórontàniómby* (attendant on cattle), a name for the Cuckoo-Shrike given by Sibree (1891a: 423). He also has *fitìlibèngy* (goat-watchman) for one of several species of ibis (1892: 118). Madagascan ibises seem to have a very strong link with goats, Sibree has yet another example – the name *manàranòsy* (goat-ibis) for an ibis species (1892: 118). One is reminded of the Zulu name **ilindankomo** (what follows cattle) for a species of egret. Another Zulu generic name for the egret is **ilanda** (the one that follows), also a reference to the way this bird follows cattle, attracted by the insects disturbed by the cattle's passage.

The Drakensberg Rockjumper is also perceived as a herder by S. Sotho speakers, but not of domestic stock, but rather of dassies: the name *molisalipela* means 'herder of dassies'. The noun *molisa* (herder) is found in the plural form *balisa* in the S. Sotho name *phakoe-ea-balisa* (falcon of herders) for the Lanner Falcon, with the name reflecting that hunters drive out birds with sticks and the falcons then catch them.

Three different kinds of behavioural patterns are seen in the Zulu names **indwazela**, **inkwalitwetwe**, and **umzwangedwa**. The first – **indwazela** – is a name for the Brown-hooded Kingfisher and it has an underlying meaning of the one that loiters around while staring vacantly into the distance.[32] The second name – **inkwalitwetwe** – literally means 'nervous pheasant' and it is assigned to the Red-legged Pheasant (Doke and Vilakazi 1958: 420).[33] The

---

31. See also an article by Nancy Jacobs on birds as cattle and goat herders – (Jacobs 2015).
32. At the July 2017 Zulu bird name workshop, this name was selected to be the generic name for kingfishers. See the Appendix on Zulu kingfisher names at the end of Chapter 12.
33. This 'Red-legged Pheasant' is likely to be the Red-necked Spurfowl of today.

third name is the name of an unidentified bird that lives in a solitary manner; the name **umzwangedwa** essentially means 'loneliness'.

I end this section with three names that are semantically unrelated: the S. Sotho name *khoholira*, the Nyabongo name *bunyiro* and the Madagascan name *fangàlimótivòay*.

Literally meaning 'chicken of the enemies', *khoholira* refers to the Spotted Thick-knee (Spotted Dikkop) and it is said that this bird gives warning of approaching enemies.

The name *bunyiro* is a generic name for owls, notoriously short-sighted in the daytime, and Hendrickx explains the name as from a verb meaning 'be blind' (1944: 196).

My last example here is the name *fangàlimótivòay*, which Sibree gives for one of several species of heron, with an underlying meaning of 'crocodile's eye-cleaner' (1892: 112).

3.2.7 Motion

Approximately half of the 'motion-names' in the database are concerned with flight. We will look at these first, and then look at other names that refer to the way in which a particular bird walks, dives, swims and so on.

Flight

We can begin with those birds that float lightly on the wind, and indeed in our first example, the name *hilembe-lembe* is derived from the verb *kulembera* 'float lightly on the wind'. Hendrickx gives this as a generic name for swallows and swifts (1944: 196). The Black-shouldered Kite is the leading hoverer on the wind, with no less than four names confirming this. Moreau gives three of these names: *kinyamhuwi* (little hoverer), *kipupwi* (little whirlwind)[34] and *msoka* (little slacker) "because it hangs about in one place in the air" (1940: 62). Hendrickx gives the fourth – *kadekere* – derived from *madekere* (leaves of the taro plant) (1944: 197). Hendrickx explains the name as "[t]he hovering of the bird is similar to the movement of the leaves of the taro plant when a light breeze moves them".

---

34. Cf. the Zulu praise name for the Yellow-billed Kite: **uNhloyile kaNgelegele** (Kite, son of Whirlwind').

The name **umathebethebana** is used generically for all the different species of kestrel found in KwaZulu-Natal.
© Adrian Koopman

Wilson gives two hovering birds from Malawi, his first name is the Tumbuka name *kamkubezi* (hoverer) for the Black-shouldered Kite (2011: 41) and the second is the Chewa name *nankapakapa* (hoverer) for the Pied Kingfisher (2012a: 53).

Other birds are far more swift and active than the hovering kites. Sibree (1891a: 425) records *firioka* (rapid, darting flight) as the name of Crossley's Warbler and the Madagascar Swallow, while Hendrickx (1944: 197) gives the name *kadurha* (from *kudurha* 'dive rapidly to ground') as a generic name for all raptors. Rapid movement is also seen in the names *kiriondànitra* (sky-rusher) and *firìotsàndro* (day-rusher), which Sibree (1891a: 430) gives for one of the Madagascan swallows. Still on the flight of swallows, we can note the Khonde name *kaweleweswa* (turning this way and that) given by Wilson as a generic name for swallows in Malawi (2012a: 58).

These names all suggest rapid, darting flight. In contrast, the Red-eyed Dove has what Wilson (2012a: 46) refers to as a "slow, gliding display flight", and for which he gives the Chewa name *chikupe* (from the verb *ku-kupa* 'to fly slowly').

Then there are birds that can fly for a considerable distance. Hendrickx (1944: 206) gives the name *nkule*, derived from *kule-kule* 'very far away', for the African Green Pigeon, saying that it can fly very far, and he gives the name *mpirahira* (from *mpira* 'far away') for the "Common Quail that flies far away when disturbed" (1944: 201).

Matjila gives the Tswana name *petleke* for the Bateleur Eagle, saying that the name relates to its mode of flight (2015: 106). He explains that the bird spreads its wings sideways just like the outstretched horns of cattle (*go di lema petleke*).

In Zulu, swifts are perceived as flying very close to cattle, and this is reflected in the generic name **ihlabankomo** (what stabs the beast). Another generic name is **ijiyankomo**, based on the verb *jiya* (be of an age to look after cattle), often found in its shortened form **ijankomo** (see Plate 27).

Finally, there is the name *sidintsidina*, which Sibree (1891a: 430) gives for one of the Madagascan swallows and one of the swifts. He explains the name as "flier [par excellence]".

When birds are not flying, they are moving in other ways. Let us look at some other 'names of motion'.

Other names expressing motion

Some names refer to ways of walking or running. Sibree gives the name *kìborànto* (far-running quail) for the Common Sandpiper and for Geoffrey's Plover (1892: 107). Geoffrey's Plover also has the name *vìkyvìky* (from a verb meaning 'to run, to leap'), and Sibree says it shares this name with the Turnstone (1892: 108). Not quite as fast-running or far-running is the Cape Raven, which Hendrickx assigns the name *tshihungwe* (from the verb *kuhunga* 'swing, dawdle, rock') in allusion to the way in which this bird walks (1944: 211). Just as slow a walker is a pipit of the genus *Anthus*, for which Moreau gives the name *kigendagenda* (little dawdler) (1940: 64). Doke and Vilakazi give the name **unondwayiza** for the "lily-trotter *Aetophilornis* [*Actophilornis*] *africanus*". The name is based on the verb *dwayiza* 'walk with long strides'. UNondwayiza is commonly found as a Zulu nickname, and is often given to white people with long legs. It is the nickname of the explorer Kingsley Holgate.[35] Waterbirds may dive, dip, plunge and float, and this is reflected in names like *hosétrika* and *hoètrika*, both meaning 'to dip', 'to plunge' and both assigned by Sibreee to the Purple Water Hen (1892: 105). The S. Sotho name *thoboloko* for the Dabchick or Little Grebe is derived from a verb meaning reappearing, (i.e. after diving). The Zulu name **inhlunuyamanzi** for the previously named Natal Kingfisher (African Pygmy Kingfisher) is a metaphor, based on *inhlunu* (a vulgar term for

---

35. See Koopman (2014b: 61).

'vulva') and *yamanzi* (of the water), referring to the bird plunging into water. Semantically similar, but without the use of vulgar metaphor, is the name that Hendrickx gives to the Lesser Pied Kingfisher: *nyamundubike* (< *kudubika* 'to plunge') (1944: 209). Floating on water, rather than plunging into it, is reflected in the generic name *nguhe* (light, buoyant) that Hendrickx gives for gulls (1944: 206).

And then a miscellany of movement: *kambanda* (little croucher) for the Flappet Lark (Moreau, 1940: 62); *kidusawarumbi* (startle hunters) for the Lemon Dove "because of its sudden rising from the forest floor" (Moreau, 1940: 56); *kumbanti* (grip-tree) for the African Harrier-Hawk or Gymnogene (Moreau 1940: 60); *lubaka* (take rapidly) for the Yellow-billed Kite (Hendrickx 1944: 200) and *tsìkoròvana* (from *rovana* 'movement en masse') for "a species of bulbul, which occurs in great numbers" (Sibree 1891a: 426).

Another type of movement is crouching and hiding. The S. Sotho name *semphoma* for Rudd's Lark means 'the one that likes to deceive' and, according to Meyer, refers to this bird as crouching down and relying on its natural camouflage when wishing to evade a predator. In similar vein, the Zulu name **isigwaca** is a generic name for quails and is derived from the ideophone *gwáca* meaning 'of lying low, of lying flat in hiding'. Nightjars also tend to lie low and flat, and Wilson (2012a: 51) gives two names from Malawi, the Chewa name *nyakalibwato* (< *ku-bwata* 'lie flat, motionless') and the Tumbuka name *bwabwalala* (< *ku-bwantalala* 'lie flat, motionless'), both used generically for nightjars.

To end this section, we note the simple name *kibikula* (wagger) for the wagtail. Moreau (1940: 69) explains that the wagtail *Motacilla aguimp* (African Pied Wagtail) is distinguished as *kibikula nyasi* (wagger of the thatch) and *Motacilla clara* (Mountain Wagtail) is *kibikula lamabwe* (wagger of the stones).

3.2.8 Belief names

These are the names given to birds according to certain cultural beliefs: beliefs that a particular bird is an evil omen, that another is a sign of impending rain, and that a third is announcing the start of spring. We start with the heralds:

Muir gives *dagbreker* (daybreaker) as the Afrikaans name for the Familiar Chat (*Cercomela familiaris*) (1940: 4), and this is echoed in Hendrickx's name *kabikabutshe* (from *kubika* 'to sing' + *butshe* '[at] dawn'). He gives this name to Heuglin's Robin, explaining that this is a bird that sings at daybreak (1944:

197). From Moreau comes the name *langazua* (call-sun) for one of the larger bee-eaters (1940: 52), and from Sibree we have the name *manàboandràno* (that which celebrates the day) for the Common Sandpiper (1892: 107).

There are many birds whose calls are believed to presage rain, and examples can be found worldwide. I discuss these in great detail in Chapter 10. The following are examples:

Langavura (call-rain) is given by Moreau (1940: 55) for the Black-and-White Cuckoo (genus *Clamator*) and he says it is "especially noisy at the break of the rains", and he gives *semchocho* (deluge chap) for the Red-chested Cuckoo, saying that this bird is "thought to call especially in very heavy rain" (1940: 56). The Zulu name **ihlolamvula** (also as **inhlolamvula**) is a generic name for both swifts and martins, and translates as 'what predicts rain'. Wilson gives the Chewa name *mvulawe* (the rain is coming) for the Striped Crested Cuckoo and the Yao name *mkoka* (drawer of rain) for the Jacobin Cuckoo (2012a: 48). He also gives the Chewa name *mwadonta* (dropping of rain) for the Red-chested Cuckoo (2012a: 49).

In almost all traditional societies there are specific taboos related to birds: there are birds that must not be eaten, birds that must not be killed, birds that must not be pointed at, and birds that may not be named. Again, I discuss these taboos in detail in Chapter 10. Here are a few examples of bird names that refer to these taboos.

Moreau says of the Red-billed Shrike that "they are birds on no account to be molested" (1940: 66). This protected status is reflected in their names *mazanamulungu* and *wanamlungu*, both of which mean 'son of God'. Hendrickx points out that in the Lake Kivu area of the Congo there is a taboo against the killing and eating of the cattle egret (1944: 210). This is seen in the name *nyange* (< *kuyanga* 'non-edible bird or thing'). Sibree tells of a species of dove which was named *tsiázotonònina* (unspeakable) "because its more common name had become tabooed or sacred through having formed part of the name of one of their chiefs" (1891b: 559). In Zulu terms, we would say that *tsiázotonònina* is a *hlonipha* word. A different kind of taboo is found in the Zulu name **ingqangqamathumba** (open abscess) sometimes given to both the hadeda and to certain larks because it is believed that a person who mocks them will break out in abscesses.

In most cultures around the world, night-birds accumulate layers of negativity, leading to them being associated with evil spirits, bad luck and death. Owls and nightjars are particularly affected in this way, and this is reflected in the names they are given.

Moreau gives two names for the nightjar that manifest these negative associations: *upapasa* "groper (in the dark)" and *mkatasanda* (cut-shroud) (1940: 63). He notes that in Tanzania the nightjar is regarded as a bird of ill-omen. His name for medium-sized owls is not quite as clear. He gives the name *babewatoto* (father of children), saying that the name is "an honorific bestowed on a bird that is much feared as an evil influence on children" (Moreau, 1940: 63) but the lexical meaning given does not tie in with the semantic explanation.

Moreau gives the name *sekikoko* for an unidentified bird, saying, "This is the name of a bad spirit, especially troublesome to women (1940: 55). The bird, which calls from the forest canopy, often at night, is known only as a voice, not by sight." A bird that is only known as a disembodied voice, especially when calling at night, almost inevitably will be connected with dark and sinister forces. A more vague allusion to such dark and sinister forces is seen in the name Sibree gives to the Fork-tailed Drongo – *andóvy* that "seems to come from a root *dóvy* 'enemy', probably from some superstition connected with it" (1891a: 422).

Finally, not all bird names linked to cultural beliefs are dark and sinister. Moreau (1940: 66) says that the call of the Puff-back Shrike (Black-backed Puffback) is a "good omen to the wayfarer", and this is reflected in the name *mkaribisha mgeni* (welcomer of the stranger).

3.2.9 Nesting

This semantic category has the fewest bird names apart from the 'Other/Miscellaneous' category that follows. Only nine names were recorded in the database.

Wilson gives the Tumbuka name *katawa* for the Hamerkop, saying that the name is derived from the verb *ku-tawa* (to weave) in reference to the bird's elaborate nest (2011: 39).

Hendrickx gives the Nyabongo name *luvungabwasi* for the Pin-tailed Whydah (1944: 201). The name, derived from *kuvunga* 'to ruffle' and *bwasi* 'type of leaf', refers to this bird laying its eggs in the nest of the waxbill, made of *bwasi*-type leaves.

Bee-eaters, as is commonly known, nest in holes they have excavated in banks. Moreau records *mchimbamchanga* (dig-soil) as a name for the smaller species of bee-eater found in Tanzania (1940: 71). I am not familiar with any raptor that excavates a hole for a nest, but Doke and Vilakazi give the Zulu name **isigumbamphalo** (what excavates a hole) for an unidentified species of hawk that catches mice.[36] This is one of two Zulu bird names found that refers to nest-building. The Cape Sparrow builds a huge, untidy nest, and is named in Zulu **undlunkulu**, derived from *indlu* (house, hut) and *[e]nkulu* (which is big) (see Plate 35).

Moreau gives the name *selukungu* "chap for *kweme* plant" (Oysternut/ *Telfairea pedata*) for one of the yellow weavers, "because the bird builds its nest with stout corkscrew tendrils typically developed by this climbing plant" (1940: 71). Moreau records two names for the Grosbeak Weaver: *shemakhome* (him with the strainer), a reference to the fine weaving of the nest, and *suwagulamilanzi* (strip-reed), a reference to the material used by this bird for building its nests (1940: 59). Still on the theme of nesting material, we note the name *topolamavi* (pick up dung). Moreau (1940: 70) says that the Morning Warbler is given this name because it is said to incorporate dung in its nest-building.

Our last two examples look at where the nest is usually found, rather than emphasising building materials, with Moreau (1940: 68) giving the name of the Black-headed [Black-crowned] Tchagra as *tagilamkoko* "lay (eggs) in dense thicket", and Wilson giving the Yao name for the African Palm Swift (*Cypsiurus parvus*) as *chiwalewale*, derived from *chiwale* (*Raphia*) (2012a: 52).

3.2.10   Other

There is inevitably a category 'other' or 'miscellaneous' whenever categorisation such as this is attempted. The following names recorded in the database either simply did not fit conveniently into one of the previous eight categories, or insufficient information was given to assign them to one. In a few cases, the name was lexically transparent, but semantically opaque. An example of this is the name *mchunga-mguruve*, given by Moreau (1940: 67) for the Stone-Curlew (the Dikkop or Thick-knee). He translates this name as 'pig-driver', so the lexical elements in the compound name are clear, but does not explain what the driving of pigs has to do with this or any other bird.[37] The only place for such a name is the category 'other'. The Zulu name for this bird (or its cousin, the Cape Dikkop), **umbangaqhwa** is lexically quite transparent – it is derived from *banga* 'cause [something to happen]' and *iqhwa* 'frost', and Doke and Vilakazi (1958: 66) stress that 'causer of frost' is the underlying meaning

---

36. Unless of course this unidentified hawk is digging a hole for some other purposes, perhaps burying the mice that it has caught for later use, in the way a dog buries a bone.
37. The second part of the name – *mguruve* – is clearly cognate with Zulu *ingulube* (pig).

of the name. They do not, however, explain why the bird has been given this name, and for many years this name was semantically opaque to me. It was only when I was working with Roger Porter in 2017 on the output of the first three of the 2013–2017 Zulu bird name workshops and told him the meanings of the elements of this name, that he made the reason for the name clear: when threatened, this bird does not fly away, but crouches low, hoping, with its 'camouflage plumage' to be mistaken for a weathered rock. And with the white flecks in its plumage, it looks just like a frost-covered rock on a wintery morning (see Plate 28).

Zulu has more bird names that are lexically transparent but semantically obscure. The name **imvunduna** for the Crested Barbet appears to be based on *imvu* 'sheep' and *induna* 'headman', which does not make much sense; the name for a small, unidentified marine bird is **intonjana** (small girl) for no apparent reason; another unidentified species of small bird that feigns injury to draw people away from its nest is named **isibulalambiza** (what breaks the clay pot). In all these examples the underlying lexical elements are clear, but why the birds should be so named is not at all clear. On the other hand, the name **unobulongwe** for Burchell's Courser, based on *ubulongwe* (fresh cattle dung) at first appears obscure, but it could be argued that this bird is attracted to the flies that themselves are attracted to fresh dung.

Further examples of the type 'lexically transparent but semantically opaque' are *kishunde mabuwa* for the Black-headed [Black-crowned] Tchagra and *lufite* for the Pin-tailed Whydah. Moreau translates the first as 'dead maize stalks' and the 'second as 'thin building pole' but for neither does he explain why the birds should be so called (1940: 68, 72).

Even more obscure, perhaps, is Sibree's name *fitosívy* for the Glossy Ibis. The name, says Sibree (1892: 118) means 'seven-nine' but the reader is left in total ignorance of why a Glossy Ibis should be 'seven-nine' in Malagasy. Not much better is *neruvuta* (oily chap), given by Moreau (1940: 68) for the Malachite Sunbird.

For our last three examples, however, the underlying meaning is clear and its link to the bird is clear.

The name *mukugwhe* (something good to eat), given by Hendrickx for the Blue-headed Coucal (1944: 203), is the only name in the database that refers to a bird as an item of food itself. Nancy Jacobs has written that in most traditional African societies, wild birds were not important as a source of food (2016: 52), and perhaps this why this role of birds vis-à-vis humans is reflected in only one name.

Sibree has the name *kitòry* (proclaimer, accuser) for the Turnstone, saying that it is a bird that warns others of approaching danger (1892: 108).

And then finally, humour is the explanatory element in the Zulu name **indlovuyenduna**, which refers to a species of waxbill similar to the common waxbill. The name means 'large male elephant'. The same humour is found in the Zulu name of the African Pygmy Kingfisher – **ungangomfula**, which means 'as large as the river'.

4 The meaning of the new Zulu bird names

4.1 INTRODUCTION

In the previous chapter I looked at the underlying meanings of a number of bird names from different parts of sub-Saharan Africa, adding names from Madagascar into the mix. A few Zulu bird names were also included. In the earlier part of the chapter, a 'Manhattan' graph showed that Zulu bird names were typical of the sub-Saharan African bird names in terms of their overall semantic profile. The Zulu names in the previous chapter were all the 'old' names, by which I mean the established names that have long been recorded in the literature and that are by and large the Zulu names given in the index of Chittenden's 2007 *Roberts Bird Guide* (first edition).

In Chapter 12 of this book I describe the various linguistic strategies adopted in the 2013–2017 Zulu bird name workshops that led to the adapting and extending of these 'old' Zulu names, as well as the coining of the new names. This chapter – Chapter 4 – continues from the previous chapter, namely to examine the underlying meanings of bird names and seeing how these meanings link directly to various aspects of the birds: their appearance, their vocalisation, their habitat, etc. The difference is that in this chapter I am looking at the new Zulu names.

Although I discuss a number of different linguistic strategies in Chapter 12, in essence there were only two strategies used: firstly, add something to an existing generic so as to distinguish a particular species from others in a cluster of species all sharing the same generic name, or, secondly, create something entirely new. The adding of something to a generic is typical of the way in which discrete English and Afrikaans names have been created for birds. A glance at any bird guide produced over the last 75 years will show sets like:

Blue Swallow – *Blouswael*
Grey-rumped Swallow – *Gryskruisswael*
Barn Swallow/European Swallow – *Europese Swael*
Angolan Swallow – *Angolaswael*
White-throated Swallow – *Witkeelswael*
Wire-tailed Swallow – *Draadstertswael*
Pearl-breasted Swallow – *Pêrelborsswael*

This type of naming individual species in a cluster of birds was indeed used in the 2013–2017 Zulu bird name workshops and a glance at Table 2.2 on page 27 (Chapter 2) shows the results for several species of eagle. Five of these are:

ukhozolumnyama (black eagle) – Black Eagle (now Verreaux's eagle)
ukhozimuhlwa (termite eagle) – Steppe Eagle
ukhozolumabala (spotted eagle) – Lesser Spotted Eagle
ukhozolusisila (tailed eagle) – Wahlberg's Eagle
ukhozolumidwayidwa (streaked eagle) – African Hawk-Eagle

It can be seen that Zulu uses **ukhozi** (eagle) in each name, just as English uses 'eagle'. Some of the Zulu names may appear to be translations of the English names, just as the Afrikaans names for the seven swallows above are translations of the English names, but that is not necessarily always the case. The Black Eagle is named **ukhozolumnyama** (black eagle) in Zulu, simply because it <u>is</u> a black eagle. Its blackness is its most obvious feature. Note that in the case of the English name Steppe Eagle, the differentiating qualifier is a habitat reference, whereas in the Zulu **ukhozimuhlwa** (termite eagle) the qualifier is a dietary reference.

In the analysis of the underlying meanings of the 'new' Zulu bird names, which constitutes the major part of this chapter, I will speak almost exclusively of the newly coined names, rather than of those names where an existing generic name has been extended by a qualifier.

It must be noted that the Zulu-speaking participants of the 2013–2017 Zulu bird name workshops were guided by their own sense of what was the most important distinguishing feature of any particular species. As with the Black Eagle above, frequently this notion of what is salient coincides with the underlying rationale of the English name, not to mention the scientific name. The shovel-shaped beak of the Cape Shoveler is clearly a distinguishing feature, hence the English name. And hence also the Zulu name **unofosholo**, based on the Zulu adoptive *ifosholo* (shovel). The bony growth on the bill of the Knob-billed Duck cannot be ignored either, and gives rise to the Zulu name **unosimila**, based on the Zulu noun *isimila* (growth).

A final point before proceeding with the semantic analysis of new Zulu names for birds: many of the previously recorded Zulu names of birds are completely opaque as regards their underlying meanings. Why is an ostrich called an **intshe** and a Blue Crane an **indwa**? Apart from referring to the birds themselves, what do these words mean? The same applies to words such as 'duck' and 'eagle': apart from referring to a loose cluster of similar birds, what do these words actually mean? Some English bird names have a well-recorded etymological history, like the word 'swallow', from the Old English *swealwe*, related to the Frisian *swale*, the Old Norse *svala*, and Old High German *swalwa*. But even so, what was the underlying meaning of these earlier forms of the word?

In the case of the names newly coined at the 2013–2017 workshops, the underlying meaning is always crystal clear, and having been present at these workshops and hearing the discussions and debates that preceded the final choice of each name, I am able to give these underlying meanings, and the reasons for creating the names, with absolute confidence.

4.2   COMPARATIVE SEMANTIC PROFILES OF AFRICAN BIRD NAMES, OLD ZULU BIRD NAMES AND THE NEW ZULU BIRD NAMES

The purpose of comparing the new Zulu names with old Zulu names (as well as sub-Saharan African names generally) is to show that even though the 2013–2017 Zulu bird name workshops were a modern phenomenon, where each participant had his or her own cell phone and laptop, and access to a data projector and the Roberts VII Multimedia program, the final choice of names, as revealed in their overall semantic profile, was typically African.

The following 'Manhattan' graphs show the distribution of the semantic categories:
Song, Appearance, Habitat, Diet, General Behaviour, Motion, Belief-orientated, Nesting and Other for the following three groups of bird names:

1. African names – representing sub-Saharan Africa: Nyabongo of the Lake Kivu area of the Congo, Malagasy of Madagascar, various Tanzanian dialects and Southern Sotho.
2. Traditional ('old') Zulu names.
3. Coined ('new') Zulu names from the 2013–2017 workshops.

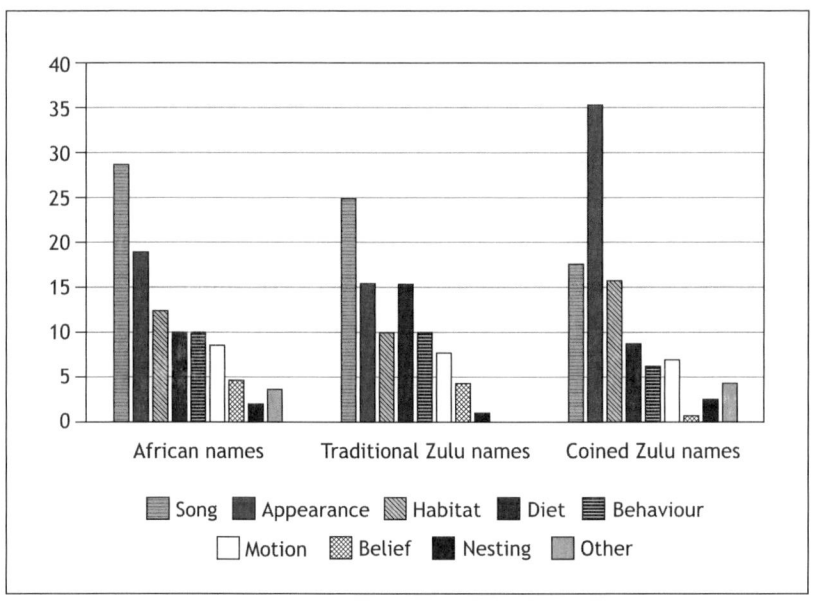

Semantic profiles of (sub-Saharan) African names, traditional, 'old' Zulu names, and coined, 'new' Zulu names.

The traditional ('old') Zulu names follow fairly closely the semantic profiles of the 'African' names represented by the four languages (or language groups) mentioned above. The only noticeable differences are:

- A greater emphasis on diet in the traditional Zulu corpus.
- A slightly smaller proportion of names linked to traditional beliefs, although the percentages involved here are so small that this is not significant.

When the modern, coined Zulu names are compared to the same category in the other two profiles, the greatest change is seen in the category 'Appearance'. Names referring to what the bird looks like show a far higher percentage in the coined Zulu names. The only conclusions that can be drawn here are:

- The Zulu-speaking participants, although well-versed in Zulu traditional bird lore, were all linked to birdwatching either in specialist avitourism safaris or as game guides in KwaZulu-Natal (KZN) game reserves. Birders today do most of their identification through visual appearance, and identifying (and ticking off) birds by song alone is sometimes frowned upon by serious birders.

- In the natural environment, many birds remain hidden while their song can still be clearly heard. In the closed environment of the five workshops of 2013–2017, each and every bird discussed was called up on screen, where they could be seen in various poses and in various lights. Their songs and calls, their habits and habitats were also called up, but each bird was revealed visually in a way that is not always possible in the field. This has surely had an influence on the high proportion of 'Appearance' names.

Also noticeable, but not as striking, is the smaller proportion of 'Belief' names. This is also explainable as bird names related to beliefs (omens, harbingers and taboos) are part and parcel of a traditional society. We would not expect the same beliefs to surface in the modern context of avitourism, where Zulu-speaking bird guides assist non-Zulu-speaking birders to tick off birds on their KZN lists.

4.3   A SEMANTIC PROFILE OF THE COINED ZULU NAMES

We follow the order of the elements in the 'Manhattan' graph above, i.e. beginning with 'Song', and then going on to 'Appearance', 'Habitat', etc.

4.3.1   The semantic category 'Song'

Twenty of the coined names had underlying meanings relating to song or call, and exactly half of these – ten names – were based on onomatopoeia, reflecting the higher proportion of onomatopoeia among 'song names' in the combined African bird name profile. There is not very much to say about onomatopoeia: the names reflect in phonemes the call of the bird as perceived by the namers. Only those who know the calls of the birds mentioned below can gauge whether or not the phoneticisation of their calls has been successful.

The workshop participants perceived the calls of the White-backed Duck and the White-faced Duck as being similar, giving the name **inzwinzwi** to the first and **inzwinzwinzwi** to the second. The Green Malkoha (previously the Green Coucal) is perceived as having a similar call, expressed in the name **incwincwincwi**. The calls of two tern species were characterised by the use of the lateral fricative /kl/, with the Lesser Crested Tern receiving the name **ukliyo**, and the Caspian Tern the name **ubhaklakliyo**.

The Common (European) Cuckoo was given the name **unokhukhuza**, which may simply be a copying of the English word 'cuckoo', or may be an onomatopoeic name *ab initio*. Cocker (2013: 266) shows that the cuckoo almost invariably has onomatopoeic names in various European languages.[1] The structure of the Zulu name is interesting: the onomatopoeic element *khukhu* has been suffixed with *–za*, creating a verb that means 'say/go cuckoo'; this has been prefixed with the name formative *–no–* giving a noun that means 'the one that goes "cuckoo"'. In other words, onomatopoeia is the base in a complex name-forming structure.

The names for two bushshrikes were found to be 'successful' instances of onomatopoeia, but, of course, as with all onomatopoeic names, it depends on how the words are said out aloud, i.e. what pitch and what rhythms and stress are used: **isicivó** (with stress and raised pitch on the final syllable) for the Black-backed Puffback (formerly Puffback Shrike), and **umabhashinhlayela** for the Orange-breasted Bushshrike.

Metaphor was used for five coinages, with three of these making comparison to musical instruments: The name **unocilongo** for the White-browed Robin-Chat likens the bird's call to a trumpet (*ichilongo*);[2] **unompempe** for the Greenshank compares the call of this bird to a whistle (*impempe*);[3] and **unonkositini** suggests that the Red-capped Robin-Chat (previously Natal Robin) sings like someone playing the concertina (*inkositini*). The two other metaphor-based names are **unonzwili** for the Melodious Lark, that suggests that this bird sings as sweetly as a Cape Canary (**umzwilili**), and **usipoki** for the Grey-headed Bushshrike, a name that is derived from the adopted noun *isipoki* (from *spook*) and mirrors the Afrikaans name for this bird – *spookvoël*.

Simile is fairly rare as a name-forming device as it requires an element meaning 'like' or 'similar to'. This is usually expressed in coined names in Zulu

---

1. His list is extensive, and includes *coucou* (French), *kukkuk* (German), *cuculo* (Italian), *kukulka* (Polish), *kukačka* (Czech), *kakkuk* (Hungarian) and *guguk* (Turkish).
2. When the new names from the first three workshops (2013–2015) were evaluated in 2016, it was felt that the call of this bird was more like a flute than a trumpet, and in the final workshop of July 2017, it was agreed to change this name to **unomtshingo**, based on the Zulu word *umtshingo* for a reed flute.
3. The word *unompempe* is also a much older Zulu coinage referring to the referee in a sporting fixture such as soccer or rugby.

by the prefix –*sa*– meaning 'something like'. This strategy was used only once in the workshops, giving **isankawu** (something like a monkey) for the Southern Pochard. The workshop participants did not know which feature of this small duck made it sufficiently distinctive from other ducks until a recording of its call was played, and someone pointed out how similar the call was to the alarm call of a vervet monkey (*inkawu*).

Two descriptive names were coined: **umagevuza** (the one always chattering) for the Sand Plover and **ihlekehleke** (continual laughter) for the Chestnut-vented Tit-babbler (Warbler). This latter name was certainly influenced by the previously recorded name **ihlekehle** (it laughs as it alights) for the Arrow-marked Babbler.

4.3.2   The semantic category 'Appearance'

As noted above, among the newly minted Zulu bird names, the semantic category 'Appearance' is the largest. We have suggested that this may have to do with the workshop participants all being involved with avitourism where bird sighting is the dominant focus area. Some of the coined names also suggest the structure of vernacular book names, which, as noted in Chapter 2, have also been artificially constructed for birdwatching. The names we refer to here are of the structure [adjective + noun + -*ed* + bird generic] where for example 'yellow' and 'bill' become 'yellow-billed' and this is then used to qualify 'duck' or 'stork' to produce a name like Yellow-billed Stork. In many cases, the Zulu equivalents of these names left out the generic, so that Yellow-billed Stork simply became **unomlomophuzi** (yellow-billed). Further examples are:

- **indebebomvu** (pink/red billed): Pink-billed Lark
- **usifubabomvu** (red chest/breasted): Scarlet-chested Sunbird
- **ukhandabomvu** (red-headed) : Red-headed Quelea
- **umqhalophuzi** (yellow-throated): Yellow-throated Woodland Warbler
- **unongilobomvu** (red-throated): Red-throated Wryneck
- **unosigqokomnyama** (black-capped): Bush Blackcap

The workshop participants did not always agree with the selected feature of the English book name. For example, the name 'Saddle-billed Stork' refers to marking on the bill, however, the workshop participants thought the red knees were a stronger diagnostic feature of this bird and so coined **umadolobomvu** (red-kneed).

Other names that refer to specific body parts include:
- **unosimila** (< *isimila* 'a growth'): Knob-billed Duck
- **unofosholo** (< *ifosholo* 'shovel'): Cape Shoveler
- **usiqhovana** (little crest): Crested Barbet[4]
- **umadevu** (moustache, whiskers): Whiskered Tern
- **umadevaphuzi** (whiskers + yellow): African Wattled Lapwing
- **uqholompunga** (grey-rumped): Fairy Flycatcher

Many coined names simply referred to the overall colour of the bird without mentioning specific body parts. These include: **unophuzana** (little yellow one) – Yellow-throated Sparrow (Petronia) and **unonsundu** (brown one) – Black Harrier. The name of the black African Oystercatcher uses an implied metaphor. The name **unozila** is based on the verb *zila* (to mourn), suggesting that the bird is dressed in widow's garb (see Plate 16).

A particularly interesting name referring to colour patterns is the name **impofana** for the Eurasian (European) Golden Oriole. Although this name literally means 'slightly dun-coloured', it is also one of the nicknames of the Kaizer Chiefs Football Club, whose uniform is a striking half-yellow/half-black, just like the plumage of this bird (see Plate 42). One is reminded of the American bird, the Baltimore Oriole, a bird that has nothing to do with the city of Baltimore in America, but everything to do with the colours of the livery of the servants of a certain Lord Baltimore.

The previous two names are decidedly metaphorical in nature, and indeed metaphor has played a strong role in the coining of new Zulu bird names. Further examples are:
- **intakanzwili** (< *intaka* 'finch' + *umzwilili* 'Cape Canary'): Chestnut-backed Sparrow-Lark
- **ivuzigazi** (what oozes blood): Grey Waxbill
- **uzibukwana** (little spectacles): Spectacled Weaver
- **ugazini** (in the blood, i.e. covered in blood): Red-headed Finch

---

4. Doke and Vilakazi's name **imvunduna** (1958: 844), given for both the "Golden-green Cuckoo, *Chrysococcyx cuprens* [*Chrysococcyx cupreus*]" as well as the Crested Barbet, has not been included in the Zulu names of any edition of *Roberts' Birds*, and the name was unknown to the workshop participants.

The name **umagumejana** for the Pink-throated Twinspot stands out among such metaphorical names. According to the workshop participants there was a young girl of the Gumede clan who was well known for her finery and beadwork. The coined name for this finely decorated bird is derived from –*ma*– 'characteristic of' + Gumede + the diminutive –*ana* (see Plate 43).[5]

There are more similes among these 'appearance' names than there are in the 'song' category, which is to say that the prefix –*sa*– (something like) has been more extensively used. A number of them define one bird in terms of another, for example:

- **isadube** (something like a zebra): Stierling's Wren-Warbler (Stierling's Barred Warbler)
- **isagundwane** (something like a mouse, i.e. mouse-coloured): Pale (Pallid) Flycatcher
- **isaqola** (something like the Southern Fiscal [formerly Fiscal Shrike]): Fiscal Flycatcher
- **usafukwe** (something like Burchell's Coucal): African Cuckoo
- **usantiyane** (something like the Common Waxbill): Melba Finch (now known as the Green-winged Pytilia)
- **usamklewu** (something like the Grey Go-away-bird (formerly Grey Lourie): Great Spotted Cuckoo

Names that refer to specific patterns, or spots and stripes, include **ucijomhlophe** (white diamond-shaped mark) for the Common Sandpiper; **unogqabhakazi** (large mark) a name coined for the Ruff; **usomthende** (father of stripes) for the African Cuckoo-Hawk; the rather predictable **usonkanyezi** (–*so*– + *inkanyezi* 'star') for the White-starred Robin; and **unosongo** (based on *isongo* 'brass ring around arm') for the Common Ringed Plover. The invader species Rose-ringed Parakeet gave the workshop participants some thought before they came up with **unocu**, based on the noun *ucu*, the bead necklace a girl gives a young man when she accepts him as a lover. This name was a brainwave of Thabile Khuzwayo,[6] and her suggestion was immediately adopted (see Plates 44 and 45).

---

5. In the morpho-phonemic process caused by the diminutive suffix –*ana*, the final 'd' of Gumede becomes /j/. This is a regular process in Zulu, seen for example in *umgojana* 'small hole' from *umgodi* 'hole' and *incwajana* 'small letter, book'; from *incwadi* 'letter, book'.
6. Participating in her first Zulu bird name workshop in 2015.

4.3.3   The semantic category 'Habitat'

At the 2013–2017 Zulu bird name workshops, it was found that many of the birds without previously recorded Zulu names were waterbirds of one kind or another, whether of pan, pond and lake, or of the marine environment of shoreline and open sea. There being few generic names like 'swallow' and 'duck' for this wide group of birds to allow for extended names of the **idadelimnyama** (black duck) type, coinages played a far great role in the creation of new names. We begin here with the broadest possible habitats and gradually work our way down to habitats that consist of a single plant species.[7]

The names that refer to the broadest habitat are surely **insukakude** and **unolwandle**, both for terns, a group of birds for which virtually no previously recorded names were found. The first name is that of the Arctic Tern, and it means 'that which comes from afar'; the second is the name given to the Swift Tern, with the name formative *–no–* prefixed to *ulwandle* (sea, ocean). Two more maritime birds are the White-winged Tern and the Kelp or Cape Gull, the former having the name **unochibi** and the latter the name **unochweba**. Both are based on names meaning 'bay' or 'lagoon'.

Moving inland, we find **inkukhuyamanzi** (chicken of the water) for the Common Moorhen, a name that mirrors the N. Sotho name *kgogomeetse* (chicken of the water) and the N. Sotho and S. Sotho name *kgogonoka* (chicken of the river). At the edge of pans and rivers we find mud, and two names refer directly to 'mud birds': **unodaka** for the Bar-tailed Godwit and **umacibudaka** for the Glossy Ibis. The first name is simply *–no–* + *udaka* 'mud', the second is a more complex name of *–ma–* 'characteristically' + *ciba* 'take pleasure in' + *udaka* 'mud'. Plants living on the fringes of water bodies are seen in the names of the Purple Gallinule (Swamp Hen) and the Sedge Warbler. The former is named **umhlangeni** (in the reed bed) and the latter **unomanduli**, derived from *–no–* + *–ma–* 'characteristically' + *induli* 'species of river grass or rush', i.e. the 'sedge' of the English name.[8]

From bodies of water, we move to rocky habitats. The name **ikhwelentabeni** (what climbs on the mountain) for the Mountain Chat follows a semantic type seen in the previously recorded traditional names like **isihlalamatsheni** (what

---

7. See section 12.5, 'A case study of shore birds and waders' (pages 351–7). There is, inevitably, some overlap between these two sections.
8. Again, a coinage that suggests that it was influenced by the English book name.

lives among rocks) for the Cape Rock Thrush and **ikhwelematsheni** (what climbs on rocks) for the Sentinel Rock Thrush. A similar coined name is **unogxumetsheni** (what jumps on a rock) for the Drakensberg (Orange-breasted) Rockjumper, where *gxuma* (leap, jump) is compounded with *etsheni* (on the rock). Also using the word *itshe* (stone, rock) is the name **umatsheni** (on the rocks), which is usually where the Freckled Nightjar is found.

Two names that reflect a forest habitat are **ujamelihlathi** and **unongoyana**. The former is the name of Barratt's Warbler and means 'what stares at the forest'. The coinage is prompted by the existing name **ujamelumhlanga,** given by Doke and Vilakazi (1958: 357) as firstly, "what stares at [the] reeds", and as "species of bird [found among reeds]", and assigned in the workshops to the Great Reed Warbler. The latter name is based on the name of an actual forest: the Green Barbet is endemic to the uNgoye Forest[9] and the name is derived from *no–* + *Ngoye* + the diminutive *–ana*, thus 'the little [bird] from the Ngoye'.

The last two names in the category 'Habitat' refer to two quite different habitats: **inqemvuma** and **ujolwane**. The name **inqemvuma** is derived from *inqe* 'vulture' + *umvuma* (Kosi palm), and so is a name that directly mirrors the English name Palm-nut Vulture. The Kosi Palm (*Raphia australis*) "occurs naturally in southern Moçambique and, in Zululand, only around Kosi Bay" (Coates Palgrave 1977: 70). Early in the twentieth century, however, a grove of these trees was planted much further south, at Mthunzini, and today they are a national monument, and of course a 'hot spot' for birders to tick off the Palm-nut Vulture.[10]

The name **ujolwane** for the House Sparrow is more metaphorical in nature, as it is derived from the noun *ujolwane*, which means 'a stay-at-home-type'.

4.3.4   The semantic category 'Diet'

Coined names reflecting diet constituted a smaller percentage than the same category among traditional Zulu names. Nonetheless, ten names referring to what particular birds customarily eat were coined in the three workshops of 2013–2017, and these are discussed as below.

We start with the insects, and the least specific of these are the two names based on *isinambuzane* (insect). These names are **usonambuzane**

---

9. And a few other isolated areas, such as a high-level plateau forest in Malawi.
10. Coates Palgrave (1977: 70); Boon (2010: 54).

for the Spotted Flycatcher and **usikhothanambuzane** for the African Dusky Flycatcher. The first name here is simply the name formative *–so–* prefixed to [*isi*]*nambuzane*; the second is a compound using the verb *khotha* (scoop [from the air]). Also using the verb *khotha* are the names **usikhothaphela** (what scoops a cockroach) and **inkothanyosi** (what scoops a bee), the former being the name of the Icterine Warbler and the second the name of the European Bee-eater.

The word *inyosi*, which means both 'bee' and 'honey', is the base of the name **umanyosini** (characteristically found among bees/honey) chosen at an early workshop for the European Honey Buzzard. Both the English names and the coined Zulu name are actually misnomers, as the so-called 'Honey' buzzard feeds mainly on wasps. A more accurate Zulu name, which would help correct the link between name and bird, would be the similar **umanyovini** based on the noun *unyovu* (wasp) and meaning 'characteristically found among wasps'. This emendation was adopted at the July 2017 workshop. Staying with insects is the name **idlantuthwane** (what eats ants) a name that mirrors the English name Ant-eating Chat as well as the scientific name *Myrmecocichla formicivora* where the generic name means 'ant thrush' and the specific epithet means 'ant-eating' (Clinning 1989: 91).

From insects to snails to snakes: the name **isigqobhamnenke** for the African Openbill is a compound of *gqobha* (peck at) and *umnenke* (snail); the word **indlanyoka** (what eats a snake) was coined as a generic for the three species of snake eagles found in the Zulu-speaking area. The word *indlanyoka* has not been used on its own, but extended with qualifiers, as in **indlanyokensundu** (brown snake eater) for the Brown Snake Eagle.

At the 2014 workshop, the Blue Waxbill was given the name **isicelankothe** (what asks for maize grain), a name that reflects the name **isicelankobe** (what requests boiled mealies) that Doke and Vilakazi (1958: 105) gloss as 'species of small bird'. At the earlier 2013 workshop this name was assigned to the Red-backed Mannikin. The result was two birds with names that suggested that they were ornithologically related, a situation that Roger Porter found misleading.[11] The name of the Red-backed Mannikin was the one that was changed at the July 2017 workshop, and Doke and Vilakazi's unassigned name **isicelankobe** was sent to the archives, and replaced with **amadojeyanabomvu**, formed by

---

11. During the several assessment meetings held in the first half of 2017.

adding –*bomvu* (red) to **amadojeyana**, the generic name for mannikins, and meaning 'the very little men'.

And then finally, a name that reflects the manner of eating rather than what is eaten: the name **indlangamandla** (what eats with strength) was coined for the Lappet-faced Vulture, presumably perceived as a more aggressive or voracious feeder than other vultures.

4.3.5   The semantic category 'General Behaviour'

Only seven names were coined that referred to the general behaviour of birds, and the first three relate to hunting activities:

The African Harrier-Hawk (previously the Gymnogene) is able to bend its legs when robbing nests, especially those nests in holes in trees. It was this unusual characteristic that led the workshop participants to coin the name **ijikanyawo** (what turns the foot).

The aggressive behaviour of the Brown Skua, especially as regards robbing other birds of their food, is what led to the coining of the name **impisiyolwandle** (hyena of the sea). The name was originally given at one of the earlier workshops to the Parasitic Jaeger (previously the Arctic Skua), but when it was pointed out that the Brown Skua was far more likely to be seen off the KZN coasts than the Parasitic Jaeger, the name 'hyena of the sea' was passed to the Brown Skua, and the Parasitic Jaeger was given the closely related name **iselayolwandle** (thief of the sea).

And then, perhaps influenced as much by the English name as the actual behaviour of the bird, is the name **umaphendula-matshe** (that which characteristically turns over stones) given to the Turnstone.

Also certainly influenced by the English name Lazy Cisticola is the coinage **unovilane** where the word *ivila* (lazy person) is surrounded by the name formatives –*no*– and –*ane*.

In the previous section, we discussed birds perceived to be cattle or goat herders because of their habits of following stock to feed on the insects disturbed by their movements. This notion surfaced again among the coined names in the name **umalusinkomo** (the herder [of] cattle) for the Eastern Nicator.

The name **umacutha** for the Squacco Heron is based on the verb *cutha* (draw oneself upright), referring to the distinct behaviour of this bird.

Quite a different kind of behaviour is seen in the name **usothathizwe** for the invader species the Common (Indian) Myna, perceived as spreading gradually all over the country. The name means 'the master of taking over the country'.

4.3.6   The semantic category 'Motion'

Only eight coined names referred to the motion of the bird. Five of these referred to the gait of the bird, the first four being of the plover/sandpiper group:

- **umaphithizela** for the Sanderling is derived from the verb *phithizela* 'move about in a haphazard, confused and disorderly manner', with the prefix –*ma*– 'behave characteristically'; a similar construction is used in
- **umathantatha** for the White-fronted Plover, from the verb *thantatha* 'deviate, walk off the path, walk to the side';
- **umatatazela** for Kittlitz's Plover is from the verb *tatazela* 'be agitated, move in disturbed fashion';
- **unothwayiza** for the Marsh Sandpiper where the verb *twayiza* 'walk with long, swinging steps' is prefixed with –*no*– rather than with –*ma*–; and
- **umatshikiza** for the Dark-capped (Yellow) Warbler, based on the verb *tshikiza* 'flick' or 'wag', as well as 'walk with sprightly gait'.

The African Finfoot was given the name **igwedlamanzi**, a straightforward descriptive name that means 'what paddles the water'. More indirect, through the use of metaphor, is the name **isicibamanzi** for the Cape Gannet, meaning 'the arrow [of the] water': the plunging dive of this bird was perceived as its most quintessential characteristic. Also using metaphor, this time referring to a different kind of flight, is the name **umasikulufu** from *ma*– 'characteristic of' and the adopted word *isikulufu* from English 'screw'. The screw-like circling courting display of the African Broadbill was seen as more diagnostic than its broad bill.

4.3.7   The semantic category 'Belief'

As pointed out above, the dearth of examples here – only one name – reflects the lesser importance of traditional cultural beliefs in modern avitourism. Our sole example here is **umvuliyeza** for the Little Swift. This name, meaning 'the rain is coming' is a clever play on the existing **ihlolamvula** (what predicts rain) as a generic name for swifts. The beliefs about the behaviour of birds presaging rain is worldwide. Cocker, who mentions many 'rainbirds', says of plovers:

> More certainly, the appearance of migrant plovers was closely related to autumn or winter conditions and specifically to rain. The word 'plover' itself is derived from *pluvia*, Latin for rain, while *Pluvialis*, the

scientific name for the four most impressive world-wanderers, means 'relating to rain' or 'bringing rain'. Today the same quartet are known as *regenpfeifer*, 'rain pipers' (2013: 201).

4.3.8   The semantic category 'Nest'

This is also a very minor category, both among African bird names generally, as well as among traditional bird names in Zulu. It is almost surprising that even three names were so coined. Two names refer to the parasitic behaviour of certain birds: **iselantaka** for the Cuckoo Finch, a compound of *isela* 'thief' and *intaka* 'finch', and **umazalashiye** for Klaas's Cuckoo, with *ma–* 'characteristically' + *zala* 'give birth, (i.e. lay eggs) + *shiya* 'leave [them] behind', thus 'the bird that characteristically lays eggs and leaves them behind for other birds to rear'. This name makes such a good generic name for all cuckoo species that the slightly variant form **unozalashiye** was adopted at the July 2017 workshop to serve this purpose.

The Sociable Weaver only occurs in a part of South Africa far from the Zulu-speaking area, and would in the normal run of things not have come up for discussion in the workshops. However, given its remarkable nesting habits, it was thought that this bird could very likely make its appearance in school biology textbooks, and it would be useful to be pre-emptive here and coin a name. The name is a fairly simple one, **unosidlekekazi**, and is based on the noun *isidleke* (nest), prefixed by *–no–* and suffixed by *–kazi*, here in the sense of 'large', thus 'the bird of the large nest'.

4.3.9   The semantic category 'Other'

Many of the 'other' names in the previous section were of the 'lexically transparent but semantically opaque' type, and we would not expect to find such examples among names that were coined for specific discussed reasons. We do, however, find one name that is semantically opaque, the name **unohlohlweni** for Montagu's Harrier. This name is clearly the prefix *–no–* followed by the locative form of '*ihlohlo*'. The problem is that the word '*ihlohlo*' cannot be traced. Also rather opaque is **unothezane** for the Sickle-winged Chat: the prefix *–no–* and the suffix *–ane* are readily recognisable here, but the core element is not clear. It could be linked to the verb *theza* meaning 'collect firewood', although that is rather unlikely. More likely is *umthezane*, a species of thorn tree. If this is so, then this name belongs to the category 'Habitat'.

The three other names in this category are all semantically transparent and their 'otherness' is of different kinds, and each is decidedly unusual.

The scientific name of the Brubru [Shrike] is *Nilaus afer*. On being told that *Nilaus* was an anagram of the shrike genus name *Lanius*, the workshop participants wished to identify Zulu with this coining style. For some reason or another, they chose the name **umnqube** (the Cape Batis) as a base and semi-anagrammed it into **ubhenqu** for the name of the Brubru. Alas! – This name did not last for more than two years. On later reflection the name '*ubhenqu*' was found to be too frivolous, and it was replaced with **usacingo** ([sounds] like a telephone).

Related to the underlying suggestions in the English name Knot, the Afrikaans name *knoet* and the Latin name *Calidris canutus*, is the Zulu name **unovimba**. Intrigued by the story of King Canute (Knut) trying to stem the incoming tide (a story they had never heard before) and how this was reflected in the various names for this bird, they came up with **unovimba**, where *no–* prefixes the verb *vimba* 'block, stem, ward off'.[12]

Our final example here is the opposite of a name referring to centuries-old traditions. The coined name **uzazu** joins the other coined name **umkholomphunga** (yellow hornbill) for the Southern Yellow-billed Hornbill. The name **uzazu** was not coined by the participants of the 2013–2017 Zulu bird name workshops, but rather by the scriptwriters of the Disney classic *The Lion King*. According to the workshop participants, through the influence of this much-loved film, many Zulu-speaking children and adults, know that the name of the hornbill character in this film is Zazu, and so apply the name to any yellow-billed hornbill (see Plate 36).

---

12. Greenoak (1997: 90) says: 'There is a tradition (mainly a literary one) that the name Knot came about because they were known as King Canute's birds. Willughby records that the old king was very fond of them . . . Poets as well as ornithologists were familiar with this association.' She goes on, however, to say that it is generally thought nowadays that the name Knot and other similar variants, (e.g. Gnat, Gnet) are imitative, (i.e. onomatopoeic) of the bird's sharp alarm note, given as 'knut, knut'. Nonetheless, it is pleasing that a modern Zulu coinage identifies with a centuries-old literary and ornithological tradition.

5 Zulu bird names used for other purposes

In the previous two chapters, I talked about the underlying meaning of bird names, showing how these are related to the birds themselves, not just as identificative tags, but directly linked to the living birds and their appearance, their songs, their nests and their diets, and other aspects of being a bird. In this chapter, I continue to talk about bird names, but no longer is the discourse about the birds themselves. This chapter in essence shows how bird names – specifically Zulu bird names – are used to refer to entities other than birds. It is still, however, a chapter about meanings and the nature of meanings. As a 'pure' bird name, the Zulu word **inkwazi** refers to the African Fish Eagle, *Haliaeetus vocifer*, and so covers the 'whole' of the bird – what it looks like, its habits and habitats, its evocative call, and so on. These different aspects of the fish eagle can be thought of as individual elements in the meaning of the word **inkwazi**. When the same word is used as a cattle colour term, or as a brand name, or as a river name, or as the name of a holiday resort, more or less of these individual elements are carried over. When the word is used as a term for a beast that is half black and half white, it is only the appearance element of the name that is carried over. When it is used as the brand name of a company that manufactures vehicle tracking systems, the overhead flight and sharp eyes of the fish eagle, just one part of the bundle of elements in the original meaning, are carried over. When the name **inkwazi** is used for a river, the riverine habitat of the fish eagle comes into play.

Because this chapter explicitly describes these transfers of pieces of meaning from one type of noun (bird names) to other types of nouns (other nouns, place names, brand names, etc.), it is very definitely a chapter on linguistics. At the same time, the chapter tries to show that birds are all around us, not just the feathered kinds in the natural environment, but their linguistic counterparts too, which can be found in the hides of Nguni cattle, on the signs at the intersections of roads and on the sides of cartons of traditional Zulu beer (see Plate 15).

5.1   BIRD NAMES AS OTHER LEXICAL ITEMS

In this section, I look at how words that commonly are taken as referring to birds may have other meanings as well. Sometimes these other meanings are entered together with the bird reference in one entry in a dictionary. When a word has more than one meaning in this way, this is known as polysemy. Look at the extract from Doke and Vilakazi's dictionary (1958: 581) below:

> **-nkwazi (inkwazi,** 2.3–8.8–3, **izinkwazi)** n.
>   1. White-headed fish-eagle, Haliaetus vocifer.
>   2. Black beast with white head.
>   3. **iziNkwazi** (pl. only): A river in Stanger distr. of Natal.

It can be seen that these two lexicographers – Professor Clement M. Doke and Dr Benedict W. Vilakazi – do not regard this as a case of three separate words each with their own meaning (in which case they would be 'homonyms', and entered separately), but as one word with three interrelated meanings. Note that they give the fish eagle as the primary meaning, the cattle colour term as the secondary meaning (derived from meaning 1) and the river name as a tertiary meaning (also derived from meaning 1). They have, in other words, treated **inkwazi** as a <u>polysemic</u> word.

I explore the notion of polysemy below, using Zulu bird names as illustration.

In some cases, the other meaning, as with **inkwazi** immediately above, is that of a cattle colour term, usually when the skin colour pattern of certain beasts is similar to that of a particular species of bird. In this chapter, I use Marguerite Poland's book, *The Abundant Herds*, to explore the relationship between Zulu bird names and Zulu cattle colour terms.

Certain bird names have also been used by other writers, notably Samuelson (1923), to create a wide range of colour terms, and this is also investigated below.

Finally, this chapter looks at how Zulu bird names are used as brand names in contemporary South Africa.

The common thread running through this whole section is metaphor: certain aspects of a particular bird (its plumage, its flight patterns, etc.) are 'copied' into another domain.

5.1.1   Polysemy as illustrated by Zulu bird names

As seen in the introduction above, the linguistic notion of polysemy is displayed when a single entry in a dictionary is given a number of different meanings, all of which are considered by the lexicographer to be related to one another. I illustrated this principle briefly with the word **inkwazi**; let us do it in more detail with the English word 'mouth'. Any English dictionary will give at least three meanings for this word: the opening in a face where food goes in and voice comes out; the open part of a jug; and the place where a river enters the sea. All are clearly different, but all are obviously linked by the notion of a place where 'things' (particularly liquid) come in or go out. Or take the word 'foot', which means a part of the body where a sock or shoe is placed, or the opposite end of the table to the head, or the bottom of a page (cf. 'footnote'), or the lowest hills of a mountain (foothills). All are linked by the notion of being at the lowest part of an entity, either physically (footnotes, foothills) or in status (foot of the table). These semantic links are not always transparent. One has to know that in earlier times the English word 'fare' meant 'travel' to understand the links between 'farewell' (equivalent of *hamba kahle*, the Zulu farewell that literally means 'go well' or 'fare well'), 'taxi fare' (the cost of travelling) and a 'bill of fare' or menu: the food available to a traveller when stopping overnight at an inn.

In the examples of polysemic bird names that follow, I look at what parts of each meaning are transferred to the next meaning, and what type of semantic process has taken place. In the examples of mouth, foot and fare above, the same process has taken place: an essential part of one meaning is transferred to a new referent, while other parts remain behind. 'Opening for liquid to enter and exit' is retained when 'mouth' is attached to 'jug' and 'river', but '[opening] in a head or face' is deleted.

We begin our examination of polysemy in Zulu by looking at a word where there are no birds present: the name of the Drakensberg Mountains, uKhahlamba in Zulu. Doke and Vilakazi (1958: 374) list the primary meaning

> **-khahlamba** ⟨**u(lu)khahlamba,** 6.6.3.9.9, **izinkahla-
> mba**⟩ n. [> loc. **okhahlambeni.**]
>   1. Rough∫ bony object; skeleton.
>   2. Tall, thin person. [cf. *u(lu)khahlambela.*]
>   3. Row of upward-pointing spears.
>   4. Broken mountain range.
>   5. **u(lu)Khahlamba:** the Drakensberg Mountains. [>
> loc. **oKhahlamba.**]

Gloss of **ukhahlamba** in Doke and Vilakazi (1958: 374).

of the noun *ukhahlamba* as "Rough, bony object; [such as a] skeleton".[1] In the second meaning "Tall, thin person", we see that 'rough' is not part of meaning (2) and that 'object' has become the new referent 'person'. But certainly a very thin person is bony and may remind us of a skeleton. The third meaning is "Row of upward-pointing spears", and this is derived directly from meaning (1), not meaning (2). A notable visual feature of a skeleton is the bony ribcage, where rows of ribs, lying parallel to each other, may point up in the air if the skeleton is lying on its back. 'Bone' is thus replaced by 'spear' but 'rows of sharp upward-pointing things' is retained. Meaning (4) is "Broken mountain range", and although this is primarily derived from meaning (1), it is meaning (3) that has led to the Drakensberg often being known as "The Barrier of Spears". The final definition, "uKhahlamba: the Drakensberg Mountains" is the easiest derivation to see: a simple onymisation (creating a proper name out of a common noun) by making the generic (any broken mountain range) into a specific (a particular named mountain range).

One can apply some of these observations and principles to bird names where the referent to a bird is only one of two or more meanings assigned to a single word. I start with the more transparent examples of polysemy and then move on to those that are more opaque.

The word **imbuzana** is derived from the word *imbuzi* 'goat' and the diminutive suffix *–ana*, thus making clear meaning (1) "Little goat". Meaning (2) is given as "Green-backed Bush-warbler,[2] *Camaroptera olivacea*, whose

---

1. Choosing this meaning as the primary out of the five offered is of course the choice of the lexicographers, and not necessarily a 'natural' feature of the language. However, one has to start somewhere.
2. Now the Green-backed Camaroptera (*Camaroptera brachyuran*).

> **dlamadoda (indlamadoda,** 2.6.3.9.9, **izindlama-
> doda)** n. [<dla+amadoda, lit. what eats men.]
>   1. A name given to the *ingqungqulu* eagle from its
> habit of eating the corpses of those slain in battle.
>   2. A serious disease, often accompanied by internal
> bleeding. [cf. *i(li)zembe.*]

Gloss of **indlamadoda** in Doke and Vilakazi (1958: 153).

> **-zembe (i(li)zembe,** 3.2.9.9, **amazembe)** n. [Ur-B.
> **-yembe.**]
>   1. Axe, hatchet. [cf. *imbazo.*]
>   2. Disease affecting the bladder, stomach, and kid-
> neys and causing bleeding.
>   3. Medicine for treating the above disease, also for
> cleansing from blood-uncleanness after battle.

Gloss of **izembe** in Doke and Vilakazi (1958: 890).

call resembles the bleating of a goat" where the semantic link is so clearly offered by the lexicographers (Doke and Vilakazi 1958: 56).[3]

The word **indlamadoda**, derived from the verb *dla* 'eat' and *amadoda* 'men' (that which eats men) is given two meanings that may at first glance appear to be easy to reconcile. Meaning (1) is "A name given to the *ingqungqulu* eagle [Bateleur Eagle] from its habit of eating the corpses of those slain in battle", and meaning (2) is "A serious disease often accompanied by internal bleeding, (cf. *izembe*)" (Doke and Vilakazi 1958: 153). Corpses slain in battle and internal bleeding somehow go together easily. But it is that "cf. *izembe*" that really completes the circle of meaning here. This word itself has three related meanings: 1) Axe or hatchet (especially a war axe or fighting axe); 2) disease affecting the bladder, stomach and kidneys and causing bleeding; and (3) medicine for treating the above disease, also for cleansing "from blood-uncleanness" after battle (Doke and Vilakazi 1958: 890). Fighting axes and the bladder/stomach/liver disease both cause bleeding that leads to death; death on the battlefield needs cleaning up, literally, in the way of putrefying corpses, and this is where the Bateleur Eagle plays a role, as well as figuratively, causing pollution in the sense of spiritual defilement, and this is where the purifying medicine *izembe* comes in. Between them these

---

3. Although they do not explain so conveniently the link to the plant *Ipomoea albivenia*, which they give as the third meaning of the word.

two words – *indlamadoda* and *izembe* – contribute to a kind of 'semantic package' of how to deal with the pollution of death, both practically (remove the corpses) and spiritually (remove the defilement). The Bateleur Eagle deals with the practical problem; I would not be in the least surprised to hear that 'muti' made from parts of this bird is used to help with the spiritual pollution.

Doke and Vilakazi assign four meanings to the word **inhlamvu** (1958: 318): 1) single seed, stone or pip of fruit; 2) bullet, pellet; 3) the honey-guide bird; and (4) bright, shining object. These meanings are not difficult to link: A bullet or a pellet is an object not dissimilar to a seed or a pip; a bullet or pellet always finds its mark as the honeyguide always finds the bees' nest; the bird leads man to a hive where there is bright, shiny golden honey.

Another word for a honeyguide is the similar word **inhlava**, again with four meanings: 1) mealie-grub; 2) place eaten out by the mealie-grub; 3) Honey-guide, *Indicator major*; and (4) scolding garrulous female (Doke and Vilakazi 1958: 323). Meanings (1) and (2) are obviously linked, but one needs a little more of a mental jump to go from the hollow place scooped out by the mealie grub to the hollow place in the tree where the bees have made a hive. Meanings (3) and (4) are again more obvious, especially if one knows how this bird flies around the head of a person, calling continuously as it guides him to a hive. We see this again in the more specific name for a honeyguide – **inhlavebizelayo** (the *inhlava* which calls) with its two meanings 'honey-guide' and 'scolding female [woman]' (Doke and Vilakazi 1958: 323). And, indeed, the word **unomtsheketshe** has exactly the same two meanings, showing that the incessant calling of this bird as it entices hunters to a beehive is one of its major distinguishing features.

Another clear example of polysemy is in the two meanings of the word **inhlazazana** (from –*hlaza* 'green' + –*azi*– 'very' + –*ana* 'small'). The first meaning is 'small, green object' and the second refers to the Collared Sunbird, *Cinnyris afra*, which, needless to say, is small and intensely green on its head and back.

The example of polysemy in the word *ukhahlamba* ended with a proper name. We can see the same in the word **impangele,** where this time Doke and Vilakazi give the bird as the primary meaning: the Crowned Guinea-Fowl, *Numida coronata* (1958: 510).[4] One of the obvious features of this bird is the white spotting on the dark body, and this leads to the other meanings; (2) 'large black bead with white spots' and (3) 'black cloth with white spots'.

---

4. Now the Helmeted Guineafowl, *Numida meleagris*.

Not as obvious is why this bird should have given its name as one of the regiments of Mzilikazi (meaning 4), but it could have had something to do with the identificatory plumes worn by this regiment to distinguish it from others, or from the active way in which this bird moves in packs through the undergrowth.

An interesting case of visual linking is seen in the three meanings given for the word *ingcungu*: 1) basket or earthenware vessel narrowing in shape at the rim, and fitted with a lid; 2) ox with horns curving in almost to meet; and (3) small light-brown bird with a long beak (Doke and Vilakazi 1958: 552). The similarity of this word to the generic names used for sunbirds, namely **incuncu** and **incwincwi** makes it fairly clear that the last reference is to a sunbird with its long curved beak like the curves of the horns of the ox and the curves of the basket or the pot.

A word for feathers rather than for a bird shows clear semantic links. Doke and Vilakazi assign to the word *isiqhova* the following three meanings, and the links are clear enough not to need discussion: 1) crest of a bird; 2) ornamental crest of feathers worn by young men; and (3) a way of wearing the hair with a bunch in front (1958: 702). The word *isiluba* has very similar meanings and very similar links: (1) crest (as on bird's head); (2) head-dress of feathers; and (3) untidy top-knot of Native [*sic*] woman (Doke and Vilakazi 1958: 466). In its plural form, this word is used as part of a praise name for the Hamerkop, the bird with a noticeable crest – **uThekwane kaZiluba** (lit. 'Mr Hamerkop, son of Mr Crests') (see Plates 8 and 9).

Owls are reputed not to be able to see very clearly in the daytime, which perhaps explains the two meanings given to the word **umandubulu**: 1) Pearl-spotted Owlet, *Glaucidium perlatum*; and (2) short-sighted person (Doke and Vilakazi 1958: 482). Similarly eagles, particularly the larger ones, could be considered aggressive birds, hence the two meanings of **isihuhwa**: 1) Crowned Hawk-Eagle, *Spizaetus coronatus* (now *Stephanoaetus coronatus*) and the Martial Hawk-Eagle, *S. bellicosus* (now *Polemaetus bellicosus*); and (2) rapacious person.

The word *inkotha* is a revealing example of polysemy. Doke and Vilakazi give the primary meaning as "Tip of tongue, esp. of an ox" and given the function of a tongue of licking at something, it is not too difficult to link this to meaning (2) "Forefinger with which food-vessel is wiped" (1958: 405). Meaning (1) is completely discarded and only the forefinger is left in the third meaning "used to express seven". The forefinger in Zulu is also known as *isikhombisa* (the pointing finger) and in a finger-based counting system the five fingers of one hand and the thumb and forefinger of the other hand make seven (5 + 2 = 7). From here to meaning (4) "Various species of bee-eaters" requires a far greater mental jump, perhaps only achieved by seeing bee-eaters catching insects in flight as being similar to the way in which anteaters eat ants with the tip of their long tongue. The fifth meaning, a *hlonipha* word for *utshani*, 'grass', seems completely unlinked.[5]

The word **unhloyile** has three meanings, which not only link to each other but also to various aspects of traditional Zulu life and culture. Doke and Vilakazi's primary meaning is that of the Yellow-billed Kite, *Milvus aegyptius* (1958: 569). The second meaning is a proper name –*uNhloyile*, one of the names given to the lunar month approximating to the period from mid-July to mid-August, which is when these birds return to South Africa from wintering abroad. The plant *Scadoxus puniceus* (Blood Lily, Snake Lily) also rises from the ground at precisely this time of year, which is why it is known as *idumbe likanhloyile* (the Yellow-billed Kite's tuber). Doke and Vilakazi's third meaning of *unhloyile* is 'whirlwind' – a clear reference to the circling flight of this raptor as it looks for prey on the ground. Curiously, a well-known praise name for this kite is uNhloyile kaGelegele, using another name (*igelegele*) for a whirlwind, and so creating a praise name that effectively means "Whirlwind, son of Whirlwind". See page 233 in Chapter 9 where the praises of the Yellow-billed Kite are given in full (see Plate 47).

5.1.2   Bird names and cattle colours

As is well known, many aspects of Zulu culture centre on the raising of cattle, and as is possibly equally well known, there are over 300 different words to describe cattle in terms of their particular cultural function, their horn shape, or the patterns of colour of the hide. We have already seen one bird name that is also a cattle term above (the ox with curving horns); our next examples are polysemic words where two of the meanings at least are that of a particular bird and that of a cattle colour term. An example is the word **unhlekwane** with a primary meaning "Common Widow-bird, black-tailed finch" and as its second meaning "Black beast with white stripes running from shoulders to sides".[6] Needless to say, this bird is black, with white stripes running from

---

5. A '*hlonipha* word' is a word used to replace another that should not be spoken out aloud if it sounds like the name of a close relative.
6. 'Beast' is the common translation of the Zulu word *inkomo*, to avoid the clumsy 'single head of cattle'.

shoulder to sides. Similar pairs are seen in the following section in Table 5.1, with the 'bird meaning' on the left, and the cattle colour term on the right.

Table 5.1   Polysemic words with bird and cattle colours.

'Bird meanings'	Cattle colour terms
inkanku: Species of cuckoo, *Coccystes cafer* and *C. Jacobinus*[†]	Brown or black beast with white stripe on belly
ilunga: Species of shrike	Black or brown beast with white stripes across stomach and legs
inkwazi: African Fish eagle, *Haliae[e]tus vocifer*	Black beast with white head
iqola: Fiscal Shrike, *Lanius collaris*	Black beast with white markings across the back and side[7]
imvunduna: Golden-green Cuckoo, *Chrysococcyx cuprens* [*cupreus*] and Crested Barbet, *Trachyphonus vaillanti*	Black beast with white spottings on the body
impofazana:[8] Wattled Starling, *Dilophus carunculatus*	Dun-coloured cow
umngquphane: Red-winged Bush-shrike, *Telophonus senegalus*	Brown beast with white above the eyes

† Doke and Vilakazi's Latin and English names may not be in current usage.

Some words given as polysemic in Doke and Vilakazi are opaque in terms of the semantic links between the various meanings. Take for example the word **isigwe**, with the meanings
 1) Red bishop bird.
 2) Flower of the pumpkin plant.
 3) Large hairless caterpillar.
 4) Green parrot.

---

7. Doke and Vilakazi (1958: 710) kindly add in a parenthesis at this point "like those of the shrike" (see Plate 55).
8. Derived from –*mpofu* 'tawny-coloured, dun-coloured' + *azana* (*azi* + *ana*) indicating 'somewhat tawny-ish or tending towards dun'.

One could reasonably suggest that a yellow pumpkin flower stands out from the surrounding greenery as a male bishop in breeding plumage does, but how does the hairless caterpillar fit in? And why are two quite different birds encapsulated in this single word, separated by pumpkin flowers and a hairless caterpillar?

Bird links in Poland's The Abundant Herds

Marguerite Poland's *The Abundant Herds* gives a fascinating account of the way metaphor links cattle hide descriptions to the natural world. Man's perception of his environment provides the basis on which terms can be created by individual herders to describe the most complex and the most individual hide patterns and colours. In the quote that follows, a wide range of potential bases of descriptive terms is noted, among them both a bird, and the eggs of another bird.

> Throughout the ages, then, the well-being of the herds and the well-being of men have been so closely connected that cattle have become part of the spiritual and aesthetic lives of people. This perception of them has given rise to a poetic and complex naming practice. Yet, it is only in experiencing the calling of the terms in the presence of the cattle that this rich aesthetic is fully appreciated: this beast is the 'eggs of the lark'; that, 'the stones of the Ngoye'; here, 'the Redwinged Starling' and 'the clouds of heaven'; there, 'the gaps between the branches of the trees silhouetted against the sky' and 'the women who cross over the river'. To watch a parade of animals, in all their variety of colour and horn shape, is to share the wellspring of admiration that has moved generation after generation of herders throughout Africa (2003: 10).

Later, Poland specifies the 'eggs of the lark' as a creamy rust-speckled colour pattern (2003: 34), the actual term being *inkomo emaqandakacilo* (beast which is eggs of the **ucilo** lark). The 'eggs of the lark' appear again a page later, in another paragraph, which both explains and illustrates how metaphor works in the use of birds and other naturally occurring entities in forming descriptive terms for cattle colour patterns:

We perceive and define reality by examining objects and observing their differences and similarities. Metaphor works by identifying things perceived as analogous and postulating an identity between them – so that the one *is* the other. Thus, a cream beast, spotted lightly with rust, is known as – *and is* – the eggs of the lark; a dun and white mottled beast is the castor oil bean; a black beast with a white head, the fish eagle. The choice of subjects for such analogies are local and contextual. Man's relationship with his own, familiar environment gives rise to such metaphors and provides insight into the shared perceptions and worldview of the people concerned (2003: 35).

Poland explains that "knowledge of the associational context is crucial to an understanding of the metaphor" (2003: 35). Indeed, one needs to be a careful observer to make the connection between a natural phenomenon and a cow or ox with a specific pattern. The term *insingizisuka* (Southern Ground Hornbill taking flight), for a black beast with a white patch below its rump, would be meaningless if one did not know that this black bird displays its white primary feathers only when taking off in flight.

Sometimes, more than one bird is utilised in creating a name, particularly when a cattle colour pattern is very unusual. Poland says that "much discussion may arise between stockmen regarding beasts with unusual or ambiguous colour patterns" and gives a number of intriguing examples collected while attending a cattle auction at Hluhluwe (2003: 38). In one particular example, a greyish-dun cow received the normal descriptive term of *inkomazi eyintenjane* (cow which is the Crowned Plover) but as it also had unusual white marks on the face and white eyebrows, it was described as [*inkomazi ey*]*intenjane emngquphane* (cow which is the Crowned Plover, which is the Black-crowned Tchagra).

Poland's descriptions of the link between birds and terms descriptive of cattle colour patterns are greatly enhanced in her book by the striking paintings of Leigh Voigt, which enable the reader to see at a glance why a speckled grey-and-red roan bull is named after the Rock Pigeon, why a black beast with a white head should be *inkomo eyinkwazi* (beast which is a fish eagle) and why other cattle should be described as the Black Eagle, the Southern Fiscal (formerly Fiscal Shrike) (see Plate 55) or the Red-winged Starling.

5.1.3 From cattle colour patterns linked to birds to general colours linked to birds

In all of the above, Poland has linked colour patterns to birds. When a particular bird such as **iqola**, **ivukuthu**, or **intenjane** is linked to a particular beast, it is the overall plumage that is taken into account, including any dots, spots, speckles or stripes. For another writer though, birds are linked to colours per se. Where the Zulu language is found wanting in colour vocabulary, birds can be enlisted to make up the shortfall, and the person to do this is R.C.A. Samuelson, the author of *Long, Long Ago*, and *The King Cetywayo Zulu Dictionary*. The following list of colour terms in Zulu, and the brief introduction that precedes them, is taken from the dictionary:

> Ulwimi lwesiZulu lubuthakathaka [weakness] mayelana nemibala, ngakhoke umlobi walengqoqelo yolwimi lwesiZulu uhlele imibala ekhona kuso isiZulu, njengoba ikho, kwathi lapho lungenayo imibala elungelene nemibala eminye esolwimini lwesiNgisi engekho kolwesiZulu, uyiqambe ngezinyoni kokunye okuphethe leyomibala. Lokhu kwenzelwa okokuba kuqiniswe isiZulu kule 'ndawo yobuthakathaka baso nceze futhi kube umumbela abazakwakhela kuwo abazolandela umlobi lo ngezikhathi ezizayo (1923: 574–5).

This brief introduction can be translated as follows:

> The Zulu language is rather weak when it comes to colours, therefore the writer of this compilation of the Zulu language has collected together the colour terms that are present in the language, just as they are, but where there are no colour terms that correspond to existing colour terms in English, he has coined such terms by using references to birds with those [missing] colours. This has been done so that Zulu is strengthened in the places of weakness, in order that there be a foundation prepared to build on for those who follow after the writer in future times.

Mnyama – black
Mnyama bhuqe – pitch black
Intengu – lamp black
Indoni yamanzi – light black

Mpmhlophe [*sic*] – white
Incombo – cream
Bomvu – red
Mgazi – deep red
Umklele – bright or light red
Iphiva – livid
Nsundu – brown
Mdaka – dark or intense brown
Bubende – liver coloured
Iphuzi – chrome yellow
Mpofu – spectrum or light yellow
Ibhonsi – orange yellow, orange chrome
Umdubu – burnt sienna
Ugolokoqo – gamboge
Ubucubu – carmine
Iminza – indigo
Amesethole – violet or spectrum violet
Umananda – scarlet
Intenjane – tan-coloured
Umthuku – mouse-coloured
Ukuphenduka kwelanga – heliotrope
Umncaka – pink
Mpunga – grey
Ukhethe – slate-coloured
Indwe – light grey [although he enters the bird in his dictionary as *indwa*]
Luhlaza – ordinary green
Uhlaza – grass green (when young)
Inyandezulu – very light and bright green
Ibhuma – sap or dark green
Incuncu eluhlaza – Hooker's green
Ithanga – Indian yellow
Umuntswi – emerald green
Ifefe – cobalt blue
Ikhwezi – Prussian blue
Isigwe – spectrum red or scarlet
Iviyo – burnt umber
Umdubu – Vandyke brown
Intsomi – intense navy blue

Ulwandle – ultramarine
Umtsintsi – scarlet lake
Uzulucwathile – azure blue
Ingxotha – gold coloured
Inhlanzi – silver coloured
Ithophi – copper coloured
Ithusi – brass coloured
Imvukuzane – fawn coloured
Isihlabathi – sand coloured
Ungcwegcwe – tin coloured
Utshani-bomile – straw coloured
Luhlaza tyoko – pure or perfectly green
Mhlophe thwa – pure or perfectly white
Bomvu tubhu – pure or perfectly red
Bomvu klele – perfectly bright red
Luhlaza cwe – perfectly ultramarine
Isiquzi – pale blue
Ithunduluka – vermilion
Uklebe – yellow ochre
Iphothwe – neutral tint
Inkankane – Payne's grey
Igwalagwala – crimson lake
Umtoto – Alizarin crimson
Ubukhwebezane – purple lake
Umgquphane – light red
Nsundukazi – Indian red
Isikhombazana – Venetian red
Ukhokhothi – raw sienna
Indhlazi [Indlazi] – brown pink
Lufipha – dust coloured

There is a certain arrogance in Samuelson's introductory words, especially when taken in conjunction with the list that follows them. The impression given is that English has a far wider colour vocabulary than Zulu, and this constitutes a 'weakness' of Zulu. However, in order to bring the Zulu 'up to strength' (that is to say, to make it correspond with English), the author is prepared to take on this task himself, and to create Zulu words for this purpose.

When one looks at the English list, though, when the straightforward colour terms such as green, red, black, white, yellow, brown and pink are removed, there remain a few curious terms such as 'heliotrope' and 'livid', and then a host of terms that refer to artists' pigments. Any artist who has painted in either oils or watercolours will recognise the pigments 'chrome yellow', 'Hooker's green', 'Prussian blue', 'Alizarin crimson' and 'burnt sienna'. But to imply that these are standard terms used by any English-speaking person as part of their normal vocabulary is an exaggeration. Had Samuelson perhaps said that the Zulu society itself was defective in some way because there weren't enough Zulu speakers painting in oils and watercolours, then we would understand this necessity to create terms for the artist's palette such as 'burnt umber', 'Payne's grey', and 'Vandyke brown'. We would be able to see why it was necessary to have three different types of black, and why we should distinguish between chrome yellow, spectrum yellow and Indian yellow, and between no less than sixteen different shades of red.

Looking back on Samuelson from this distance and comparing his analysis of existing Zulu colour terminology with that of Marguerite Poland some 80 years later, we can notice both fundamental differences as well as some similarities. The most obvious difference is that where Samuelson found it necessary to create colour terms because "[t]he Zulu language is rather weak when it comes to colours", Poland's work is all about discovering the wealth of Zulu colour terms that exist in the Zulu language to describe a wide variety of cattle hides. She recognises the creative ability of the Zulu stockmen themselves to create ever more original terms when they come across a particular configuration for which no suitable terms exist.

In a way, one can understand Samuelson: he was very likely a water-colour painter, and his sister Letitia Samuelson was almost certainly one too. He was also a fluent Zulu linguist, and it possibly bothered him that the pigments he was using had no Zulu words to describe them. And certainly if he himself did not create such terms, no one else was going to do so.

Where the two systems of colour terminology resemble each other is the way that each draws from nature. Poland has described for us exactly how the Zulu language draws images from natural phenomena such as clouds and shadows between the trees, and particularly how the plumage of birds (and sometime the colouring of their eggs) are used to describe a particular hide coloration or pattern. Samuelson has drawn his azure blue from a cloudless sky (*uzulucwathile*), his purple lake from the wild lantana bush (*ubukhwebezane*), and his chrome orange from the fruit of the *Salacia alternifolia* shrub (*ibhonsi*). His alizarin crimson comes from the same *umtoto* roots that we will meet again in the praise of the **inqomfi** longclaw in Chapter 9. But most of all, Samuelson has drawn his colour terms from the plumage of birds. In his list we find 'lamp black' drawn from the Fork-tailed Drongo (**intengu**), 'light grey' from the Blue Crane (**indwe**), 'Hooker's green' from one of the brilliant green sunbirds (**incuncu**), 'spectrum red' from the Red Bishop (**isigwe**), 'intense navy blue' from the Red-winged Starling (**intsomi**), and several more.

One wonders whether Samuelson's carefully constructed vocabulary of colour terms has ever been used. Perhaps some art teacher at a mission school in the 1920s or 1930s was grateful for this opportunity to help her Zulu-speaking pupils select the colours for their palettes in terms from their own language, drawn from the natural world.

5.2   BIRD NAMES IN SECONDARY LEXEMES

5.2.1   Onymisation

In the photograph below we see a sign marking Mngcelu Road.[9] This road name – quite obviously – is derived from the noun **umngcelu** (pipit). The roads all around it have similar origins: for example Nkwazi Road (< **inkwazi** 'African Fish Eagle'), Hlalanyathi Road (< **ihlalanyathi** 'Oxpecker') and Ntinginono Road (< **intinginono** 'Secretarybird').

---

9. Photograph taken in Stage 2 of Imbali suburb in the city of Pietermaritzburg, KZN.

The linguistic process whereby words for birds have become names for streets is known as onymisation: the process by which 'common nouns' become 'proper names'. In this book we been referring to words such as **uthekwane, inkwazi** and **ihlalanyathi** as the 'names' of birds, but, as pointed out in Chapter 1, these are not really proper names at all, and it is merely a non-linguistic convention to call them 'bird names'. On the other hand, the names of streets (and suburbs and towns and cities) are all regarded as 'proper names' (street names are known as 'odonyms' and the names of towns and cities are 'urbonyms').

The linguistic process of onymisation implies three different changes:[10]
   1. As we see above, there is a change in onomastic status: a 'common noun' such as **uthekwane** becomes the proper name Thekwane Road,

---

10. The International Council of Onomastic Sciences (ICOS) has a number of standing committees, one of which is the Terminology Committee. In the ICOS Terminology list, 'onymisation' is defined as "the transfer of a linguistic unit (including common nouns, adjectives, verbs, interjections, phrases, etc.) to the class of proper names".

with a consequent change in capitalisation, and dropping of the noun class prefix.

2. There is a change in meaning: the meaning of the noun **uthekwane** is 'the bird known to scientists as *Scopus umbretta*, i.e. the Hamerkop'; the meaning of the name Thekwane Road is 'a short stretch of road running south-west to north-east between Hlalanyathi Road and Ntshingizi Road in the suburb Imbali, in Pietermaritzburg, KwaZulu-Natal [KZN].'[11]

3. There is a change in lexical status: **umngcelu** is a primary lexeme (in other words, this is the original form of the word with its original meaning), whereas Mngcelu Road is a secondary lexeme (in other words, derived from the original word, with a new form and new meaning, but with clear links to the original word). In non-linguistic speak we would say, "This road has been named after the pipit in Zulu."

In this section, I look at a variety of Zulu bird-origin secondary lexemes: the names of people, places and brands that are derived from the Zulu names for birds. This is a huge topic in itself, and so cannot be discussed in any great detail. The intention of this sub-section is to give an idea of the range of entities that originate in words resembling **umngcelu**, **inkwazi** and **ihlalanyathi**.

We will first look at the names of people and then at a variety of place names: natural geographical features such as mountains and rivers, and man-made locations such as towns, suburbs and streets, also farms, hotels and holiday lodges.

5.2.2   Names of people

In a previous publication, I have collected together, from various sources, a number of nicknames of colonial whites in the Xhosa-speaking and Zulu-speaking areas of eastern South Africa.[12] A number of these are bird-derived, ranging from the general such as uNyonende (tall bird), a nickname often given to a tall person, to the individual, such as uBusobendlazi (face of a mousebird), given to a particular individual with a round face. The birds themselves range

---

11. This holds true in this particular suburb, at least. In other places, where there is also a 'Thekwane Road', each would be in its own locational situation.

12. See Koopman (2014b).

from the taller ones, as in the name uNtinginono (Secretarybird) given for a tall, thin man with a long stride, to the smaller ones, as in the Xhosa name uNgqatyana (sparrow), given to a certain Dr Henderson of Lovedale who was very small in stature.

The links between the characteristics of certain birds and the characteristics of certain persons are various:

The name uThekwane (Hamerkop) was recorded several times for individuals perceived as vain, from the Hamerkop's habit of staring for long periods at its reflection in the surface of still pools (see Plate 8).

The name uGwalagwala (< *igwalagwala* 'turaco') was given to Henry Fynn, son of early Natal settler Henry Francis Fynn, because he always wore a turaco feather in his hat.[13]

The name uNondwayiza (the jacana or 'lily-trotter') has on occasions been given to individuals with long legs.

Early magistrates give a number of examples of nicknames derived from the names of birds: John Bird, resident magistrate in Pietermaritzburg from 1859 to 1876, was known as uNyoniyentaba (bird of the mountain). Another magistrate from the same era was named uMpangele (< **impangele** 'guinea-fowl'), while yet another was uNgqungqulu (< **ingqungqulu** 'Bateleur Eagle'). Magistrate 'Bateleur' was probably named for the fear he induced, but the man called uKhozi (< **ukhozi** 'eagle') was given this name because he had forward sloping shoulders. Another magistrate was nicknamed uMngqangendlela (Rufous-naped Lark) because he always took the straightest and most direct route to anything.[14] A pipit, the early-rising **umngcelu**, gave rise to the name uMngcelu for magistrate G. Walker Wilson, well known for his habit of rising early.

In 2015, I worked with Dr George Hughes, at one time the head of the former Natal Parks Board (now KZN Wildlife), to collect all the Zulu

---

13. See Koopman (2014b: 58)
14. This bird is known for running straight along the path in front of people. See Chapter 9, page 245.

nicknames of staff who had worked for the NPB over the years. Of the 103 nicknames collected, six were bird-based:

Hugh Dent – uThekwane (Hamerkop)
Matt Jackson – uNyonezinde ('tall birds' – refers to a tall man)
Keith Meicklejohn – uNondwayiza ('jacana' – normally given to a man with long legs and a big stride)
Peter Ruddle – uPhuphulendlazi ('mouse-bird chick' – "he resembled a mouse-bird chick")
David Skead – uNtungunono ('Secretarybird' – usually given to tall person with big stride)
'Tonk' Tomkinson – uNsingizi ('Southern Ground Hornbill' – walked with the gait of this bird).

As can be seen from these examples, bird-derived nicknames are not just chosen arbitrarily, but as with all Zulu nicknames, are chosen because there is some perceived resemblance between the bird and the person nicknamed: Hamerkops do seem to look at their reflections in water, as vain people do in mirrors;[15] the **umngcelu** (pipit) does rise early in the morning as magistrate Walker was wont to do. In such cases part of the meaning of the bird name is carried forward to the meaning of the new nickname.

5.2.3   Names of geographical features

This section looks specifically at the names of mountains and hills, on the one hand, (officially oronyms,), and at rivers and streams, on the other hand, (hydronyms, a word that encompasses names for all bodies of water: rivers, ponds and streams, lakes and lagoons, seas and oceans).

I have not been able to find many mountains and hills named after birds. The few I have are listed here:

- Majuba Mountain near Newcastle, KZN, is perhaps the best-known (< **amajuba** 'pigeons').
- In Ithala Game Reserve are the hills iKholwane Hill[s?] (< **ikholwane** or **umkholwane** 'hornbill'). Other examples are iTshelamasomi

---

15. I do not know Mr Hugh Dent personally: there could well have been many other reasons why he was nicknamed uThekwane.

(rock of the Red-winged Starlings) and iTshelenkonjane (rock of the swallow).

- In the Tugela Ferry area is the smaller hill iNyonyane (< *inyoni* 'bird' + *–ane* 'small species') and further to the east lies the highest mountain in the area eQhudeni Mountain (< **iqhude** 'rooster').
- In the greater Pietermaritzburg area lies the iNyonithwele Mountain (< *inyoni* 'bird' + *ithwele* 'it is carrying [something]').

Although there are few hill and mountain names, there are more examples of rivers and streams with bird-derived names, and I have benefited here from a study done on Zulu hydronyms by T.J.R. Botha, published in 1977, in Afrikaans, as *Watername in Natal*.

The Boboyi River is a tributary of the uMpenjathi River on the KZN South Coast. Botha speculates that this Xhosa word for a hoopoe has been used here (1977: 25). Also in the South Coast/southern KZN area is the Nomyayi River that Botha translates as 'Swartkraai-Spruit' (black crow stream), suggesting that this name is derived from the Xhosa word *unomyayi* for *Corvus capensis* (1977: 187).

Botha erroneously translates the name of the Gqumusheni River in the Greytown area as 'Janfiskaalspruit' (fiscal shrike stream) (1977: 79), because the Doke and Vilakazi gloss (1958: 268) for the bird **igqumusha** that he relies on says "Bush shrike, Butcher-bird".[16]

The **igqumusha** is the bird that features in the proverb *insele yasulela ngegqumusha* (the genet wiped [its feet] on those of the bushshrike).[17]

In the Hlabisa district is the minor river the Mankankaneni, the locative form of the plural *amankankane* (< **inkankane** 'hadeda'). Botha explains: "*Die spruit is 'n gewilde hadida-oord*" (the stream is a popular hadeda resort) (1977: 107).

The Ngagalu River, in the Nongoma district, takes its name from **ingagalu** (korhaan)', with Botha explaining: "*Dié voëlsort is in hierdie en in die aangrensende gebied verbrei*" (this bird species is widespread in this and the surrounding areas) (1977: 174).

---

16. Doke and Vilakazi also say that this name is "applied to several species of Oriole". The name **igqumusha** is used today for the Southern Boubou, a type of bushshrike.
17. See Chapter 9, section 9.2.5, 'Proverbs and folk tales', where the role of this bushshrike is explained.

The Zinkwazi (< **izinkwazi** 'fish eagles'), on the KZN North Coast, is well known. Perhaps less well known is the Nkwazi River in the Ixopo district, with Botha explaining: "*Soos aan andere Natalse riviere kom die* inkwazi *ook aan hierdie spruit voor*" (as with other Natal rivers the fish eagle is also found near this stream) (1977: 184).

Botha's comments on two particular river names (eNkwalini and Mbuzana) are worth quoting in full because they are general comments about onymisation and about the difficulties of reconstructing the etymologies of geographical names.

The place name eNkwalini refers to the valley that lies between Melmoth and Eshowe in Zululand, as well as the river at the base of the valley and a small village along this river. It is derived from **inkwali**, variously interpreted over the years as 'pheasant', 'quail' and 'partridge'. Botha cites Bulpin (1952: 286) as naming this valley "Place of the Red-legged Pheasant" and says:

Patryse was en is egte besonder volop in die gebied, sodat dit eintlik 'n onbegonne taak is om to probeer rekonstrueer watter voëlsoort die oorspronklike naamgewers presies in gedagte gehad het (1977: 184).

[Partridges [and related gamefowl] were and are rather common in this area, so it is actually a thankless task to try to reconstruct exactly what bird species the original name givers had in mind.]

The Green-backed Camaroptera is named **imbuzana** (little goat) because of its bleating call. Botha's explanation for the Mbuzana River in the Lower Umfolozi district is worth quoting in full, as it provides a rationale for naming rivers after birds, which can be applied generally:

Omdat hierdie voëlsoort in die betrokke streek voorkom, by voorkeur in die boomryke gebiede aan riviere waar hy gesien en, weens sy onderskeidende roep, ook gehoor kan word, is daar 'n assosiasie tussen dié voëlsoort en die rivier vasgelê, sodat die voëlnaam op die rivier oorgedra is (1977: 123).

[Because this bird species is found in this particular region, by preference in the well-wooded areas on rivers, where it is seen and, because of its distinctive call, also heard, an association between this bird species and the river is established, so that the bird name is transferred to the river.]

5.2.4   Street names

This is a particularly rich field for examples of onymisation, as I discovered when I entered the words 'Iphothwe Road' into Google Maps, thinking that this name for the well-known 'toppie' (Dark-capped Bulbul, *Pycnonotus tricolor* (previously *P. barbatus*) must be reflected in a street name <u>somewhere</u> in KZN. Google Maps immediately asked me whether I meant the Iphothwe Road in Pinetown, KwaMashu, Inanda, Ntuzuma, Umlazi, Empangeni or Pietermaritzburg. Checking all of these in turn showed that municipal officials charged with assigning names to streets in newly laid out areas frequently turned to Zulu avian nomenclature for inspiration. No matter in which area I sought Iphothwe Road, I found it surrounded by street names reflecting dozens of other birds. The area in 'J' section, KwaMashu,[18] is known locally as 'eZinyonini' (among the birds), as can be seen in the following, image, scanned from page 33 of the 2005 publication, *Map Studio: Street Guide, Durban: Includes Towns of KwaZulu-Natal*:

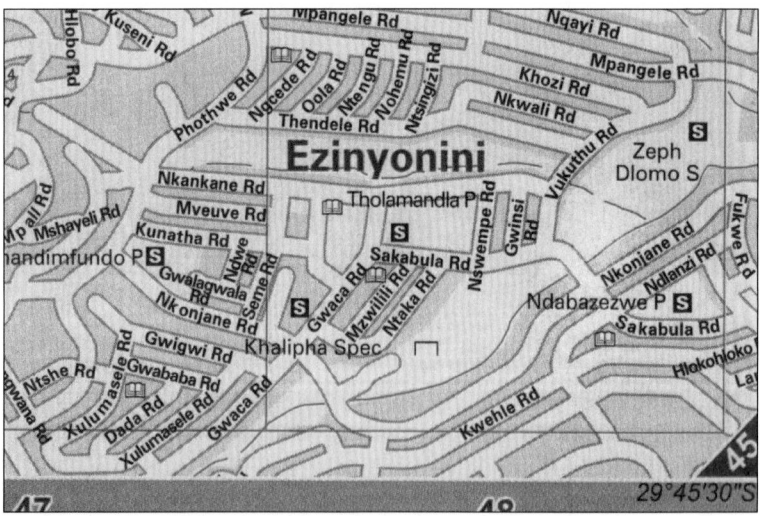

[i]Phothwe Road can be seen just to the left of the 'E' of 'Ezinyonini'.

---

18. Where, ironically, I worked as a municipal official in the mid-1970s, without noticing any street names derived from birds.

As it was a little difficult for me to visit 'The Place of Birds' in KwaMashu, I decided rather to pay a visit to Stage 2 of the suburb Imbali, a lot closer to my home in Pietermaritzburg, to see if I could find street signs that related to the names on the map shown below, taken from page 117 of the same *Map Studio: Street Guide, Durban: Includes Towns of KwaZulu-Natal*: showing the map of Pietermaritzburg's streets.

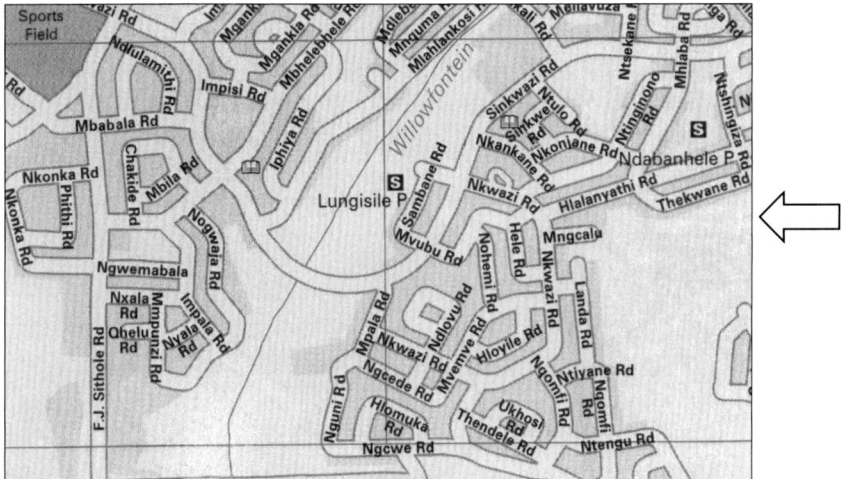

The 'bird section' is shown to the south-east of the Willowfontein River.

I entered this area from the Slangspruit area to the south-east (not shown on the map) and soon found myself on the intersection of Ntshingiza Road (< **insingizi** 'Southern Ground Hornbill') marked by an arrow on the map above, Thekwane Road (< **uthekwane** 'Hamerkop') and Hlalanyathi Road (< **ihlalanyathi** 'oxpecker'). There were no street signs to tell me this; the information was conveyed to me by three older men sitting on the kerb on the side of Ntshingiza Road. When I asked them how they knew the names of the intersecting streets, they laughed and pointed out that they were long-time residents of the area – why should they not know the street names? Driving a little further into the area, where Mvemve Road (< **umvemve** 'wagtail') and Nkwazi Road (< **inkwazi** 'fish eagle') intersected, I was able to find some rather battered concrete street signs, two of which have already been shown above, with three more shown below:

The three roads shown here are Nqomfi Road (< **inqomfi** 'Yellow-throated Longclaw'), Ntiyane Road (< **intiyane** 'Common Waxbill') and Landa Road (< **ilanda** 'Cattle Egret').

5.2.5 Brand names and logos

One of the best-known Zulu bird names serving as a brand name is surely **ijuba,** the generic term for a dove that has become synonymous with factory-produced traditional beer sold in a waxed carton (see Plate 15). A web search of some of the more well-known Zulu bird names such as **igwalagwala** (turaco), **impangele** (guineafowl), **indwe** (Blue Crane), **intendele** (partridge) and **ukhozi** (eagle) turned up scores of different businesses, institutions, and products using these Zulu bird names as brand names, usually with an accompanying logo based on the same bird. Two birds that are particular productive in this way are the Hamerkop and the African Fish Eagle.

In the corporate world, the Hamerkop's Zulu name suffers from spelling problems. While I was able to find one firm (Uthekwane Projects and Services) that uses the correct noun prefix, most prefer to drop it off (Thekwane Lodge, Thekwane Projects, Thekwane Holdings and Construction, The Thekwane's Nest and Thekwane Logistics Holdings). Also very popular is the aberrant form 'Tegwaan' (Tegwaan Country Getaway, Tegwaan Builders, Tegwaan Investments, Tegwaan Guest Farm and Tekwani Sawmills).

The African Fish Eagle does not suffer such orthographic indignities: almost every use of the name **inkwazi** gives it with the noun prefix, with only a few dropping it off. The following are a few examples: Inkwazi Country Hotel, Inkwazi Travel and Tours, Inkwazi Software Solutions, Inkwazi

Shopping Centre, Inkwazi Video Vehicle Tracking Solutions, Inkwazi Self-catering Cottage, Inkwazi Flyfishing Safaris, Inkwazi Learning Network and Inkwazi Glass and Aluminium. Those without the prefix include the Nkwazi Lake Lodge, and the Nkwazi Football Club in Lusaka, Zambia.

I do not wish to go into a detailed analysis of the use of the brand name 'inkwazi' among this different institutions and products, beyond the observation that different aspects of the fish eagle are used for different marketing intentions. Going by information on the various websites of these country hotels, glass and aluminium firms and learning networks, we can observe:

- Almost every country hotel, lodge, bed-and-breakfast, rest camp and other holiday accommodation stresses the fish eagle as a symbol of freedom from the hustle and bustle of city life (for example, "wake up to the sound of the fish-eagle" – a refrain repeated in various ways).
- The Video Vehicle Tracking form stresses the eagle as the 'eye in the sky'.
- Travel and tour firms stress the far-ranging flights of eagles generally.
- The fly-fishing safari firm that uses the fish eagle stresses (as one might expect) its prowess in catching fish.

For scholars focusing on this particular sub-discipline of onomastics, brand names (or *ergonyms* or *chrematonyms* as they are known to some European onomasticians) provide much interesting data. One needs, perhaps, a devoted ornithological chrematonymist to follow up on Zulu bird names used in this way.

6 The morphology of bird names

6.1 INTRODUCTION

The term 'morphology' refers to structure, and can be used both for birds and bird names. When used for birds, it is a matter of wings and toes, bones and feathers, in other words the constituent elements of the body of a bird and how these differ from other species, genera or families. When used of bird names, it is a matter of word roots and word stems, of prefixes and suffixes, in other words the constituent elements of words as grammatical items. In previous chapters, I have dealt with the meanings of bird names; now I look at how bird names are constructed.

In this chapter I deal mainly with Zulu bird names. As all names are nouns, no matter what the language, and as Zulu is a Bantu language, it is necessary to understand the basic nature of the Bantu noun class system. All nouns in Bantu languages are divided into a number of noun classes, each with a distinguishing prefix that indicates whether the noun is singular or plural. The relationship between singular and plural, and the use and function of the noun classes is reasonably uniform across all 400 or so Bantu languages. For example, all the languages use class 1 for humans, with a plural in class 2, as in Zulu *umuntu/abantu*, S. Sotho *motho/batho*, Shona *munhu/vanhu*, etc. All the languages use class 7 to indicate languages, so *isiZulu, seSotho, tshiVenda, xiTsonga, ciShona, chiChewa* and *kiSwahili*. All the languages use class 15 to indicate infinitives or gerunds, so *ukufa* in Zulu, *hufa* in S. Sotho and *ku-fa* in Nyanja all mean 'death', 'to die' and 'dying'.

The Zulu bird names that follow illustrate the noun classes:

Noun class 1 with the prefix *um–* or *umu–* is used only for humans, and there are no bird names in this class. The plural, in class 2, with the prefix *aba–* is also only used for humans, and there are no bird names, with the sole exception of the name **abayeni**, used only in this plural form, for the White-crested Helmetshrike.

Noun class 1a, with the prefix *u–*, is also only used for humans, and there are no birds in this noun class. Class 3a, also with the prefix *u–*, does contain bird names, and examples are **ufukwe** (Burchell's Coucal), **ujojo** (Pin-tailed Whydah), **ubhamukwe** (Wattled Crane) and **unohhemu** (Crowned Crane). In the singular these names mean either a single specimen of each bird, or the species as a whole. In other words, **ufukwe** means a single Burchell's Coucal, or the species as a whole, and **ujojo** means a single Pin-tailed Whydah or the species *Vidua macroura*. The plural of class 1a is found in class 2a, with the prefix *o–,* so several Pin-tailed Whydahs are **ojojo**, and **obhamukwe** means two or more Wattled Cranes.

Noun class 3, with the prefix *umu–* or *um–*, and with the corresponding plural in class 4, with the prefix *imi–*, contain a number of birds. Examples are **umngcelu** (used generically for larks) with the plural **imingcelu**, **umunswi** (used generically for thrushes) with the plural **iminswi**, and **umvemve** (used generically for wagtails) with the plural **imivemve**.

Noun class 5 (prefix *i–*), with its corresponding plural in class 6 (prefix *ama–*), contains many birds. The following are examples: **ijuba/amajuba** (used generically for doves and pigeons); **ifefe/amafefe** (used generically for rollers), **iqola/amaqola** (Southern Fiscal, formerly called Fiscal Shrike) and **ikhwezi/amakhwezi** (used generically for glossy starlings).

Classes 7 (singular, with prefix *is–* or *isi–*) and 8 (plural, with prefix *iz–* or *izi–*) are equally productive for bird names, including: **isakabuli/izakabuli** (Long-tailed Widowbird), **isihuhwa/izihuhwa** (Martial and Crowned Eagles), **isikhwehle/izikhwehle** (used generically for some francolins or partridges) and **isikhova/izikhova** (used generically for owls).

Classes 9 (prefix *in–* or *im–*) and 10 (prefix *izin–* or *izim–*) contain more birds that any of the other noun classes: well-known birds in these classes are **indlazi/izindlazi** (Speckled Mousebird), **inkwali/izinkwali** (used generically for some francolins or partridges), **intengu/izintengu** (drongo), **inkwazi/izinkwazi** (African Fish Eagle) and **intshe/izintshe** (ostrich).

A few birds are found in class 11 (prefix *u–* or *ulu–*), and the plural is in class 10 (prefix *izin–* or *izim–*). Some examples are **uluve/izimve** (Paradise Flycatcher), **uheshe/izinkeshe** or **iziheshe** (used generically for the smaller hawks) and **ukhozi/izinkozi** (used generically for eagles).

Classes 12 and 13, used to form diminutives in Bantu languages north of the Limpopo River, are no longer used in the Bantu languages of South Africa.

These have the prefix *ka–* in most of the languages, and we saw a number of smaller birds in this class in Chapter 3: the Tanzanian *kahuji* (dove) and the Nyabongo *kadekere* (Black-winged Kite) for example.

Class 14, with the prefix *ubu–*, is used to form abstract nouns in the Bantu languages. The best-known example is the word *ubuntu* (humanity). Only one single bird name occurs in this class – the name **ubucubu** (African Firefinch, previously the Blue-billed Firefinch) – and there is no plural for this class.

Class 15 is only used for 'verbal nouns' (infinitives, gerunds), and there are no birds in this class.

6.2   CLASSIFICATION OF NOUN STEMS

In all the examples shown above, the bird names consist of a noun prefix and a noun stem. The noun class prefix indicates whether the noun is singular or plural and also indicates grammatical agreements with other words in the sentence. The noun stem gives the meaning of the noun, in this chapter indicating the folk genus, or the species of bird.

Thus '*is*' in **isakabuli** indicates singularity, while '*akabuli*' indicates Long-tailed Widowbird; '*izi*' in **izikhova** indicates plurality and '*khova*' indicates owl.

Noun stems contain lexical meaning, sometimes described as 'real life meaning': owls, eagles and widowbirds are 'real things' – biological entities existing in nature.

Noun prefixes, however, (and all other prefixes and suffixes that attach themselves to nouns), contain grammatical meaning. Concepts such as singularity, plurality, tense, mood, etc., are 'notional' entities – they exist in linguistic theory and grammatical description.

In this chapter, I classify bird names according to the structure of their noun stems. These may be divided into four main categories:
1. Names with simple stems
2. Names with reduplicated stems
3. Names with complex stems
4. Names with compound stems

6.3   NAMES WITH SIMPLE STEMS

Simple stems are stems that cannot be broken down into smaller meaningful units: they have one single, lexical meaning and no grammatical meaning. In single-stemmed words such as **in-gududu** (Southern Ground Hornbill),

i-titihoye (plover), **in-taka** (finch), and **isi-phungumangathi** (Long-crested Eagle), the stems '*gugudu*', '*titihoye*', '*taka*' and '*phungumangathi*' carry the identity of the birds mentioned and nothing else. They may indeed be broken down into syllables (*gu-du-du* and *phu-ngu-ma-nga-thi*) but syllables on their own carry no meaning unless they are a monosyllabic stem such as **ulu-ve** (Paradise Flycatcher) and **in-tshe** (ostrich), or are a grammatical affix.

The following are all examples of bird names with simple stems:

uzavolo: used generically for the nightjars
umqoqongo: Black-headed Oriole
ingqungqulu: Bateleur Eagle
indwe: Blue Crane
inqomfi: used generically for the long-claws
inkanku: Jacobin Cuckoo
intinginono: Secretarybird
umjenenengu: Narina Trogon
umxhwagele: Bald Ibis
unhloyile: Yellow-billed Kite
iseme: used generically for the bustards or korhaans
isikhobothi: Forest Buzzard
insingizi: Southern Ground Hornbill

Bird names with simple stems are generally monosemic, which is to say they carry only one meaning, that of the bird they refer to. For example **umjenenengu** means 'Narina Trogon' (either as a single specimen of the bird, or as the species as a whole). There is no other meaning. Compare this situation to a complex-stemmed name such as **isagwaca** (a generic name for quails) which consists of –*sa*– 'something like' and the ideophone *gwáca* 'of hiding', giving the underlying meaning of a bird that habitually hides in long grass. And then take a compound noun such as **ihlalanyathi** (oxpecker), where the underlying meanings of *hlala* (sit) and *inyathi* (buffalo) tell you that this is a bird that sits on a buffalo.

The only time single-stemmed nouns can give further information is when they are based on onomatopoeia, and thus give an idea of the bird's call. The name **umklewu** – does this, as does its English name Grey Go-away-bird[1]

---

1. Now the official book name (Chittenden, Davies and Weiersbye 2016: 266).

(formerly Grey Lourie [Turaco]) as well as the Afrikaans name *kwêvoël*. Other onomatopoeic names are **ititihoye** (plover), **ingududu** (Southern Ground Hornbill) and **inzwece** (Paradise Flycatcher).

It is noteworthy that the only simple-stemmed nouns coined in the 2013–2017 Zulu bird name workshops were those onomatopoeic names that referred to bird calls. Far more is said about this in the chapter on bird names and bird calls, but we can give a few examples here: **umcwicwicwi** (Green Malkoha, formerly Green Coucal), **ukliyo** (Lesser Crested Tern) and **usibó** (Black-headed Oriole).

6.4   NAMES WITH REDUPLICATED STEMS

Reduplication of stems is used for both nouns and verbs in Zulu, with opposite meaning – in nouns it is a kind of augmentation or pluralisation; in verbs it reflects diminution of the action. For example:

Nouns:
imifula 'rivers' > *imifulamfula* 'lots of little rivers', as in a delta
amafu 'clouds' > *amafumafu* 'lots of clouds all over the sky'
imimoya 'winds' > *imimoyamoya* 'lots of gusts of wind all over the place'

Verbs:
phuza 'drink' > *phuzaphuza* 'take little sips'
dla 'eat' > *dlayidla* 'nibble at something'
gijima 'run' > *gijigijima* 'jog along'

The use of reduplication of stems in the formation of Zulu bird names, is probably much closer to the way in which certain nouns in Zulu are formed from ideophones, as in the following examples:

isiphalaphala 'woman with lovely eyes' < *phála* 'of searching with the eyes'
isintwanguntwangu 'slovenly, untidy person' < *ntwángu* 'of slovenly dressing'
isinqandunqandu 'person who is always talking' < *qándu* 'of making a noise'

It is difficult to say whether a name such as **igwalagwala** (Purple-crested Turaco, previously Purple-crested Lourie) has been derived in a similar way. There is no ideophone '*gwála*' in Zulu, so has this bird name been derived from *igwala* 'coward' or perhaps from *ugwala* 'type of musical instrument similar to the Khoi *gora*'? If there is any derivation involved at all, it is probably from the word for a musical instrument, and a reference to the musicality of the bird's call.[2]

Reduplication is definitely linked to bird calls in the three previously recorded names for the Pied Starling: **ingwangwa**, **ikhwikhwi** and **igwayigwayi**. Maclean, among other verbalisations, records this bird as calling *skwee-skwee*, *skweer-skweer*, *skik-skik* and *chirrup-chirrup* – exact reduplications of syllable each time (1984: 668).

The following are examples of Zulu bird names with reduplicated stems:

igwigwi: species of waterbird
isiwelewele: species of marsh bird which utters a shrill, screeching cry (< the ideophone *wéle* 'of screeching' – also as *isiwekeweke*)
umjekejeke: Corn Crake
ibhoyibhoyi: Cloud Cisticola (*Cisticola textrix*)
ihlokohloko: generic for the yellow weaver birds (and surely derived from the ideophone *hlóko* 'of noise')
ingwangwa: generic for glossy starlings

This method of coining a new name was not popular at the 2013–2017 workshops and only three names were coined in this way, all onomatopoeia-based:

- **umcwicwicwi**: Green Malkoha (formerly Green Coucal)
- **isishishi**: Southern Black Tit
- **inswinswi**: Orange Ground Thrush

6.5 NAMES WITH COMPLEX STEMS

Names with complex stems are those that have one lexical element and one or more grammatical affixes (prefixes and suffixes). The most basic kind of complex names are those where a name is derived from a noun of a different noun class, but retains part of its original noun class prefix when it becomes a name. The workshop-coined name for the Grey-headed Bushshrike – **usipoki**

---

2. Although not everybody will agree that this bird has a musical call.

– is an example. Here the class 7 noun *isipoki* (from English 'spook') has been moved into class 3a with the prefix *u–*, but retains part of its original *isi–* prefix. The noun stem can thus be divided into the grammatical element *–si–* and the lexical element *–poki*. The name refers to the ghostly, hooting whistle of the bird, which has also given rise to the Afrikaans name *spookvoël*.

Another coined example is **umadevu** for the Whiskered Tern: the original class 6 noun *amadevu* 'whiskers' is moved into class 3a.

Slightly more complex examples are seen in the three names **ubantwanyana** (Emerald Cuckoo), **umehlwane** (Cape White-eye), and **usiqhovana** (Crested Barbet). In each of these cases, the original noun class prefix is retained in part, followed by the noun stem and then a diminutive suffix, as seen in the breakdown below. The noun stems are given in upper case:

u-bantwanyana: (*a*)*ba* + *NTU* + *ana* + *ana*[3]
u-mehlwane: (*a*)*me*[4] + *HLO* + *ana*[5]
u-siqhovana: (*i*)*si* + *QHOVA* + *–ana*

The name **ubantwanyana**, for reasons unknown, thus means literally 'the very little children'; **umehlwane** means 'the little one with [notable] eyes'; and the coined name **usiqhovana** means 'the little one with a crest' or 'the one with a little crest'.

In the next three coined names, a noun has again been moved from its original noun class into class 3a, and then further extended with a colour adjective:

usifubabomvu: Scarlet-chested Sunbird (< *u–* + [*i*]*sifuba* 'chest' + *–bomvu* 'red')
ukhandabomvu: Red-headed Quelea (< *u–* + [*i*]*khanda* 'head' + *–bomvu* 'red')
umadevaphuzi: African Wattled Lapwing (< *u–* + [*a*]*madev*[*u*] 'whiskers' + *aphuzi* 'which are yellow')

---

3. Phonological changes that have taken place when each *–ana* suffix adapts to the previous syllable have given rise to the final form *–bantwanyana*.
4. The word for 'eyes' in Zulu (*amehlo*) is exceptional in having *ame–* for a prefix instead of the standard *ama–*.
5. In Zulu biological nomenclature *–ane* and *–ana* are interchangeable. The *–ane* suffix is described in greater detail later in this chapter. Again the addition of the diminutive suffix has caused phonological changes, the noun stem *–hlo–* becoming *–hlw–*.

In the following three coined bird names, all indicating a place where the bird is commonly found, the base is a locative (in the place of . . .), and the locative has been moved into class 3a, with the prefix *u–* replacing the initial vowel of the locative:

> **umatsheni**: Freckled Nightjar (< *ematsheni* 'among/on the stones')
> **umalaleni**: Lemon-breasted Canary (< *emalaleni* 'among the lala palms')
> **ugazini**: Red-headed Finch (< *egazini* '[covered] in blood')

We go on now to discuss the use of the name-forming prefixes *–no–*, *–ma–*, *–so–* and *–sa–*.

6.5.1   The use of the prefix *–no–*

The name formative *–no–* is very productive in the forming of Zulu bird names. Originally used for forming both male and female personal names, but now only used in the names of girls,[6] it is also used extensively in the forming of names of biological entities. In an earlier publication (Koopman 2015: 57–9), I gave examples of plant names such as *unobebe* (*Eugenia albanensis*) and *unokleshe* (*Habenaria epipactidea*) and *–no–* can be used in the names of a number of other living entities such as *unogwaja* (hare, rabbit), *unozimponjwana* (type of beetle with horns), and *unohhohha* (onomatopoeic praise name for a baboon).

Bird names using this formative include **unondwayiza** (African Jacana), **unobulongwe** (both Burchell's and Temminck's Coursers), and the onomatopoeic names **unohhemu** (Crowned Crane) and **unogilonki** (the Grey and the Black-headed Herons). In the 2013–2017 Zulu bird naming workshops, using this formative was one of the most popular ways of coining new bird names, for example:

- **unokukhukhuza**: Common Cuckoo, formerly European Cuckoo (< *–no–* + (*u*)*ku-khukhuza* 'make the sound "cuckoo"')
- **unongoyana**: Green Barbet (< *–no–* + Ngoye[7] + *–ana*)
- **unongilobomvu**: Red-throated Wryneck (< *–no–* + (*i*)*ngilo* 'throat', 'neck' + *–bomvu* 'red')

---

6. See Koopman, 'Gender shift in the use of the formative *–no–* in Zulu given names', in T. Meyiwa and M. Cekiso (eds), *Names Fashioned by Gender* [forthcoming].
7. The Green Barbet is endemic to the Ngoye Forest in Zululand, and a few other isolated spots.

- **unonkositini**: Red-capped Robin-Chat, formerly Natal Robin (<–*no*– + (*i*)*nkositini* 'concertina')
- **unovilane**: Lazy Cisticola (< –*no*– + (*i*)*vila* 'lazy person' + –*ane*)
- **unophuzana**: Yellow-throated Petronia /Sparrow (< –*no*– + –*phuzi* 'yellow' + –*ane*)

6.5.2   The use of the prefix –*ma*–

The name formative –*ma*– usually indicates a characteristic behaviour. It is seldom used in the giving of first names to children, but is used with the maiden surname of married women, thus a Mrs Mhlongo, born Dlamini, may be known as MaDlamini, and someone who has married into the Mkhize clan, but was born in the Cele clan, may be known as MaCele. It is used to form nouns, often compound nouns, such as *umahlekehlathini* (lit. he who laughs in the bush), for a heavily-bearded man and *umagoningane* (he who marries a child), for a man married to a woman considerably younger than himself. Koopman shows the use in the formation of plant names, as in *umakuphole* (lit. 'let it cool down') for the plant *Pentanisia prunelloides* and *umayime* (lit. 'let it stop') for the well-known Clivia (2015: 56–7).

This formative is found in bird names such as **umahube** and **umangube** (both names for the Fan-tailed Widowbird (formerly Red-shouldered Widowbird), **umabhengwane** (the male of the Wood Owl) and **umabhelwane** (Bar-throated Apalis).

The participants of the 2013–2017 workshops used –*ma*– in the coining of a number of bird names, including the following examples:

- **umacutha**: Squacco Heron (< –*ma*– + *cutha* 'draw body tense')
- **umadletshana**: African Scops Owl (< –*ma*– + (*i*)*dlebe* 'ear' + –*ana*, i.e. 'the little-eared one')
- **umanyovini**: European Honey Buzzard (< –*ma*– + (*e*)*nyovini* 'among the wasps')[8]
- **umaphithizela**: Sanderling (< –*ma*– + *phithizela* 'move about in a haphazard, disorderly and confused manner')

---

8. The English name 'European Honey Buzzard' is a misnomer, as this bird feeds on wasps (see also page 371).

- **umasikulufu**: African Broadbill (< –*ma*– + (*i*)*sikulufu* 'screw' – a reference to the circular display flight of this bird)
- **umaphendulamatshe**: Ruddy Turnstone (< –*ma*– + *phendula* 'turn over' + *amatshe* 'stones')
- **umazalashiye**: Klaas's Cuckoo (<–*ma*–+*zala* 'lay eggs' + *ashiye* 'and then leave them behind')

6.5.3   The use of the prefix –*so*–

The name-forming prefix –*so*– is not nearly as productive in name-forming as the prefixes –*no*– and –*ma*–, but there are a few bird names that use this prefix. An example is **usomheshe** (African Goshawk), where –*so*– is prefixed to (**u**)**heshe**, a generic name for the smaller hawks.[9] As –*so*– usually carries the meaning of 'father of' or 'master of', this suggests an ethnobiological perception of the African Goshawk as the 'master' of other smaller raptors.

It is not clear whether the participants of the 2013–2017 workshops had the same idea in mind when they used –*so*– to coin the following bird names:

- **usomthende**: African Cuckoo-Hawk (<–*so*– + (*u*)*mthende* 'stripe')
- **usonkanyezi**: White-starred Robin (<–*so*– + (*i*)*nkanyezi* 'star')
- **usonambuzane**: Spotted Flycatcher (<–*so*–+ (*isi*)*nambuzane* 'insect')
- **usothathizwe**: Common (Indian) Myna (<–*so*– + *thatha* 'take [over]' + *izwe* '[the] country')

6.5.4   The use of the prefix –*sa*–

Also not as productive a name formative as –*no*– and –*ma*–, but still found in bird names, is the prefix –*sa*–, which means 'something like'. It is found in words such as *isamuntu* ('ghost', from –*sa*– + *umuntu* 'person') and *isambane* ('antbear' < –*sa*–+ *mba* 'dig' + –*ane*, i.e. 'something like a creature that digs').

Examples of bird names using this formative are **isagwaca** ('quail' <–*sa*–+ *gwáca* 'of hiding'), **isanxa** (Common Buzzard) (<–*sa*–+ (*i*)*nxa* 'side', 'edge', i.e. 'like something found on the edge', a reference to this bird being found on the margins of forests), and **usangqwashi** ('species of lark-like bird' <–*sa*– + (*i*)*ngqwashi* 'lark').

The 2013–2017 workshop participants found this formative useful in coining bird names, and the following are examples:

---

9. Now assigned specifically to the Lanner Falcon.

- **isankawu**: Southern Pochard (< *–sa–* + (*i*)*nkawu* 'monkey', a reference to the call of this duck, which sounds something like the alarm call of a vervet monkey)
- **usafukwe**: African Cuckoo (< *–sa–* + (*u*)*fukwe* 'Burchell's Coucal')
- **usamklewu**: Great Spotted Cuckoo (< *–sa–* + (*i*)*klewu* 'Grey Go-away-bird')
- **isaqola**: Fiscal Flycatcher (< *–sa–* + (*i*)*qola* 'Fiscal Shrike', now Southern Fiscal) (see Plates 32 and 33)
- **isangulube**: Gorgeous Bush Shrike (< *–sa–* + (*i*)*ngulube* 'pig', a reference to their grunting "*graak-graak*" alarm call)
- **usantiyane**: Green-winged Pytilia (< *–sa–* + (*i*)*ntiyane* 'Common Waxbill') (see Plates 32 and 33)

6.5.5   The use of the suffixes *–ana* and *–ane*

The previous sections dealt with the use of various prefixes in the formation of Zulu bird names.

We now go on to look at various suffixes found in bird names and, without any doubt, the most common here are the suffixes *–ane* and *–ana*. We treat these under one heading as they are frequently interchangeable and if a Zulu-speaking source is asked "Is the name of this bird '**umehlwane**' or '**umehlwana**'?" the response is usually bewilderment, as the person being asked perceives these as the same name. It is my personal perception, however, that where *–ana* is purely diminutive (as seen below), *–ane* also implies the designation of a biological species. The White-eye may be called **umehlwane** or **umehlwana**, both derived from *amehlo* (eyes), but if one were to talk of small human eyes, one would always use 'amehlwana', never 'amehlwane'

The suffix *–ana* is the diminutive suffix, found in Zulu words such as *umntwana* ('child' < *umuntu* 'person' + *–ana*) and *indodana* ('son' < *indoda* 'man' + *–ana*). This suffix is found in all South African Bantu languages, in contrast to the Bantu languages found north of the Limpopo River, where the nouns classes 12 (plural) and 13 (singular) are used to form diminutives. As an example, when the Zulu language wishes to indicate a small goat, *–ana* is suffixed to *imbuzi* (goat) to produce *imbuzana*, with the plural *izimbuzana*. However, when the Shona language wishes to do the same thing, the noun *mbuzdi* (goat) is moved into class 13 (with the noun class prefix *ka–*) to form *kambudzi* (small goat) with the plural found in class 12 (with the noun class prefix *tu–*) as *tumbudzi* (small goats).

The diminutive noun classes account for the names of a number of small birds in Bantu languages, such as Nyanja of Malawi and Nyabongo of the Congo, as seen in the following examples:

From Nyanja:[10]
kadzidzi: a very small species of owl, with a large head
kakozi: a small species of hawk[11]
kansire: a small grey bird
kanchenge: a small species of kingfisher

From Nyabongo:[12]
kafunzi: Common Waxbill
kakukhwe: small Blue-headed Coucal < *mukukhwe*
kanyamarhaza: Green Sandpiper and Wood Sandpiper
kashuge: Bully Canary

There are a considerable number of Zulu bird names where the written forms reflect *–ane*, and among these we can include:

imbuzane: Green-backed Camaroptera   **inkankane**: Hadeda Ibis
iklebedwane: Grey Cuckooshrike   **inqwathane**: Collared Sunbird
udemezane: Black-winged Kite   **umbhukwane**: Blue Korhaan
uphalane: Egyptian Vulture   **inkonjane**: generic for
                                            swallows

The *–ane* suffix was used occasionally in the coining of new names at the 2013–2017 workshops for example:
- **intuntwane**: Red-knobbed Coot
- **umkhololwane**: Crowned Hornbill (cf. **umkholwane**: Southern Red-billed Hornbill)
- **unonklilwane**: Little Tern (cf. **unonkliyo**: Sandwich Tern and **ukliyo**: Lesser Crested Tern)

---

10. These examples are taken from Scott (1929).
11. The early Bantu root *–kozi* can be distinguished here, found in Zulu as **ukhozi**, a generic name for 'eagle'.
12. These examples are taken from Hendrickx (1944).

Bird names with *–ana* suffixes as written forms are equally common, and the following are a few examples:

ibomvana: Southern Red Bishop (the little red one)
uhlazanyana: Malachite Sunbird (the little green one)
isiqhanazana: Black Crake
inkovana: Pearl-spotted Owlet (< *isikhova*: generic for owls)
impofazana: Wattled Starling (< *–mpofu* 'tawny-coloured' + *–azi* + *–ana*)

The 2013–2017 workshop participants found this suffix to be very useful in the coining of new names for birds, especially where the English name included the designation 'lesser' as in 'Lesser Honeyguide' and in the following examples:

- **ingedana**: Lesser Honeyguide (< **ingede**: generic for honeyguides)
- **unongoyana**: Green Barbet (< Ngoye, i.e. 'the little one from Ngoye Forest')
- **isizinzana**: Baillon's Crake (< **isizinzi**: generic for crakes)
- **inkothana**: Little Bee-eater (< **inkotha**: generic for bee-eaters)
- **usiqhovana**: Crested Barbet (< *isiqhova* 'crest')

6.5.6   The use of the suffixes *–azi* and *–kazi*

In the name for the Wattled Starling, given above as **impofazana**, we noted the suffix *–azi* was used as well as *–ana*. This suffix, also found in the form *–kazi*, can indicate either the feminine or the augmentative. Thus in *umfazi* ('woman' < *umfo* 'human') and *indlovukazi* ('cow elephant' < *indlovu* 'elephant'), the suffix is expressing the notion of 'female'. In words like *ifukazi* ('large cloud' < *ifu* 'cloud') and *itshekazi* ('boulder' < *itshe* 'stone'), the suffix is expressing the notion of large size.

The following were words coined during the 2013–2017 workshops:

- **unogqabhakazi**: Ruff (< *no–* + *gqaba* 'mark the face' + *–kazi* 'greater')
- **unosidlekekazi**: Sociable Weaver (< *–no–* + (i)*sidleke* 'nest' + *–kazi* 'large')
- **umfelokazi**: Dusky Indigobird (formerly Black Widowfinch): here the Zulu word *umfelokazi* 'widow' (< *umfelo* 'one who is bereaved'),[13] was simply given new life as a bird name, influenced no doubt by both the earlier English name for the bird, as well as the specific epithet in *Vidua funerea*.

---

13. Itself a derivation from the verb *fa*, 'to die'.

6.5.7   Various other suffixes

Various other suffixes may be found in Zulu bird names, for example:

ukholwase: flamingo (here the archaic suffix –*se*, found in clan names such as Mpungose, Nyambose and Shangase, is added to the verb *kholwa* 'believe').

unkombose: Namaqua Dove (here the same suffix –se is added to the noun *inkombo*, derived from the verb *khomba* 'point out').

unukani: species of woodhoopoe (here the interrogative suffix –*ni*? (what?) is added to the verb *nuka* 'smell').

imbuyelelo: Woodland Kingfisher (the noun *imbuyelelo* is derived from the verb *buyelela* 'persistently return', in turn derived from the verb *buya* 'return'. In coining this name, the workshop participants noted how this kingfisher consistently returned to the same perch after catching something).

isiphikeleli: Brown-hooded Kingfisher (the noun *isiphikeleli* is derived from the verb *phikelela* 'persist in doing something', where the same verbal suffix –*elela* is used).

umbicini: Cape White-eye (the noun *umbicini* uses the locative suffix –*ini*).

umamhlangeni: African Marsh Harrier (the noun *umhlanga* 'reed bed', is suffixed with the locative –*ini*, causing the final 'a' to change to 'e'. The bird's name means 'that which is characteristically found in a reed bed').

The locative form of the noun, as seen in **umbicini** and **umamhlangeni** above, was used in a number of names coined in the 2103–2017 workshops, and the following examples have all been given elsewhere in the book:
- **umanyovini**: Honey Buzzard (< ma– + (*e*)*nyovini* 'among the wasps')
- **umatsheni**: Freckled Nightjar (< (*e*)*matsheni* 'among the stones')
- **ugazini**: Red-headed Finch (< (*e*)*gazini* 'in the blood', the bird looks as though its head has been drenched in blood)

6.6   NAMES WITH COMPOUND STEMS

Compound nouns are nouns where there are at least two lexical elements in the stem, and usually with one or more grammatical affixes. Take for example the name **ubhavuzilomidwayidwa** for the Striped Flufftail, at the heart of this name are the ideophone *bhávu* (of making a noise on a paraffin tin) and the noun *umudwa* (stripe). The ideophone has been made into a verb by adding the suffix –*za*, and this verb has been put into the perfect tense with the suffix –*ile*, creating a generic name for flufftails that essentially means, 'the one which has made a resonating sound'. The noun stem –*dwa* has been duplicated to form the plural *imidwayidwa* (a great number of stripes). The two words are joined together with –*o*– filling the function of the English 'which'. We thus have a name meaning 'the flufftail which is heavily striped', and using upper case for the lexical roots and lower case for the grammatical morphemes, we can show the structure of this bird name as

u–BHAVU–z(a)–ile–o–mi–DWA–yi–DWA

I start with those that have a noun as the first lexical element, and sub-divide them according to whether they are NOUN + NOUN, NOUN + ADJECTIVE, NOUN + VERB; and so on.

6.6.1   Noun + noun

We start with four bird names, to illustrate how two nouns placed 'side by side' in a compound noun may relate to each other semantically in the same way, but are morphologically different. In each of the four Zulu names below, the second noun is used to qualify the first noun, in exactly the same way as an adjective would.[14]

itithoyelimlotha: Grey Plover
itithoyenomqhele: Crowned Plover
ukhozilwentshebe: Bearded Vulture or Lammergeier
isigqobhamithintshebe: Bearded Woodpecker

The first name (**itithoyelimlotha**), is compounded from the nouns **itithoye** (plover) and *umlotha* (ashes). The basic form is [**itithoye** 'plover' + *eli* 'which

---

14. In Zulu, adjectives come after the noun, not before the noun as in English.

is' + (*u*)*mlotha* 'ashes']. This construction, where a noun (without its initial vowel) is used to qualify another noun, is usually found in Zulu *izibongo* (praise poems), and normally occurs as a three-word phrase such as *umuntu ozitho zinde* (person [who is] limbs long) and *isitulo esimilenze mithathu* (stool [which is] legs three). These constructions equate to the English constructions, 'long-limbed person' and 'three-legged stool'. The workshop participants showed that they were familiar with this type of grammatical construction, and used it often if a bird name would otherwise have been clumsily overlong. The name for the Grey Plover then is literally 'plover [which is] ashes [in colour]'.

The second name (**ititihoyenomqhele**), is also based on two nouns – **ititihoye** (plover) and *umqhele* (crown). These two nouns are joined by –*na*– (and, have) in the construction **ititihoye** + *na* + *umqhele*, giving a name that literally means, 'plover [which] has a crown'.

The third name (**ukhozilwentshebe**), uses a possessive concord to join the two nouns **ukhozi** (eagle) and *intshebe* (beard), so the basic construction is [*ukhozi* + *lwa*– 'of' + *intshebe*].

The fourth name (**isigqobhamithintshebe**), simply juxtaposes two nouns: *isigqobhamithi* (itself a compound of the verb *gqobha* 'peck' and the noun *imithi* 'trees'), and the noun (*i*)*ntshebe*, beard. The literal meaning of the name is thus simply 'peck-trees beard'.

Several examples are given below of these different constructions. Those coined at the 2013–2017 workshops are marked with [#]:

inkonjenesibhakabhaka[#]: Blue Swallow (< **inkonjane** 'swallow' + (*i*)*sibhakabhaka* 'sky, firmament')

inkonjanesiloside[#]: Wire-tailed Swallow (< **inkonjane** 'swallow' + (*e*)*silo* [which is] tail + *eside* 'which is long', i.e. 'long-tailed swallow')

uncedoselesele[#]: Croaking Cisticola (< **uncede** 'cisticola' + (*i*)*selesele* 'frog')

ukhozolumabala[#]: Lesser Spotted Eagle (< **ukhozi** 'eagle' + *olumabala* 'which has spots')

inqelendlovu: Marabou Stork (< **inqe** 'vulture' + *la* 'of' + *indlovu* 'elephant', i.e. 'the elephant-sized vulture')

impisiyolwandle[#]: Brown Skua (< *impisi* 'hyena' + *ya* 'of' + *ulwandle* 'sea')

inkukhuyamanzi[#]: Common Moorhen (< **inkukhu** 'chicken', 'domestic fowl' + *ya* 'of' + *amanzi* 'water')

imbuzane yomnqawe: Burnt-necked Eremomela (< **imbuzane** 'species of warbler' + *ya* 'of' + *umnqawe* 'mimosa tree')

The Zulu name **inqelendlovu**
(lit. 'elephant vulture') refers to both
the large size and the bald head
and neck of the Marabou Stork.
© Adrian Koopman

ifefemidwa[#]: Purple Roller (< **ifefe** 'roller' + (*i*)*midwa* 'stripes')
inqemvuma[#]: Palm-nut Vulture (< **inqe** 'vulture' + (*i*)*mvuma* 'Kosi palm')
usikhothamlotha[#]: Ashy Flycatcher (< **isikhotha** 'flycatcher' +
(*u*)*mlotha* 'ashes')
isikhovanhlanzi[#]: Pel's Fishing Owl (< **isikhova** 'owl' + (*i*)*nhlanzi* 'fish')[15]
intakansinsi: Southern Red Bishop (< **intaka** 'finch' + (*i*)*nsinsi* 'red and
black seed of the *umsinsi* [Erythrina] tree")
umvemventaba[#]: Mountain Wagtail (< **umvemve** 'wagtail' + (*i*)*ntaba*
'mountain')
isikhovampondo: Cape Eagle-Owl (< **isikhova** 'owl' + (*izi*)*mpondo*
'horns')

6.6.2   Noun + adjective

This is another very common way of constructing bird names, and almost
every example shown below was coined at the 2013–2017 workshops. Some

---

15. After this name was coined at the 2013–2017 workshops, it was discovered that
Doke and Vilakazi's dictionary had the entry **ifukezi** for this owl species. None of
the workshop participants had heard this name, however, and, as it might have been
an earlier misprint for the well-known name **ifubesi** for the Giant Eagle Owl, it was
decided to leave the coinage **isikhovanhlanzi** in place until further fieldwork shows
'*ifukezi*' to be an extant name or an error.

of the names have been written as two-word phrases, rather than as compound nouns, mainly because if written as one word they would have produced an awkwardly long compound. For example, '*ubhavuzilomidwayidwa*' and '*ujamelumhlangomncane*' are better written as **ubhavuzile omidwayidwa** and **ujamelumhlanga omncane**. These are the names of the Striped Flufftail and the Lesser Swamp Warbler respectively.

As there are over 80 examples of names with this structure, only a handful can be shown. It is worth noting, that qualifying a noun with an adjective is precisely what vernacular names committees did with English generics in the earlier years of the twentieth century, when 'eagle' became 'Martial Eagle', 'Booted Eagle', 'Wahlberg's Eagle', 'Southern Banded Snake Eagle', etc., and the generic term 'swallow' separated into 'White-throated Swallow', 'Greater Striped Swallow' and several others. A large proportion of the 80-plus newly coined Zulu bird names, follow exactly the same pattern, as can be seen in a few of the names given to eagles and swallows respectively:

African Hawk-Eagle: **ukhozolumidwayidwa** (eagle which has many streaks)
Booted Eagle: **ukhozolumadladla** (eagle which has shaggy growth of hair)
Wahlberg's Eagle: **ukhozolunsundu** (eagle which is brown)
Lesser Spotted Eagle: **ukhozolumabala** (eagle which has spots)

White-throated Swallow: **inkonjane emqalomhlophe** (swallow which is white-throated)
Red-breasted Swallow: **inkonjanesifubabomvu** (swallow which is red-chested)
Lesser Striped Swallow: **inkonjanencane** (swallow which is small)
South African Cliff Swallow: **inkonjane yamawa** (swallow of the cliffs)

The most common adjectives used after a noun, in forming this type of compound noun, are –*khulu* (big) and –*ncane* (small), separating such pairs for example as the Greater Flamingo (**ukholwase omkhulu**) and the Lesser Flamingo (**ukholwase omncane**). There is the same distinction between the kestrels, **umathebethebana omkhulu** (Greater Kestrel) and **umathebethebana omncane** (Lesser Kestrel). In addition to **inkothana** (honeyguide which is little) we have **inkotha enkulu** (Greater Honeyguide); we have the swallows **inkonjanenkulu** as well as **inkonjanencane**; the Pink-backed Pelican is known as **ivubelincane** (as opposed to the Great White Pelican, which is simply

ivuba); and among all the different ducks (**amadada**), is the little Hottentot
Teal (**idadelincane**).

After the adjectives indicating big and small, the most common adjectives
are the colour adjectives, and a whole rainbow of avian compound nouns:

> **isagwacesibomvu**: Harlequin Quail (< **isagwaca** 'quail' + –*bomvu* 'red')[16]
> **indlanyokephuzi**: Southern Banded Snake Eagle (< *dla* 'eat' + (*i*)*nyoka*
> 'snake' + –*phuzi* 'yellow')
> **umhlanebomvu**: Swee Waxbill (< *umhlane* 'back' + –*bomvu* 'red')
> **igwalagwala eliluhlaza**: Knysna Turaco (< **igwalagwala** 'lourie' +
> –*luhlaza* 'green')[17]
> **unomlomophuzi**: Yellow-billed Stork (< –*no*– + (*u*)*mlomo* 'mouth', 'bill'
> + –*phuzi* 'yellow')
> **unobulongwonsundu**: Bronze-winged Courser (< **unobulongwe** 'courser'
> + –*nsundu* 'brown')
> **ubhavuzilobomvana**: Red-chested Flufftail (< **ubhavuzile** 'flufftail' +
> –*bomvu* 'red' + –*ana*)[18]
> **uqolompunga**: Fairy Flycatcher (< *uqolo* 'rump' + –*mpunga* 'grey')
> **incwincwemphunga**: Grey Sunbird (< **incwincwi** 'sunbird' + –*phunga*
> 'grey')

A combination of the previous two compound noun categories (NOUN +
NOUN and NOUN + ADJECTIVE) can be seen in compounds with the three
elements NOUN + NOUN + ADJECTIVE, sometimes written as one Zulu
word (leading to very long compounds) and sometimes as two-word phrases:

> **intakemahlombabomvu**: Fan-tailed Widowbird (< **intaka** 'finch' + (*a*)
> *mahlombe* 'shoulders + –*bomvu* 'red)

---

16. Zulu –*bomvu* is not red in the sense of crimson, carmine or 'pillar-box' red, but more
    the colour English would describe as 'russet'.
17. There is in fact no distinction between green and blue in the Zulu language: –*luhlaza*
    means both green and blue, or blue-green. This lack of distinction is surprisingly
    common in languages, leading writers about colour terminology to coin the term 'grue'.
18. The addition of the diminutive suffix –*ana* to an adjective, is the equivalent of adding
    –*ish* to an English adjective, thus –*bomvana* 'reddish' (or even pink), –*luhlazana*
    'greenish', –*nsundwana* 'brownish', etc.

umfelokazi omlenzemhlophe: Purple Indigobird (< *umfelokazi* 'widow' + (*u*)*mlenze* 'leg' + –*mhlophe* 'white')

isikhwenene esikhandansundu: Brown-headed Parrot (< **isikhwenene** 'parrot' + *esi* 'which is' + (*i*)*khanda* 'head' + –*nsundu* 'brown')

ijubelintamemhlophe: Eastern Bronze-naped Pigeon (< **ijuba** 'dove', 'pigeon' + (*i*)*ntamo* 'neck' + –*mhlophe* 'white')

umbalane okhandampisholo: Black-headed Canary (< **umbalane** 'canary' + (*i*)*khanda* 'head' + –*mpisholo* 'dark', 'black')

6.6.3   Noun + possessive locative

These are the Zulu equivalents of scientific specific epithets, in names such as *abyssinica* in *Cecropis abyssinica* (Lesser Striped Swallow), *namaquus* in *Dendropicos namaquus* (Bearded Woodpecker), and *capensis* in *Morus capensis* (Cape Gannet). They indicate the place where the bird is commonly found.[19]

Only three names in the previously existing literature displayed this structure, these being **ivukuthu lehlathi** (African Olive [formerly Rameron] Pigeon, < **ivukuthu** 'pigeon' + *lehlathi* 'of the forest'), **isikhombazane sehlathi** (Tambourine Dove, < **isikhombazane** 'dove' + *sehlathi* 'of the forest'), and **isikhombazane sehlanze** (Emerald- [Green-] spotted Wood Dove, < **isikhombazane** 'dove' + *sehlanze* 'of the bushveld'). For the 2013– 2017 workshop participants, however, it proved to be a very popular way of distinguishing various individual species in a folk genus. The following are just a few of the names that were coined at the workshops:

- **igwalagwala logu**: Livingstone's Turaco (< **igwalagwala** 'lourie', 'turaco' + *la* 'of' + *ugu* 'coast')
- **ijubalaphansi**: Lemon Dove/ Cinnamon Dove (< **ijuba** 'dove' + *la* 'of' + *phansi* 'on the ground')
- **isigqobhamithi saseningizimu**: Knysna Woodpecker (< **isigqobhamithi** 'woodpecker' + *sa* 'of' + *iningizimu* 'the south')
- **isikhova sexhaphozi**: Marsh Owl (< **isikhova** 'owl' + *sa* 'of' + *ixhaphozi* 'marsh')
- **iseme lasentshonalanga**: Ludwig's Bustard (< **iseme** 'bustard' + *la* 'of' + *intshonalanga* 'west')
- **iwamba lasenyakatho**: Collared Pratincole/Red-winged Pratincole (< **iwamba** 'pratincole' + *la* 'of' + *inyakatho* 'the north')

---

19. Or where it was first found.

- **ujojekhaya**: Pin-tailed Whydah (< **ujojo** 'long-tailed finch' + *ekhaya* 'at home')

6.6.4   Noun + verb

There are only three bird names with the structure NOUN + VERB, two in the existing literature and one coined at the 2013–2017 workshops. The earlier names are:

> **undodosibona**: Black Cuckoo – derived from *indoda* 'man' and *osibona* 'who sees us', explained by Zulu-speaking bird experts as "this bird hides in thick foliage when it calls – we can't see the bird, but it can see us"; and **inhlavebizelayo**: Greater Honeyguide – derived from **inhlava** 'honeyguide' + *ebizelayo* 'which is calling to someone'. The behaviour of honeyguides is too well known to need an explanation here.

The coined name is **umvuliyeza**: Little Swift – derived from *imvula* 'rain' and *iyeza* 'it is coming'. The name is a clever play on the generic term for swifts – **inhlolamvula** (what predicts the rain), a reference to the Zulu belief that the swift is one of dozens of birds worldwide that are weather portents.

We turn now to those compound nouns where a verb is the first element. By far the most common is where the verb is followed by a noun, usually in the object position.

6.6.5   Verb + noun

We start here with a few examples from the previously existing literature and then go on to give a few examples with coined names:

> **isixulamasele**: African Spoonbill (< *xula* 'catch up', 'snatch up' + *amasele* 'frogs')
> **unogolantethe**: White Stork (< –*no*– + *gola* 'catch' + (*i*)*ntethe* 'locust')
> **unozalizingwenyana**: Goliath Heron (< –*no*– + *zala* 'give birth to' + *izingwenyana* 'little crocodiles')[20]

---

20. From its habit of standing in the shallow water of pans, where young crocodiles are basking.

The African Jacana is known as **unondwayiza** (a name given to many birds with long legs) and **umathandaluzibu** ('the one that likes lily pads').
© Adrian Koopman

umambathingubo: African Hoopoe (< *–ma–* + *mbatha* 'enfold', 'wrap up in' + *ingubo* 'blanket') (see Plate 20)[21]

inhlekabafazi: Green (formerly Red-billed) Wood Hoopoe (< *hleka* 'laugh' + *abafazi* 'women')[22]

isithandamanzi: Woolly-necked Stork (< *thanda* 'like' + *amanzi* 'water')

The following are names coined at the 2013–2017 workshops:

- **isicibamanzi**: Cape Gannet (< *ciba* 'move like an arrow' + *amanzi* 'water')
- **ivuzigazi**: Grey Waxbill (< *vuza* 'leak', 'ooze' + *igazi* 'blood' – a reference to the red markings on the side of the bird)
- **umathandaluzibu**: African Jacana (< *–ma–* + *thanda* 'like' + *izibu* 'water-lily')
- **indlantuthwane**: Ant-eating Chat (< *dla* 'eat' + *intuthwane* 'ant')
- **isigqobhamnenke**: African Openbill (< *gqobha* 'extract' + *umnenke* 'snail')
- **inkothanyosi**: European Bee-eater (< *khotha* 'lick up' + *inyosi* 'bees')

---

21. This name has only ever been seen in print in a very brief piece in the journal *Ostrich*, written by a certain Chadwick, in 1947 and entitled, 'Zulu names for birds'. The workshop participants were intrigued by the idea of this bird wrapping itself in a colourful blanket, and it was decided to retain the name.

22. In this very well-known Zulu name, the women are the object of the verb *hleka*; the women being laughed at by these birds, as they alight.

6.6.6   Other verb-based names

There are a few bird names, both previously recorded as well as newly coined, which have a verb as their first element, but where the second part of the concord is not a noun as in the names given above. The following are examples, with the coinages marked with a hash symbol:

> **indlangamandla**: Lappet-faced Vulture (< *dla* 'eat' + *ngamandla* 'with strength')
> **umazalushiye**[#]: Klaas's Cuckoo (< *zala* 'lay eggs' + *ushiye* 'leave [them] behind')
> **insukakude**[#]: Arctic Tern (< *suka* 'come from' + *kude* 'far away')
> **isihlalamatsheni**: Cape Rock Thrush (< *hlala* 'sit', 'live' + (*e*)*matsheni* 'on the rocks')
> **ikhwelentabeni**[#]: Mountain Wheatear (< *khwela* 'climb up onto' + (*e*)*ntabeni* 'on the mountain')

6.7   THE VALUE OF A DERIVATIONAL SYSTEM

Zulu, like all the Bantu languages, is known for its extended derivational system. The number of different affixes for both nouns and verbs allows the language to create new words from existing ones without much effort. The ease with which the Zulu-speaking participants in the 2013–2017 Zulu bird name workshops were able to weave their way through the complexities of Zulu grammar and, from the same base form, create new names is a sign of how the oral poetic tradition is still very much alive in Zulu culture today. In traditional Zulu oral culture, children from a young age were encouraged to create nicknames for each other and to compose brief *izibongo* (praise poems) in their own honour as well as in honour of others. That knowledge, of how to manipulate the derivational system of grammatical affixes, goes hand in hand with the other aspects of an oral poetic culture: familiarity with metaphor and simile; with symbolism and personification, and with onomatopoeia and other phonetic adaptations. All of these are described in other chapters, but in this particular chapter on structure (the morphology of names), we can give a few examples of how one base form can be adapted and extended to give rise to many different names. In the following groups of birds, an existing Zulu name that has been given to a loosely connected cluster of birds has been morphologically adapted to create various new names to distinguish between species in this cluster:

- The '*zwili*' group: from the names *umzwili* and *umzwilili*, both given by Doke and Vilakazi as a name for the Cape Canary (1958: 903). The name **umzwilili** was affirmed for the Cape Canary. For the Melodious Lark (previously Singing Bush Lark), the prefix *–no–* was added to **umzwili** to create the name **unonzwili**. This was then extended with the adjective *–bomvu* 'red', to create **unonzwilobomvu**, a name for the Eastern Long-billed Lark. For the Chestnut-backed Sparrow Lark, the generic name for a finch was prefixed to (*i*)*nzwili* to form **intakanzwili**.

- The '*swi/shwili*' group: the generic name **umunswi** for a thrush lies at the base of the names for this group. This name was confirmed as a generic name for thrushes, and then amended with an extra syllable to create **umunswili** for the Olive Thrush. The single syllable stem /swi/ in *umunswi* was reduplicated to form **inswinswi** as a name for the Orange Ground Thrush. Changing the /s/ in *umunswili* to the closely related sound /sh/, and prefixing *–ma–*, produced **umashwili** as a name for the Sombre Greenbul, and then, finally, extending this with the noun *imidwa* (stripes) the name **umashwilomidwa** was created for the Yellow-streaked Greenbul.

- The '*heshe*' group: Doke and Vilakazi gloss **uheshe** as "Species of hawk, *Falco biarmicus*", i.e. the Lanner Falcon (1958: 301). The previously recorded Zulu names for the smaller raptors are confusing, with names such as **uheshe** and **uklebe** used for a variety of falcons, buzzards, hawks, goshawks, harriers and sparrowhawks. At the July 2017 workshop it was decided to use **uklebe** for falcons and buzzards, **umamhlangeni** for harriers, **uheshe** for hobbies and goshawks and **uheshane** for sparrowhawks. This last name – **uheshane** – is an adaptation of **uheshe**. Further similar skilled manipulation of derivational morphology also produced the following names, all with the base form '*heshe*': **usomheshe** (African Goshawk), **uheshanobomvu** (Rufous-breasted Sparrowhawk), **uheshanomncane** (Little Sparrowhawk) and **uheshomlotha** (Sooty Falcon).

- The '*kliyo*' group: Even onomatopoeic bases can be extended to form new names. The name **ukliyo** was coined during the 2013–2017 workshops to name the Lesser Crested Tern. This name was prefixed by *–no–* to create **unonkliyo**, a name assigned to the Sandwich Tern. The diminutive *–ana* (in the form of *–ane*) was added to this last name to create the name **unonklilwane** for the Little Tern. In a last example from this group, the workshop participants wanted to add the rasping sound of the Caspian Tern – verbalised by them as '*bhakla*' – and they prefixed this to the name **ukliyo** to form the name **ubhaklakliyo**, a wonderfully evocative piece of verbalisation.

A chapter on the structure of nouns – effectively the grammar of nouns – can easily be dismissed as boring and unnecessary linguistic background. I hope I have shown that forming and re-forming of new words out of the same bases, using grammatical affixes, shows the creative adaptability of the Zulu language itself and of the people who use it. Showing the mechanics – the nuts and bolts – of the coining processes (both the age-old process that created the original bird names, as well as the more recent coining processes executed in the 2013–2017 naming workshops) is yet another way of underlining the three-way link between birds, language, and the people who use language to talk about birds.

7 Verbalisation of birdsong in Zulu

7.1 INTRODUCTION

7.1.1 Preamble

In the thickets and the long grass of the bushveld, in north-eastern South Africa, the sound of a single Arrow-marked Babbler (*Turdoides jardineii*, **ihlekehle**) can frequently be heard, followed by others, as they start up with their characteristic calls. This vocalisation has been described by Prozesky as:

> A chattering, squabbling sound uttered by all members of a party, working up to a climax and then subsiding again as individual birds stop calling (1974: 217).

Maclean describes the call of this bird as:

> Nasal whirring *ra-ra-ra-ra-ra*, usually in chorus; harsh *chak-chak-chak*; 1 or 2 birds start calling, others join in crescendo, then calling dies away (1984: 492).

Both authors <u>describe</u> the vocalisation of these birds, using words such as 'chattering', 'nasal' and 'whirring'. Prozesky uses the anthropomorphic word 'squabbling'. Maclean adds to his descriptive words with imitative phrases: *ra-ra-ra-ra-ra* and *chak-chak-chak*. Both authors mention how the sound rises to a climax or crescendo and then dies down again. These are all typical ways in which authors of field guides for birdwatchers describe bird vocalisation. But notice that neither of these authors mention how the <u>names</u> of this bird species also remark on the vocalisation. In this case, the names in English, Zulu and Afrikaans all comment in one way or another on the "chattering, squabbling" call of this bird.

The English name 'babbler' itself, used for five different species in South Africa, refers to a bird that babbles – obviously. The Zulu name **ihlekehle** is derived from two verbs: *hleka* (laugh) and *ehla* (descend), so it is a bird that laughs as it descends or alights. Here too, is a touch of anthropomorphism. The Afrikaans *katlagter* provides us with the interesting metaphor of a cat that laughs (or possibly a bird that laughs at cats).

The Green Wood Hoopoe, formerly Red-billed Woodhoopoe, is another bird that calls in a chorus that rises to a crescendo and then dies down again, and while the English name says nothing about this, the Zulu and Afrikaans names again provide commentary on the vocalisation. Afrikaans *kakelaar* (*rooibekkakelaar*) simply means 'chatterer' so it is similar to the 'babbler' of the previous species. The Zulu name **inhlekabafazi** is based on the same verb *hleka* (laugh) as in the name of the previous species and, used with the noun *abafazi* (women), it means the bird that laughs at women or, perhaps, as it is commonly interpreted, 'the laughter of women'.

In this chapter, we intend to explore further Zulu names for birds that make reference to call, song or vocalisation. We will compare such Zulu bird names with names from other South African or African languages.[1]

7.1.2   Birdsong renderings in field guides for birdwatchers

Birdsong (bird 'vocalisation') is turned into human language in three ways (which overlap and intersect):[2]

1) as aids in bird guides;
2) as bird names;
3) as 'interpretations' in various languages.

We will look at the aids in bird guides first, then turn our attention to names and 'interpretations'.

Mynott refers to a number of different ways in which birdsong has been rendered by humans, either for entertainment or for the purposes of assisting

---

1. See Koopman (1990) for more detail.
2. See Koopman (2018a).

in the identification of birds (2009: 145ff.). For pure entertainment, we can note that at the beginning of the twentieth century, people who were expert at mimicking bird calls were popular as music hall 'turns', and occasionally composers included phrases of birdsong into their compositions.[3] Writers of early bird guides also used musical notation to render bird calls for their readers. But most renderings of bird calls were, and are, verbal, in the sense that the calls are rendered in words. These verbal renderings can be subdivided into:

- those that are descriptive in nature, which may include descriptions via metaphor and/or simile, and
- those that are imitative in nature, where the notes and the phrasing of a bird call is, as far as possible, rendered in the phonemes of a particular language.

We will look briefly at these two categories of verbalisations, see how they are echoed in similar categories in bird names themselves, and then look more specifically into Zulu bird names.

Descriptive rendering

In the examples given above of the rendering of the call of the Arrow-marked Babbler, we noted Prozesky and Maclean used terms such as 'chattering', 'nasal', 'whirring' and 'squabbling', and noted that the word 'babbler' itself is descriptive of a type of call. A very superficial search of different bird guides shows phrases such as "a shrill, whistling cry", "a sort of twittering cry", "a wild, rattling, crowing note that can be heard at a great distance" and, "loud, shrill screaming . . . also a variety of whistles and chattering notes".

Descriptions via simile and metaphor

These renderings, again, describe rather than imitate, but do so via comparison: "a deep, frog-like croak", "harsh, ratchet-like alarm note", "the song [is] flute-like", "like a squeaky wheel", "like a rusty hinge", "characteristic *klink, klink klink* hammer on anvil", and referring to birds as mewing, barking or

---

3. For example, Mynott (2009: 164) refers to Vaughan Williams's composition, *Lark Ascending*, where "the solo violin traces in sound the upward, spiralling flight song of the skylark".

neighing, in other words, describing their calls by using descriptive terms usually associated with various other animals, such as cats, dogs and horses.

Imitative renderings

This involves trying to find phonemes, which, as closely as possible, approximate the sounds made by the birds. The hundreds of examples in various bird guides show how, in this effort, phoneme combinations are given that would never be found in English:

> "*tlop-tlop-tlop*", "*tuckle-tuckle-tuckle kara-karrikarra-keek gurk gurk gurk*", "*tzzee switty-sweety-tsweep-sweepy-tsweep*", "*peet-achuke-achuke-achuke . . . pheeeeeoooooo*", "*tee-too-kzzzzzrrreeee*", and so on.

Imitative rendering fully verbalised

These are the same as above, except that the phoneticisation of the bird sound has been extended to create recognisable English words. They do not necessarily combine into fully meaningful phrases, however: "*it's-up-to-you it's-up-to-you IT'S-UP-TO-YOU*", "*think-of-it think-of-it THINK-OF-IT*", "*how's fa-a-a-ther how's fa-a-a-ther*" or "*werk sta-a-a-dig werk sta-a-a-dig*", "*Pretty Georgie*", "*don't-you-do-it*", and "*Willie, come-out-and-fight sca-a-a-red*".

Bird guides, bird calls and bird names

Occasionally the writer of a bird guide indicates an awareness that the call of the bird has led to the name of the bird in one or another language. Here, for example, is McLachlan and Liversidge on the call of the Red-chested Cuckoo (1978: 244): "A loud, ringing, descending 'whip . . . whip . . . *wheeeooo*' . . . giving rise to the popular name 'Piet my vrou'." And McLachlan and Liversidge again on the 'Grey Loerie' (1978: 242): "The most characteristic call is a loud drawn out 'go-away' or 'kweh' hence the popular name 'Goaway Bird'."[4] In another example, Maclean says of the call of the Chestnut-vented Titbabbler, (Afrikaans name *bosveld-tjeriktik*, i.e. 'bushveld tjeriktik'), that

---

4. They do not mention that the Afrikaans name, *kwêvoël*, is derived from this call as well, as is the Zulu *umklewu*. Nor would they have known, in 1978, that in 2015 the formal book name Grey Loerie [Lourie], would be changed to Grey Go-away-bird.

its "commonest callnote [is a] sharp *cheriktik* or *cheriktiktik* (hence Afrikaans name)" (1984: 543).

In 1978, occasionally reference was made to what was still called the "native" name, as in McLachlan and Liversidge on the Natal Francolin (1978: 130):[5] "Most often to be heard at sunrise and sunset, sounding like 'kwaali, Kwaali, kwaali', whence the native name." This is a little confusing, because the only "native" name they give for this bird is the Zulu **isikwehle**. The Zulu word **inkwali** is indeed a bird name, used generically of francolins, so it is not wrong to make the link between name and bird call here.

Still on the subject of francolins, we may note the Coqui Francolin, with the Afrikaans name *swempie* and the Latin name *Peliperdix coqui*. Maclean has the bird calling "2-syllabled *kwee-kit, kwee-kit* or *coqui, coqui* much repeated (from which name derived)" (1984: 170). McLachlan and Liversidge, on the other hand, say that the name '*swempie*' is onomatopoeic, i.e. is derived from the call (1978: 122). Neither remark on the Zulu name that McLachlan and Liversidge give as '*i-Swempie*', and Maclean as '*iNswempe*'. As neither source gives any evidence of any element of the call of this bird that resembles 'swempe' my guess is that the Afrikaans name is derived from the Zulu name **inswempe**.

7.1.3   Birdsong references in bird names

Bird names in all languages reflect the linguistic categories given above: descriptive names, descriptive names using metaphor and simile, and imitative names, which are usually called onomatopoeic names, or, by some earlier authors, 'onomatopes'.

Descriptive names can be seen in English in generic names, such as 'babbler' and 'warbler', or in descriptive epithets, such as with the 'Zitting', 'Wailing', 'Rattling' and 'Croaking' Cisticolas. Such descriptive references may be overt and obvious, as in 'Melodious Lark', or they may be hidden and covert, as in 'nightjar', derived originally from 'night churr'.

Jean Branford says of the *korhaan* that "its name derives from its noisy call, esp. if disturbed" (1980: 143, 146). She explains that this Afrikaans name, for any of the larger bustards, is a variant of *knorhaan*, derived from the Dutch *knorren* (to scold) with *haan* (cock). One could perhaps construct a new English vernacular name to replace *korhaan*, which is still in use for a number

---

5. Now officially the Natal Spurfowl, *Pternistis natalensis*.

of species. A new name 'scoldcock' would be in line with the British name 'woodcock', and with older English vernacular names, such as 'stormcock' for the Green Woodpecker and the Fieldfare, and 'sandcock' for the Redshank (Greenoak 1997).

The generic 'woodpecker' refers to noise as well as habits, as does the Afrikaans *houtkapper*. Many African language generics for woodpecker mean simply 'knocker', as in the Tanzanian word *nkong'ota*. From Madagascar, Sibree gives the name *gàdragàdra* (harshness or roughness of voice) for a type of sandgrouse, and the name *jìjỳ* (well recited) as a generic for three related species of warblers (1891b: 562). For another group of warblers, Sibree records the name *fitatra* (expanded, drawn out). Two descriptive Tanzanian names from Moreau are *mlangilambago* (chatterer of the forest) for the Green Coucal and *mguna* (grunter) for a species of large bustard. Hendrickx gives the Nyabongo *mwanana* for the Hadeda Ibis, a name derived from the verb *kuyanana* (cry in a loud or harsh voice).

An interesting, descriptive name from Tanzania is *chemalango*, given by Moreau for Heuglin's Robin (1942: 65).[6] Here the class 7 prefix *che–* is added to *malango*, the instructions given to initiation candidates. Moreau finds this very suitable for the variable phrases[7] uttered by this bird, which he describes as "a bird which sings with much *empressement*".

Another interesting descriptive name from Tanzania is *komandugu*, which Moreau gives for the Rufous Chatterer, another bird that like the Arrow-marked Babbler and the Green Wood Hoopoe travels in parties and sings in chorus (1940: 54). The name literally means 'slay-brother', and Moreau explains that the name is given because the birds "have excessively bad-tempered voices". This name starts to lean toward metaphor, and, like the renderings of bird calls in bird guides, metaphorical descriptions may be found in bird names as well.

Given that song-reference in bird names was discussed in chapters 3 and 4, there is inevitably some overlap. Such names have been discussed twice, and while the emphasis was on underlying meanings in those previous chapters, in this chapter the focus is on the relationship between human and avian language.

---

6. Now the White-browed Robin-Chat.
7. Often compared to the reading of a shopping list.

7.2 DESCRIPTIVE NAMES, INCLUDING THE USE OF METAPHOR AND SIMILE

A good example of a previously recorded descriptive Zulu name for a bird is that of the Greater Honeyguide, **inhlavebizelayo**. It is derived from **inhlava**, (a generic name for honeyguides), and *ebizelayo* (which is calling to [someone]). And a good example of a coined descriptive Zulu bird name is **umagevuza** for the Greater Sand Plover, derived from –*ma*– (behave characteristically), and the verb *gevuza* (chatter, talk incessantly): yet another chatterer such as the *kakelaar* (*rooibekkakelaar*).

Another descriptive Zulu name is **unogandilanga** for the Red-fronted Tinker Barbet (now the Red-fronted Tinkerbird). Doke and Vilakazi give the underlying meaning of this name as 'what pounds the sun', but *ilanga* means 'day' as well as sun, and this bird, with its monotonous non-stop clinking is certainly a bird that 'pounds all day long' (1958: 583). Godfrey records this bird from the Eastern Cape, where it was known as the 'Anvil Bird' or simply as 'the tinker' (with Godfrey noting that the East London 'boys' referred to it as 'Johnny Blacksmith') (1941: 69). The Xhosa name is *unoqand'ilanga*, which Godfrey translates as 'sun-chipper', saying "from his monotonous metallic note."

Three Zulu names coined at the 2013–2017 workshops are **isankawu**, **isangulube** and **unonzwili**. The first refers to the Southern Pochard. It literally means 'like a monkey', and refers to the bird's call, which sounds exactly like the alarm call of a vervet monkey. The second name means 'something like a pig', and was coined for the Gorgeous Bushshrike, which has a grunting alarm call. The third name is that of the Melodious Lark (also known in South Africa as the Singing Bush Lark), and is derived from **umzwili**, the name of the Cape Canary. It would seem that the English expression 'sings like a canary' resonates in Zulu as well!

The metaphoric name Bleating Warbler,[8] is echoed in the two Zulu names for the same bird: **umbuzana**, from a diminutive form of *imbuzi* (goat), and **imbuzi yehlathi** (forest goat). Hendrickx gives the similar example of the Nyabongo name *kahene* ('little goat', from *hene* 'goat'), for an unidentified bird that cries like a goat (1944: 198).

The Grey-headed Bushshrike is well known for the ghostly sound of its call, leading to the Afrikaans name *spookvoël*, as well as the coined Zulu name

---

8. The metaphor has gone in the current name Green-backed Camaroptera.

usipoki. Prozesky describes the call as "An eerie, protracted mournful whistle '*hooo-whoeee*' . . ." (1974: 270). McLachlan and Liversidge say: "A loud, long ghostly and monotonous whistle . . ." (1978: 517).

Many descriptions of bird calls in bird guides refer to the calls as being melodious or musical, which leads very easily to the notion of 'music in bird names'. For example, the Afrikaans name for the Forest Weaver (now Dark-backed Weaver) is *bosmusikant*, literally 'bush musician'. Godfrey (1941: 120) notes:

> [T]he Forest Weaver . . . [with the] Colonial name of Bush Musician reminds us that the bird when singing seems to be playing on an instrument . . . Eastwards its name is *ingilinkingci*, a very good attempt to reproduce the musical note.

(We might note here that the Zulu word for a guitar is *isigingci*.) Godfrey goes on to say that "[t]he Zulu name is *ithilongo* or Bugle". Doke and Vilakazi record the name **itilongo** (without the 'h'), for the "Forest weaver-bird", (1958: 815), but make no mention of trumpets or bugles. During the 2013–2017 workshops, the Zulu name **unocilongo** was coined for the White-browed Robin-Chat (previously Heuglin's Robin), derived from *icilongo*, which Doke and Vilakazi define as a "Native trumpet (made of a long reed with ox horn fixed to the end" (1958: 121). The word is also used for a modern trumpet or bugle. In the July 2017 workshop, it was felt that the call of this bird was more flute-like than trumpet-like, and so the newly minted name **unocilongo** was discarded in favour of **unomtshingo**, based on the noun *umtshingo*. Doke and Vilakazi describe this as a reed-pipe made of an oblique flute of reed, saying that "a great variety of individual tunes can be played on this" (1958: 821). They give the secondary meaning of the word *umtshingo* as "sweet, well-trained woman's voice", which seems equally suitable for this particular robin.[9]

In fact, there are many bird names, in all languages, that use musical instruments as metaphors.[10] Sometime these instruments are descriptive prefixes attached to a generic name, as in 'Trumpeter Hornbill'. In other names, the musical instrument constitutes the base of the name, as in the coined Zulu

---

9. Doke and Vilakazi's third meaning, 'trombone', a modern usage, can safely be discarded here, together with the trumpets and bugles of **unocilongo**.
10. Not all bird names incorporating musical instruments refer to the bird's call. The name 'lyrebird' refers to the shape of its tail.

name **ivevenyane**, for the Pygmy Goose, which is derived from the noun *uveve* (horn, trumpet), suffixed with *–nyane*.[11] Another robin-chat with a coined named derived from a musical instrument is **unonkositini**, from *inkositini* (concertina). The name has been given to the Red-capped Robin-Chat, formerly known as the Natal Robin.

For the Common Greenshank, the 2013–2017 workshops took the already coined word for a referee **unompempe**, and applied it to the bird. The name is derived from *impempe* (a whistle or flute), and refers to the bird's "loud ringing *tew-tew-tew* on take-off" (Maclean 1984: 243).

Musical instruments are found in bird names from further afield in Africa: Moreau gives *kimpululu zeze*, for the Button Quail, *kimpululu* being a generic for quails and *zeze* being a "stringed instrument with a note like the bird's" (1940: 66). And Hendrickx suggests that the name *mushumbiza-ngoma* for the Red-eyed Turtle Dove is "perhaps an allusion to the bird's cooing" (1944: 204). The name is derived from *kushumbiza* 'strike' and *ngoma* 'drum'.

7.3   IMITATIVE NAMES: THE USE OF ONOMATOPOEIA

7.3.1   Onomatopoeia

As mentioned above, by far the most productive linguistic technique for producing bird names that refer to birdsong is onomatopoeia. When Mynott asks himself the question, "How best can we represent and remember [bird calls]?", he supposes that rendering them in human language seems the most common way, and notes:

> We can look at some of the ways this conversion has been attempted. One of the commonest is to use *transcriptions* into what seem the nearest equivalent human sounds in our own spoken language. We all learn this at a very early age: rooks say *caw*, ducks say *quack*, owls say *tu-whit tu-whoo*, and cuckoos say, well, *cuckoo* . . . Lots of birds have onomatopoeic names; some very obviously so, such as the cuckoo, chikadee, towhee, curlew, jackdaw, and chiffchaff (2009: 147).

Before we look at the numerous examples of onomatopoeic bird names in Zulu, let us consider a few examples from other African languages.

---

11.   Which combines the double diminutive *–nyana*, with the species-forming suffix *–ane*.

Moreau gives the two onomatopoeic names *kipulipuli* and *mkubwamkubwa* for the Black Boubou Shrike, and *kokoko* for the Banded Harrier Eagle (1940: 66). Sibree, too, has a number of onomatopoeic Malagasy names from Madagascar, and the two names *trìotrìo* and *trìotrìotsa* for a species of yellow-bellied wagtail immediately give an idea of this bird's call (1891a, 1891b, 1892). No less effective is *toitòy*, which Sibree gives for both the Curlew and the Common Sandpiper, saying it is "imitative of their cry". South African readers may be forgiven for associating the name of this bird with a kind of walking/dancing motion, linked to political protest.

Hendrickx contributes a number of what he calls 'onomatopes' from his Congo bird names, such as *kurukuru* for the White-faced Owlet, and *tshibiribiri* for the Blue-headed Coucal (1944).

7.3.2   The call and the names of the Hadeda Ibis

The strident call of the Hadeda Ibis (hadeda) is well known to all South Africans, (if not necessarily always well loved), and it is worth looking in some detail at how this call has given rise to names in South African languages, as well as in other African languages. At the same time, it is of interest to note how earlier travellers from abroad reacted to the sound of this bird when they first heard it.

One of the earlier travellers in South Africa to write about the hadeda, was the Frenchman, Adulphe Delegorgue, who on his arrival in Port Natal in the 1830s, immediately identified the hadeda as a quick and easy source of meat. He writes about "bringing down a couple of *addidas* which invariably formed the basis of my meagre meal", and says:

> [I]t was almost always one or two gunshots which determined the noisy departure of the flock. The contrast with the silence was almost amusing, the uproar surpassed that of all the tin kettle bands in the world. Once again, man must learn from the animals if he wishes to obtain perfection (1990: 57).

Note here the "tin kettle bands", showing that even the raucous calls of the hadeda can be associated with music.

The brothers Richard and John Woodward, ostensibly in colonial Natal as missionaries, but really the forerunners of later generations of serious 'birders',

The name **inkankane** for the Hadeda Ibis is onomatopoeic. © Adrian Koopman

travelled in colonial Natal and in neighbouring Zululand in the 1890s.[12] They also found the hadeda a convenient, if noisy, source of food:

> There was a large roosting-place of the 'Hadidah' or Hagedash Ibis (*Geronticus hagedash*), near our encampment, and when we were short of meat we had only to go there after dark and shoot a brace. This bird derives its name from its peculiar cry of '*ha-ha-hadadah*', with which it makes the woods resound. Its voice can be heard a long distance off, and harmonizes well with the grand scenery amongst which it dwells (1897: 403).

Delegorgue recognised the onomatopoeia in the name of the hadeda. For a short while in 1840, before he travelled extensively through Zululand and further into the interior, he had a small dwelling in the Berea bush, overlooking the bay of Port Natal. Here is part of a description of the delights of living amongst all the animals and birds of this coastal bush:

> There were *addidas, Ibis addidas*[13] filling the air with their cries in the early morning and at twilight, and there were the calaos, *Buceros buccinator*, imitating the sound of the trumpet and looking like carnival characters (1990: 128).

---

12. For details of their travels, see Koopman (2017d).
13. A footnote at this point in the original text says: "Addidas: name given to this type because of its cry."

There are two points to be made about this brief extract: first is the reference in the footnote that the name 'addidas' was given to the hadeda because of its call; the second brings back the metaphor of the trumpet again. Delegorgue's 'calaos' are Trumpeter Hornbills, as we might expect, and his scientific name contains another trumpeter in *buccinator*, which is a Latin word for a trumpeter. In fact, the current scientific name is *Bycanistes bucinator*, with the genus name containing yet another trumpeter: *bykane* is Greek for 'trumpet' and *bykanistes* means 'trumpeter' (Clinning 1989: 25).

To return to the hadeda, the nasal quality of this bird's call is expressed in its name in several South African languages: in Zulu **inkankane** and **ihahane**, for instance, and in Xhosa *ng'ang'ane*. Godfrey says, that in Pondoland, this is shortened to *haan* (1941: 19). The Southern Sotho name *lengangane* and the Northern Sotho names *le-nkagata* and *le-ngao* likewise reflect the harsh, nasal cry. Maclean gives the Tsonga name *man'an'ani* (1984: 74).

Moving further afield, Hendrickx says that the Nyabongo name *mwanana* for the hadeda is derived from the verb *ku-nana* (cry like a child) but it is easy to see the same nasal sounds and the vowel /a/ as in the southern African names (1944: 204). Moreau's four names from the eastern side of the continent, in Tanzania, are less nasal, but are still full of /a/ vowels: *kwarara, nyawawa, khako, haha* – all these names for the hadeda are described by Moreau as onomatopoeic. Wilson gives the Chewa name *mng'ang'a* for this bird, the Tumbuku names *chihaha* and *mwanawawa*, and the Nkhone name *mwalala*. The last three he has marked as 'O+', his shorthand for a "particularly apt onomatope" (2011: 40).

7.3.3   Previously recorded Zulu onomatopoeic names

Having looked at onomatopoeic bird names from languages other than Zulu (apart from the hadeda) let us now go on to look more specifically at how this name-forming strategy applies to Zulu bird names.

The Zulu bird name **ititihoye** for the Black-winged Lapwing (formerly Plover), is well known, if only from the famous lines in Alan Paton's *Cry, The Beloved Country* (1948: 1).

About you there is grass and bracken and you may hear the forlorn crying of the titihoya, one of the birds of the veld.

McLachlan and Liversidge say of the call of this bird that it "has a fine vocabulary of curses of varied pitch and intensity rising to high screams, '*che-che-che-chereck*' and '*titihoya*'" (1978: 174).

Three Zulu names have been recorded for the Pied Starling: **ingwangwa**, **ikhwikhwi** and **igwayigwayi**. The last of these is seen most clearly in Prozesky's rendering of the call as "a rather weak '*gwah-i, gwah-i*' repeated several times" (1974: 279). Less clear is McLachlan and Liversidge's "a rather soft 'squeer, squeer' or 'squeerky-week'" (1978: 533). Maclean gives the call as (among others): "skwee-skwee . . . skweer-skweer, skik-skik, chirrup-chirrup . . .". Maclean's 'skwee-skwee' is certainly the closest to the Zulu name **ikhwikhwi**, but all the descriptions of the call of this bird emphasise the repeated syllables found in **ingwangwa**, **ikhwikhwi** and **igwayigwayi**. Note that Godfrey, in talking about the Pied Starling, says: "The northern Natal name of *ingwi-gwi* . . . also appears to be derived from the cry" (1941: 113).

Maclean's rendering of the call of the Sombre Bulbul (now the Sombre Greenbul) includes the phrase "Willie, come out and fight . . . sc a-a-a-a-red." The 'willie' in this song-phrase is also seen in the Zulu name **iwili** and the Afrikaans name *willie*. Prozesky (1974: 221) has the bird calling in the London Cockney dialect, where /w/ is pronounced as /v/: "A loud, clear '*villi'* . . . At a distance only the '*villi'* is heard . . . Alarm-call a series of '*villi*'s' . . .". He gives the Afrikaans name as *willietiptol*, turning the onomatopoeia into a descriptive epithet.

The Zulu name of the Grey Crowned Crane is **unohemu [unohhemu]**, where the onomatopoeic element *hemu* is prefixed by –*no*–. The Afrikaans name is *mahem*, and we can see the origin of this onomatopoeia when McLachlan and Liversidge describe the bird's call (if not all that clearly . . .) as "A two-syllabled trumpet hence the name Mahem" (1978: 151). Prozesky, who gives the Afrikaans name as *mahemkraanvoël* (*mahem*-crane-bird), is clearer with "A not very loud bisyllabic trumpeting '*ma-hém*', with all the stress on the last syllable" (1974: 101). Xhosa *ihem* and Southern Sotho *lehehemu* complete the onomastic profile of this bird.

The Zulu name **inkwali** has some interesting echoes in East Africa. This name of the Natal Spurfowl clearly has its origin in the call, described by McLachlan and Liversidge as "Most often to be heard at sunrise and sunset, sounding like 'kwaali, Kwaali, kwaali', whence the native name" (1978: 130).[14]

---

14. As mentioned above.

From Tanzania, Moreau has *kwale kwechi* or *kikwele kwechi* for the Crested Francolin, marking both names as onomatopoeic (1940: 59). The Zulu words **inkwali** and **isikwehle** are recognisable in *kwale* and *ki-kwele,* so presumably *kwechi* is an onomatopoeic epithet for these generics. And then given that *kwale* is obviously cognate with Zulu **inkwali**, the name *kwale kwechi* is doubly onomatopoeic.

Maclean's description of the call of the Gorgeous Bushshrike as a "loud bell-like kong-kowit-kowit" immediately points to the Afrikaans name *konkoit* (1984: 657). Perhaps less obvious is the Zulu name **ingongoni**, recorded by Maclean. There is a problem here, though, because Doke and Vilakazi record the name **ingongoni** as belonging to the Black-collared Barbet (where of course it might also be onomatopoeic) (1958: 257). Perhaps this is why the 2013–2017 workshop participants preferred the coined name **isangulube** (like a pig), referring, as we noted above, to the grunting alarm call of this bird.

The Speckled Pigeon or Rock Pigeon has the Zulu name **ivukuthu**, clearly an onomatopoeic name, as seen in Maclean's rendering "emphatic Vukutu-kooo, accented on first syllable, drawn out on last, repeated several times . . ." (1984: 303). Prozesky gives the call of this pigeon as "a hollow, ringing, '*voo-voo-voo-voo*'; courting call a trisyllabic '*voo-goo-too*' " (1974: 141).

Birds create other sounds apart from vocalisation, and these can occasionally produce onomatopoeic names as well. Clapping of wings, for example, gives the descriptive epithet to the name of the Flappet Lark, with its loud wing-rattle at the top of a steep climb (Maclean 1984: 435), and the knocking or tapping of the beaks of barbets, woodpeckers and similar birds produces a number of Zulu names, for example, **inqondanqonda** for the Cardinal Woodpecker and **isinqonqotho** for the Black-collared Barbet. As is generally known, in traditional Zulu society one does not knock on doors with the knuckles of the hand – one stands outside the door and says "*Nqo nqo nqo*".

Whether or not the names for the two KwaZulu-Natal (KZN) mousebird species are onomatopoeic or not is debatable. Godfrey thinks the Xhosa name *untshili* and the Zulu name **umtshivovo** for the Red-faced Mousebird, and the Xhosa and Zulu name **indlazi** for the Speckled Mousebird, are all derived from the cry. For the name **indlazi**, there is no indication in the field guide literature that suggests onomatopoeia, but Godfrey (1941: 62), describes the Xhosa verbalisation of the Speckled Mousebird's call as "*dlatsi dlatsi dlatsi*". For the name **umtshivovo**, we can only quote McLachlan and Liversidge as saying of the call:

Rendered as 'tree-ree-ree' and responsible for several onomatopoeic names such as 'Tshivovo' which matches it pretty closely (1978: 281).

Given the large number of onomatopoeic names for birds in southern African languages, and in other African languages, it is not surprising that this linguistic strategy was employed at the 2013–2017 workshops to form names for those birds with no previously recorded names.

7.3.4   Zulu workshop-coined onomatopoeic names

During the 2013–2017 bird naming workshops, onomatopoeia was frequently considered as an option when basing the choice of a new name on the call of a particular bird. For example, Maclean notes of the White-backed Duck that it has a "quiet musical 2-syllabled whistle *curwee curwee* . . ." (1984: 81–2), and of the White-faced Duck that it has a "characteristic 3-syllabled whistle *swee-swee-swee* . . ." (1984: 80). For the former, with its two syllabled call, the name **inzwinzwi** was coined, and for the latter, with its three syllabled call, the name **inzwinzwinzwi** was coined. In similar fashion, the name **umcwicwicwi** was coined for the Green Malkoha (Green Coucal) to reflect its call, described by Maclean (1984: 338) as a "Sharp, metallic *tsik-tsik, tsik, tsik-tsik* . . .".

More reduplication of syllables is seen in the coined name **igwigwi** for the Three-banded Plover. Prozesky describes the call: "Alarm-call a loud, high-pitched whistle: '*twi-twi*' . . ." (1974: 114).

The Caspian Tern has a harsh, rasping call, described by Prozesky (1974: 135) as "a harsh, raucous, heron-like '*kraark*' or '*kwarkwa*'", and by McLachlan and Liversidge as "a loud and raucous 'krake-kraah' or 'kraark'" (1978: 210). To capture the essence of this harsh, rasping sound, the workshop participants chose the harsh velar fricative sound /kl/ and came up with **ubhaklakliyo**. The same phoneme /kl/ was chosen in coining the name **iklosi** for the Grey Penduline Tit, but it is difficult to reconcile this name with Maclean's description of the call as "Raspy *chiZEE-chiZEE-chiZEE* or *chikiZEE-chikiZEE-chikiZEE*" (1984: 490).

7.3.5   Onomatopoeia that becomes generic

Godfrey, perhaps unwittingly, gives an example of a bird name that is onomatopoeic, and is then extended to become a generic, thus including in the denotation those birds whose calls do not match the original onomatopoeia:

Other names applied to herons can hardly be regarded as strictly specific. The name *ugilonko*, evidently an attempt to reproduce the *kronk* of the herons has a wide range ... at the Gordon Memorial Mission in Natal [Pomeroy] it is in use for the black-headed heron (1941: 9).

Godfrey's '**ugilonko**' is one of a number of different versions attempting to "reproduce the *kronk* of the herons": Doke and Vilakazi (1958: 248) have '**ugilonci**' for a "Blue Heron"[15] – whatever that might be. They did not inherit this word from Bryant's 1905 dictionary. Maclean has '**uNokilonki**' for the Grey Heron (1984: 45), but he too did not inherit this name from previous editions of *Roberts*. At one of the earlier 2013–2017 workshops, Maclean's name was adopted for the Grey Heron, but at the July 2017 workshop it was decided to act on what Godfrey had suggested, more than 75 years previously, and adopt **unokilonki** as a generic name for herons. I am not sure if what Godfrey also suggested is true, namely that all herons go *kronk*, but having adopted this apparent onomatopoeic name as a generic, they now all do indeed do so, onomastically at least.[16]

What Godfrey has identified is a phenomenon that revealed itself in the coining of new names at the 2013–2017 workshops, and it is illustrated by the names of the Hadeda Ibis and the Sacred Ibis. As we saw above, in the extended narrative of the hadeda, its resonant nasal cries are reflected in the Zulu names **inkankane** and **ihahane**. Earlier versions of *Roberts Birds* gave the Zulu name **umxwagele** to both the Sacred Ibis and the Bald Ibis (McLachlan and Liversidge 1957, 1970, 1978). However, Bryant (1905: 706) and Doke and Vilakazi (1958: 870) only assign the name **umxwagele** to the Bald Ibis, and this was ratified at the 2013–2017 workshops. This meant that a name had to be found for the Sacred Ibis, and the workshop participants chose to take the two onomatopoeic names of the Hadeda Ibis and extend both of them with –*mhlophe* (white) to produce **inkankanemhlophe** and **ihahanemhlophe** for

---

15. They do not say so, but this might be from *ugilo* 'neck', 'throat' + *nci* 'small' 'thin'. My Zulu bird name workshop colleagues, Noleen Turner and Roger Porter, are firmly in favour of this interpretation, with Porter saying that 'kronk' is certainly not a typical noise of herons.

16. Having adopted **unogilonki** as a generic name for herons, the workshop participants had no problem constructing **unokilonki elikhandamnyama** (black-headed kronker), for the Black-headed Heron and **unokilonkomnyama** (black kronker), for the Black Heron.

the Sacred Ibis. In so doing, as always happens when an onomatopoeic name becomes a generic in this way, the original song-reference become irrelevant. The call of the Sacred Ibis is given in various field guides as a harsh croak, nothing like the resonant nasal calls that gave rise to the names of the hadeda in various languages. However, in July 2018, after considerable discussion, it was decided to coin the name **inkankanelunga** for the Sacred Ibis, with the effective meaning of 'pied hadeda'.

We see the same occurring a number of times in Zulu bird names, mostly with the newly coined names, but also with previously recorded names, as with the name **ivukuthu lehlathi** of the African Olive Pigeon (previously the Rameron Pigeon). Bryant, Doke and Vilakazi, and earlier versions of *Roberts* all assign the name **ivukuthu** to both the Speckled Pigeon and the African Olive Pigeon (presumably as these two species are the only speckled pigeons in KZN), but Maclean records the differentiated **ivukuthu lehlathi** (vukuthu of the forest) for the African Olive Pigeon (1984: 304). This bird, needless to say, has no '*vu-ku-tu-vooo*' in its vocalisation, again rendering this name an 'ex-onomatope'.

The element 'willie' in the vocalisation of the Sombre Greenbul, and how this has contributed to its Zulu name **iwili**, has already been discussed above. In the 2013–2017 workshops, the regional variant **umashwili** was also accepted as a name for this bird. Note that this variant still contains the element *wili*. This variant was then extended with –*omidwa* (which has stripes), to coin **umashwilomidwa** for the Yellow-streaked Greenbul. Maclean records the "noisy, penetrating nasal notes" of this bird, as "*winky-wink, winky-wink chink CHANK chow*", among other partial verbalisations, ending with a remark that part of this bird's vocalisation sounds like the call-note of the Black-eyed Bulbul (1984: 500).[17] But there is absolutely no mention of the "willie" element that led to the names *iwili* and *umashwili* in the first place.

The names **imbuzi yehlathi** (goat of the forest) and **imbuzane** (little goat) for the Bleating Warbler (now the Green-backed Camaroptera) have already been mentioned. At the 2013–2017 workshops, the name **imbuzane** became a generic and was qualified by two colour-reference adjectives to give the names **imbuzanephuzi** (yellow *mbuzane*) for the Yellow-bellied Eremomela and **imbuzaneluhlaza** (green *mbuzane*) for the Green-capped Eremomela. Needless to say, neither bleats.

---

17. But not the Sombre Greenbul which is the 'real' Willie.

The coined name **umabhashinhlayela** for the Orange-breasted Bushshrike was based entirely on the rhythms and stresses of this bird's call, as explained above. These rhythms are completely missing in the vocalisation of the Olive Bushshrike, for which the name **umabhashinhlayela ohlaza** (green *bhashinhlanyela*) was coined.

The underlying meaning of song-reference is not <u>always</u> lost when an onomatopoeic name becomes a generic, to be further qualified for other birds. For example, take the onomatopoeic name **umklewu** for the Grey Go-away-bird, already mentioned above. The name **usamklewu** (like an *mklewu*) was coined at the 2013–2017 workshop for the Great Spotted Cuckoo and, as this bird does not at all resemble the Grey Go-away-bird in appearance, this must clearly be a reference to its "loud, rasping *keow*".

The same can be said for three species of terns, none of which had previously recorded names when the birds came up for discussion at the workshops. The Lesser Crested Tern came up first, and its sharp call of *chirruk* was captured in the coinage **ukliyo**.[18] The Sandwich Tern, with its "loud, grating *kirrik*" was given the adapted form **unonkliyo**, and the Little Tern, with a call described as "sharp *kreekl* or *kree-uk*", received the name **unonklilwane**, using the diminutive suffix *–ane*. There are three separate names for three separate species here, but each name is based on the same onomatopoeia, for calls that are very similar.

* * *

Returning once again to the strident nasal cries of the hadeda, I note that several writers (Dunning 1946; Bulpin 1969; Lugg 1970) have said that when the hadeda utters these cries, it is saying, "*ngahamba, ngahamba, ngahamba*" (I went and I went and I went). This is obviously a verbalisation of the call, in the same way as we saw bird calls above rendered as '*it's-up-to-you*', '*think-of-it*', '*how's fa-a-a-ther*' and '*Pretty Georgie*'. Another way, though, of looking at "*ngahamba, ngahamba, ngahamba*" of the hadeda is to see this as an <u>interpretation</u> of the bird's call. This is what the bird is saying, not as a meaningless <u>imitation</u>, but as a translation of bird language into human language. The bird utters this call as it flies off, so it is sensible to interpret the cries as referring to its leaving and going away. This notion of <u>interpretation</u>, rather than <u>imitation</u>, is explored in the next section.

---

18. Details of vocalisation from Maclean, (1984: 285, 286, 292).

7.4   INTERPRETED CALLS: WHAT THE BIRD IS REALLY SAYING

7.4.1   The notion of interpretation

In this section, we move away from descriptions and imitations of the bird calls (whether in field guides as an aid to identification, or in the names of the birds themselves), and move more towards a 're-invention' and a development of the bird calls as speech acts in human language. Let us take the Southern Ground Hornbill as an example. The purely denotative name of this bird in Zulu is **insingizi**, but when the bird is calling the imitative (onomatopoeic) name **ingududu** is used. The call itself can then be interpreted. Lugg is one of a number of writers to do this:

> In their wanderings during the mating season, the female [Southern Ground Hornbill] is often heard to address her mate with a deep booming note – '*Ngiyemuka, Ngiyemuka, Ngiyakwabakithi*' 'I am going, I am going back to my people,' to which the male replies with a caustic '*Hamba, Hamba, kad' usho*' 'Go, Go, Long have you said so' (1970: Appendix p. 16).

In such interpretations, the vocalisations of the birds suggest human situations, here a wife, unhappy with her marriage, tells the husband that she is returning to her kinfolk, and the husband, clearly tired of his wife, tells her to go by all means. In such recasting of bird calls as human scenarios, anthropomorphic notions such as the female bird 'addressing' the male, and the male being 'caustic' in reply, are not out of place.

As with all oral narratives recast in writing, the actual wording is not fixed and Bryant says:

> The cry of the female hornbill is said to be *Ngiyamuka, ngiyamuka, ngiya kwabetu* (I am going, I am going, off to my people)! to which the male-bird replies in an undertone *Hamba, hamba: kad'usho* (Go, for goodness sake; you've been saying so long enough) (1905: 654)!

Bulpin takes his Zulu version from Bryant, but adds his own interpretative layers to the Zulu interpretation of the call. The anthropomorphism is strongly evident:

> The foolish-looking iNsingizi (ground hornbill) always amused the tribes-people with his apparent pomposity. There seemed no doubt

that the female had good reason for her cry: '*Ngiyemuka, ngiyemuka, ngiya kwabethu*' (I am going away, going away to my people). To this the male always replied gruffly: '*Hamba, hamba, kad'usho*' (Go, go, you have been saying so for a long time) (1969: 27).

The Xhosa people, as recorded by Godfrey, take this dialogue between the female and the male hornbill much further. Godfrey gives the following for the duet, but it is not clear whether each male-call/female-response couplet constitutes a single variant, or whether the dialogue recorded below is a single version:

Male: *Iph' impi* (Where is the enemy?)
Female: *Naantsiya!* (Yonder he is!) or, *naants es'apha!* (Just over the hill!)
Male: *Uph' umhlakulo?* (Where is the hoe?)
Female: *Usekoyeni!* (It's in the maize-crib!) Or *Awukh' ekoyeni* (It's not in the maize-crib!)
Male: *Aphi amakhwenkwe?* (Where are the boys?)
Female: *Ases'apha!* (They're over the hill!)
Female [again]: *Ndiyemka, ndiyemka, ndiya kowethu!* (I'm off, off to my father's place!)
Male: *Hamba ke, kad' usitsho!* (Off you go then; you've talked about it long enough!)
Or, hitting off pat the hollow boom,
Male: *Awumki, awumki, kad' usitsho* (You'll not go: that's your old threat!)
Female: *Ndiyemka, ndiyemka emhlabeni!* (I am going away, I am going away from the earth!)
Male: *Mus' ukut fho! mus' ukut fho!* (Don't say so! Don't say so!) *Yithi!* (Do it!)
Mus' ukuthi (Don't do it!) (1941: 65).

Rather than the verbalisation of a bird call, this has the feel of a children's oral game along the lines of extended riddles, something similar to the bird-name game played by Zulu children called '*Bhula 'ntsentse bo!*'[19]

---

19. See Chapter 9, section 9.3, 'Birds in riddles and children's games', for further details.

7.4.2 Contributors to the interpretations of Zulu bird calls

In this section, I look at five of the main contributors of material on the interpretation of bird calls. Taking them chronologically, these are:

- A.T. Bryant: In his 1905 *Zulu-English Dictionary*, Bryant was careful to include these interpretations with his gloss for certain bird names, one of the reasons that makes Bryant so important a contributor to the lore of birds, as well as their identity.

- Rev. Robert Godfrey: Godfrey's 1941 book, *Bird-lore of the Eastern Cape Province*, focuses most strongly on Xhosa bird names, but as these frequently overlap with Zulu bird names, and Godfrey often has much to say about Zulu bird names and interpretations of their calls, (even when these differ from Xhosa), he is included in this list.

- R.G. Dunning: In 1946, Dunning produced his splendidly named *Two Hundred and Sixty-four Zulu proverbs, Idioms, etc., and the Cries of Thirty Seven Birds (Fully Translated)*. At 37 interpretations, he is well ahead of the other four authors and he also has far more to say about each interpretation than they do.

- T.V. Bulpin: Bulpin's 1969 *Natal and the Zulu Country* is a 'popular' history of this region. For some reason, he has chosen to open his Chapter 3, 'The Door of Night', with two pages of interpretations of Zulu bird calls, perhaps to give a 'folkloristic flavour' to the Zululand region. The opening sentence – "To the mind of the African, nature is full of prophets" – certainly seems to suggest this (1969: 26).

- H.C. Lugg: Lugg's 1970 *A Natal Family Looks Back* is the latest chronologically of these five sources, but this personal history was written at the end of a long life, and the several extended vignettes he provides of Zulu culture, including the section on the interpretations of bird calls, would have been collected far earlier than the publication date of the book. Lugg was born in 1882, and the Zulu cultural material in this book would have been collected in the earlier decades of the twentieth century (perhaps around 1910 or 1920) when he was already an established linguist, working in the colonial civil service).

Before going on to compare and contrast their offerings of the interpretations of selected birds, there are some general remarks to make about the collections of these five authors, taken as a whole.

In all these contributions, the authors have been involved in two simultaneous transformational processes.

First, they have converted oral sources into written material, which in itself is not without its difficulties. Oral versions of anything: praise poems, clan praises, children's songs, proverbs and riddles, folk tales – none of these have a fixed form: they vary from teller to teller and from performance to performance. These interpretations of bird calls are undoubtedly part of the oral cultural heritage of Africa; by transforming them into written forms their versatility has been restricted, and they are removed from the context of performance, where teller or singer is, together with the audience, part of the same performance.

Second, these creations of oral cultural heritage have all been recorded by white authors who, despite often having a fluent command of Zulu or Xhosa, are not members of the cultural groups that 'created' the interpretations. These white authors have often coloured the versions with their English translations and with their own cultural perceptions. Bulpin differs from the other four in that it is clear that his versions are taken second hand from the writings of others. This has not, as we have seen above, prevented him from adding his own layers of interpretation on top of those who have gone before him.

It is worth noting that although the authors are extracting material from what we might assume to be a single pool of cultural notions regarding bird calls and their interpretations, they do not all talk about the same species of birds. While most of the time the interpretations offered vary only slightly, sometimes, however, the interpretations differ significantly. As regards them not all talking of the same birds, despite Dunning offering the interpretations of 37 birds[20] and Lugg giving interpretations of the calls of 21 birds, only eight of these birds are common to both their lists. As regards similarities and differences, we will give examples of both under the heading 'Selected Zulu interpretations of bird calls' below.

A final point, which comes under the more general rubric of bird names in the wider sense, we may note the extraordinary difficulty of simply identifying some of the birds whose calls are interpreted. Dunning, for example, interprets the calls of the birds for which he only gives the Zulu names **uqotshane**, **insindaphi**, and **mlele**. This last named is described as a "very small dove smaller than preceding", but, as the preceding entry is identified only as "one of the smaller-ring-necked doves", this is not very useful. These three Zulu names are recorded by no other source in the literature and are unknown to our Zulu-speaking workshop participants. Dunning gives us no Latin names, and no English vernacular names, so it will forever remain a mystery as to

---

20. Or so the title claims: I was only able to count 34 in his book.

which exact species of bird it is that tells us (in translation) "I was asleep at Sokhulu's" and, "I come from Cekwana's place". Nor will we ever find out which bird proclaims to the world (again in translation), "I would say *nti nti nti*, [so] let them be satisfied."

Not that giving an English vernacular name would necessarily have been any use. When talking about the Hamerkop (**uthekwane**), Godfrey refers to it as the "hammerhead", Bryant as the "Hammerhead or Mud Lark", Bulpin as the "hammerhead crane", Lugg as the "hammerhead", and Dunning as '*uthekwane kaZiLuba*', with the English vernacular names "Hammerhead, tufted Umber, Mud-lark, [and] Heron".

When it comes to the interpretation of the call of the Fiery-necked Nightjar (**uzavolo**), Godfrey refers to this bird as the "South African Nightjar", Bryant as the "Goat-sucker or Nightjar", Bulpin as the "nighthawk", Lugg as the "night jar", and Dunning as the "Nightjar, Whip-poor-will [or] Goat-sucker".

Our old friend the hadeda, whose call has featured so often in this chapter alone, is Bryant's "Common or Hadadah Ibis", Bulpin's "black ibis or hadadah", Luggs's "hadadah", Godfrey's "Hadadah Ibis", and Dunning's "Black or Common Ibis, Hadadah, Au-to-dor". If this last one – 'au-to-dor' – were a bird, it would be a 'lifer' for me, as I have never come across it before.[21] It is pleasant to know that despite recent changes in the English vernacular names of southern African birds, this bird is still officially known as the Hadada Ibis,[22] showing that Bryant was correct when, in 1905 (more than 110 years ago), he decided that this bird was the 'Hadadah Ibis'. It is also pleasing to see that in the tight confines of their 2016 *Roberts Bird Guide*, where space is at a premium, Chittenden, Davies and Weiersbye still find place for "Loud, raucous *ha-ha-haaa* (hence common name)".

7.4.3   Selected Zulu interpretations of bird calls

There is complete unanimity among our five contributors in the interpretation of the call of the Emerald Cuckoo. Dunning, Lugg, Bulpin and Bryant all agree that this bird says *Bantwanyana, ningendi!* (Little children, don't get married!). Godfrey says that among the Xhosa, at different places in the Eastern

---

21. Early Natal Colony settler, Henry Francis Fynn, had his own uniquely spelt name for this bird, as seen in, 'The birds known in the Colony as Haw-di-das rousing their voice [*sic*] on the approach of the army . . . [is an omen of ill-fortune]' (Fynn 1969: 317). At least this is recognisably a variant spelling of 'hadadah'.

22. Chittenden, Davies and Weiersbye (2016: 106).

Cape the interpretation is variously *Ziph' iintombi?* (Where are the girls?) and *Helen! Ntombi!* (Helen! Girl!), but that the Zulu rendering of the song is *Bantwanyana! ning'endi!* Of interest is Godfrey's mention of Moreau, who, in a 1932 contribution to *Ibis* on the birds of the erstwhile Tanganyika (which I have not seen), says:

> A native who interpreted its call (in [the] kiZigua [dialect]) as *kulwa tuoge* (let us go and bathe) gave the best impression of the sound I have ever heard (cited in Godfrey 1941: 57).

The Hamerkop is generally considered by Zulu people to be an ugly bird, with its overlarge bill and its big crest at the back of the head. The hours that this bird spends at the edge of the still water of ponds, waiting for frogs and other aquatic prey to appear, is interpreted as an unhappy bird peering at its reflection in the mirror of the water (see Plate 8). Lugg, Bulpin and Bryant all agree on variants of *Ngangimuhle, ngangimuhle, kodwa ngoniwe yilokhu nalokhu* (I used to be good looking, but I have been spoilt by this [the bill] and that [the crest]). Godfrey shows that the Xhosa have the same interpretation when he gives a Xhosa version as:

> *Ndimhle ngapha, ndimbi ngapha, ndoniwe yile ndawo* or *yile nkobonkobo.*

> (I am pretty on this side [looking at its face], I am ugly on this side [looking at the back of its head], I am quite spoilt by this affair [referring to its crest]) (1941: 11).

Dunning is the odd one out here. He does not have the bird peering into the mirror of the water, complaining about its appearance. Instead, Dunning's Hamerkop says, "*Ngasho, Mina, Thekwane, onesicholo unozulele emlanjeni. Mina Mathanga Zimbombo*" (In my opinion, me, the Hamerkop, [the one] with the top-knot who is Wanderer-about-the-River.[23] I [am] Thighs [like a] long range of mountains).[24] The translation I have given here is my own, and is as close as I can get to the Zulu original. Compare this translation to the one Dunning himself gives for this Zulu interpretation of the bird's call:

---

23. *Isicholo* is any tuft or topknot, but usually refers to the high headdress of a woman.
24. [*I*]*zimbombo* is the plural of the noun *u*(*lu*)*bombo* (range of mountains), the base of the name of the Lubombo Mountains in northern KZN.

I myself, have often said:– Thekwane! you, with your (fine-looking) crest (and) your leisurely (and graceful way of) strolling when frequenting the spring, at the time it has been (dug around) opened up – mark you as a very fine fellow – (Besides this) You have (such) large (and shapely) thighs (1946: 44).

Authorial imagination is clearly evident here. It is unfortunate that for many English-speaking readers of Dunning's book – if they have no command of Zulu – this will be what they think the bird is 'really saying'. (Although they may struggle to understand how Dunning's 50-word translation comes from a mere nine-word Zulu original).

It is clear from the Hamerkop interpretations that we have moved some distance from imitative verbalisations. These narratives are more a way to explain certain behavioural patterns of the birds, recast, as mentioned above, in typically human situations. As regards the Hamerkop looking at himself in water, it has been noted that uThekwane is often given as a nickname for people who are considered to be vain or pompous. [25]

A similar type of human situation explanation is offered for the call of the **isikhombazane**. Bulpin refers to this bird as the "bush dove"; the name has traditionally been used of both the Emerald-spotted Wood Dove and the Tambourine Dove, both of which have calls that end with *tu-tu-tu-tu-tu-tu-tu-tu*. The interpretation applies to both doves. It is important to note that some authors of field guides, for example, Maclachlan and Liversidge, hear the call of these birds as "mournful" (1978: 312). Bulpin's rendering of the call is "*Ngibe ngiyazalele lapho, ngithathelwe; Ngibe ngiyazalele lapho, ngithathelwe; ngize ngizwe inhliziyo yami ithi to-to-to-to-to-to-to*" (Whenever I lay eggs here, I am robbed of them; Whenever I lay eggs here, I am robbed of them; Until I feel my heart go *to-to-to-to-to-to-to*.) (1969: 27).

The wording is slightly different but the sentiment the same in Dunning's interpretation of the call of the "Small Natal Bush Dove" with the same name **isikhombazane**: "*Ngabengazalele, babathatha abantabami, ngezwa inhliziyo ithi ndo, ndo, ndo, ndo, ndo, ndo . . .*" (I once gave birth, they took my children, I felt my heart go *ndo, ndo, ndo, ndo, ndo*) (1946: 51).

Relevant here is McLachlan and Liversidge's description of the call of the Emerald-spotted Wood Dove:

---

25. Koopman (2014b: 61); Turner (1997: 56).

One of the most characteristic and monotonous sounds of the bush.
Consists of a series of coos, 'du, du . . . du; du . . du . . du, du . . dudu,
du, du, du, du, du, du, du, du, du', the final run descending quickly.
Likened by various native tribes to 'My mother is dead! My father is
dead! All my relations are dead! Oh, oh, oh, oh, oh, . . .' (1978: 231).

It can be seen here that it is not important what the actual situation is: loss of
eggs, loss of children, loss of parents – what is important is that a situation has
to be created to explain the 'mournful' series of notes. Incidentally, McLachlan
and Liversidge's assigning the mournfulness of the bird's call to the loss of
its parents is echoed in Lugg, who gives a similar background to the bird that
he names as **isibhelu** but wrongly identifies as the 'Namaqua Bush Dove".
The Zulu name **isibhelu** refers to the Tambourine Dove (White-fronted/
White-breasted Dove), another bird as noted above with a "mournful series
of 'dus' . . . but ending more rapidly" (McLachlan and Liversidge 1978: 230).
Lugg interprets the call of the Tambourine Dove as "*Ubaba no mama bafile,
ngingedwa. Inhliziyo ithi du, du, du, du, du, du, du . . .*" (My father and mother
are dead, I am alone. My heart goes *du, du, du, du, du, du, du . . .*).

Occasionally, what I am calling "interpretations" of bird calls are, in fact,
explanations of why the bird is calling in the manner it does. As an example
of this, let us look at the case of the Black Cuckoo:

Maclean gives *unomntanofayo* (the one with a dying child) as the Xhosa
name for the Black Cuckoo (1984: 329), and, according to Godfrey, this name
is derived from the verbalisation of its 'plaintive' song, which he gives in the
form of a brief poem:

Ndina mntan' ufayo,	I have a sick child,
Ndiba ndiya mbika,	I think I am reporting him,
Kanti akabikeki	But he is ignored (1941: 57).

It is difficult to see how this can be a straightforward verbalisation of the 2–3
syllable whistle that Maclean gives as sounding like *hoop-hoo whooo*.[26] It can
only be assumed that in traditional Xhosa lore, there must be a reason why the
bird sounds so sad, perhaps because it has a sick child. I know from personal
experience that this bird calls non-stop for hours on end, so therefore, although

---

26. Popularly verbalised in English as '*I'm so sca-a-a-red*'.

it has reported that the child is sick, no one is listening, so it has to keep on reporting.

We bring this chapter to a close by looking at just a few more interpretations (or perhaps explanations) of bird calls, beginning with owls.

Owls are often considered to have human-like voices, so their calls lend themselves easily to human interpretations. Bulpin suggests that the **umabhengwane** owl (unidentified in Bulpin, but always used of the male of the African Wood Owl)[27] suffers from similar personal misfortune as the dove above, calling "*Maye, maye, maye babo!*" (Alas! Alas! ALAS!), to which the female responds "*Yini, yini, yini, nje?*" (what is it, what is it, what is it, then?) (1969: 27).

For this same owl, Lugg has the male calling "*Woza, woza, woza Nobathekeli!*" (Come, come, come Nobathekeli!), to which the female responds, "*Woza, woza, woza, Mabhengwane!*" (Come, come, come, Mabhengwane!). Dunning and Bryant both have the same interpretation as Lugg.

Still on owls, we note Bulpin's rendering of the call of **umandubulu** (a name normally used to refer to the Southern White-faced Owl) as "*Vuka, vuka sekusile!*" (Wake up, wake up, dawn has come!) (1969: 27). And Bryant says of an unspecified owl (although almost certainly the Giant Eagle Owl – now known as Verreaux's Eagle-Owl) that it calls "*Vuk' ungibhule!*" (Wake up and prophesy for me), using the verb *bhula*, which is often translated as 'throw the bones', and links this bird to notions of witchcraft at night.

The chatty Dark-capped Bulbul (*toppie, tiptol*) is not just indulging in idle chatter. According to Dunning, the bird is asking "*Bafana, bafana, izinkomo ziyobuya nini?*" (Boys, boys, when will the cattle return?) (1946: 51), while according to Lugg, the same bird is saying "*Ngihlezi lapha nje, ngihlezi lapha nje, ngihlezi phezu kwendlu ka nyoko*"[28] (I am just sitting here, I am just sitting here, I am just sitting here on top of your mother's house) (1970: Appendix p. 18).

There are many more such examples, but there is not enough space to record everything that Dunning, Lugg, Bulpin, Bryant and Godfrey have to offer on this

---

27. The female is named **unobathekeli**. This name is based on the noun *umthekeli*, for which Doke and Vilakazi give the primary meaning of, "One who solicits food from friends in times of scarcity", while the secondary meaning simply as "woman" (1958: 789).

28. The noun *unyoko* (your mother), is considered highly offensive in Zulu, equivalent to the Afrikaans "*jou ma se . . .*" (your mother's . . .).

topic. The important thing to note is that these interpretations of bird calls fall under the general rubric of 'verbalisations of vocalisations': the expression of bird language in human language. An examination of this transformation from avian 'speech' to human speech has taken us from descriptions of bird calls in bird field guides, through the realms of metaphor and simile, to exploring the roles of human imitation of bird sounds in language (in the way of onomatopoeic names), and then, finally, to these interpretations that cast the birds into human situations, which explain, among other things, their mournful notes and the plaintiveness of their songs.

I end this chapter with a brief summary of the way in which different authors have turned the call of the Emerald Cuckoo (see Plate 34) into words.

The Emerald Cuckoo and interpretations of its call

Godfrey says that the usual rendering of the cry of the Emerald Cuckoo is *ziph'iitombi* (where are the girls?) and that this is the base of the name *uziph'iintombi* recorded at Clarkebury (1941: 57). At Pirie, the call of the same bird is rendered as *Helen! Ntombi!* (Helen! Girl!). Bryant (1905: 23) says that the call of this bird is rendered as *Bantwanyana! Ningendi!* (Little children! Don't get married!) and clearly this is where the name *ubantwanyana* for this bird comes from. For comparative purposes, Maclean verbalises the call of this bird's "highly characteristic sweet whistled 4-syllabled phrase" as *Pretty Georgie* or *hullo Aunt Bet*, and Godfrey quotes the Tanganyikan ornithologist R.E. Moreau as saying that the Zigua language interpretation *kulwa tuoge* (let us go and bathe) was the best impression of the sound he had ever heard. Somehow these interpretations, in their translated forms anyway, together form a sort of fanciful monologue:

"Where are the girls? Helen! Girls! Oh, there you are. Listen, girls – just don't get married. Let us rather go and bathe. And then you'll be pretty again, Georgie."

I just can't figure out where Aunt Bet comes into it!

8 Birds in Zulu praise poetry

8.1 INTRODUCTION

Traditional Zulu oral poetry falls into the genre known as 'panegyric' or 'praise poetry'. In this chapter I will look into the three sub-genres: *izibongo zamakhosi* (praises of kings and chiefs), *izihhasho* or *izibongo zabantukazane* (the praises of commoners) and *izithakazelo* (clan praises). Also included under the broader ambit of 'poetry' are the oral forms of proverbs, popular or idiomatic sayings and riddles and children's games, which will be dealt with in the following chapter.

Although it is not within the scope of this book to include the role played by birds in the poetry of modern published Zulu poets, I do, however, refer to the poem 'Inqomfi' by the Zulu poet B.W. Vilakazi, and this is done as a literary foil to the oral praises of the same bird, the longclaw *inqomfi*, which is explored further in the next chapter.[1]

This chapter is structured not by literary sub-genre, but rather in terms of loose groups of birds and their symbolic value, with swallows representing travel over the seas, birds with red feathers symbolising blood spilt in battle, eagles and other raptors symbolising power and majesty, and so on. First, though, some background information is necessary about the different genres of praise poetry, and about the historical figures who are invariably the subject of praise poetry.

---

1. See section 9.1.3, pages 234–42.

8.1.1   Different genres of praise poetry

This section covers three different genres of praises:

- *izibongo zamakhosi*: the praises of kings and chiefs. These are composed by professional *izimbongi* (bards), and they employ a number of highly stylised literary structural features.
- *izibongo zabantukazane*: the praises of commoners. These are not composed by professionals and do not have such a high literary style. The term *izihhasho* is often used for this genre.
- *izithakazelo*: clan praises: these praises, individual to each clan, are recited by family heads at important rites and ceremonies.

Izibongo zamakhosi

These are composed by professional *izimbongi* (bards) and they employ a number of highly stylised literary structural features such as cross-linking, and parallelisms of various kinds such as antithetic parallelism. The former is illustrated by these lines from the praises of Sotobe kaMpangalala:[2]

UMkhunjini wolwandle	Great ship of the ocean
Ulwandle kaluwelwa	The uncrossable sea
Luwelwa yizinkonjane	Which is crossed only by swallows
nabelungu	and white people

and the latter illustrated by

Joj' onamagomb' ebusika[3]	Widow bird with a long tail in winter
Ojojo banamagomb' ehlobo[4]	Whereas other widow birds have a long tail in summer

Antithetic parallelism always reflects a contrast of one sort or another. The one shown here is typical of the contrast 'X does this, while Y does that', a

---

2. Cope (1968: 180–1).
3. Doke and Vilakazi (1958: 255), give <u>umgomba</u> (*imi-*) as the long tail plume of a bird, often worn as a headdress.
4. From the *izibongo* of King Sobhuza II (Nyembezi 1958: 149).

typical feature of *izithakazelo*. Here we see that Swazi King Sobhuza II is 'Mr Widow Bird', who has long tail feathers in winter, whereas [real] finches have long tail feathers in summer.

The main sources used for these praises are Cope (1968) and Nyembezi (1958).

Izibongo zabantakazane

The praises of commoners. These are composed by members of a family, about themselves and about each other. While including a number of literary features found in *izibongo zamakhosi*, they are not as lofty in style, and are characterised more by earthy humour. The main source for these is Gunner and Gwala (1994).

Izithakazelo

Each and every Zulu clan has a number of praises attached to it. Single names from clan *izithakazelo* are used in greeting; the full praises are reserved for important ceremonies. They include the literary features found in *izibongo zamakhosi*. The main source for these is Sithole (1982).

8.1.2   Historical figures in Zulu praise poetry

The following figures occur repeatedly in the praises that are quoted in this section. Rather than explain them later and singly, it would be better to introduce them now. Most, but not all of the praises from which extracts are given below, are of members of the Zulu royal line, generally considered to have started with Malandela. The dates of the first four are unknown – they are generally considered to be pre-1727. Gumede is placed in square brackets as not all sources include him in the line-up of chiefs of the Zulu clan. Phunga and Mageba are listed here as succeeding one another, but some sources see them as brothers. The dates given are dates of birth and death, not of reign.

Malandela
Zulu
[Gumede]
Phunga
Mageba (*c*.1667–*c*.1745)
Ndaba (*c*.1745–*c*.1763)

Jama (*c*.1752 –1781)
Senzangakhona (*c*.1762–1816)
Shaka (*c*.1787–1828)
then his brother Dingane (*c*.1795–1840)
then his brother Mpande (1798–1872)
Cetshwayo (1834–1884)
Dinuzulu (1868–1913)
Nkayishana (Solomon) (1891–1933)
Bhezuzulu (Cyprian) (1924–1968)
Zwelithini (Goodwill) (1948–)

In Zulu oral poetry it is Ndaba who is the 'father' of the Zulu royal line, not only of his son Jama but of all Zulu kings since then. Thus Shaka is *uDlungwana kaNdaba* (Wild Rager of Ndaba) and *uSikhukhula sikaNdaba* (Raging Floods of Ndaba); Mpande is *Inzingelezi kaNdaba* (circular marks of Ndaba) and *Intakansinsi kaNdaba* (Red Bishop Bird of Ndaba); and Dinuzulu is *Ukhozi lukaNdaba* (Eagle of Ndaba).

Two other people from whose praises extracts are given include Mangosuthu Buthelezi and Isaiah Shembe.

Buthelezi (1928–) is the hereditary chief of the Buthelezi clan, whose chiefs have been hereditary advisers and senior councillors to Zulu kings since the time of Shaka. He is a royal prince of the Zulu ruling family through his mother Princess Magogo Constance, daughter of Dinuzulu, and he is the founder of the Inkatha Freedom Party, an opposition party in the South African Parliament, and which he has lead since 1975.

Isaiah Shembe (1865–1935) is the founder of *Ibandla lamaNazaretha* (the Nazareth Church), the largest African-based Christian church in South Africa.

8.1.3   The copying of bird-related memes

Dingiswayo, son of Jobe, was the king of the large Mthethwa clan at the time the young Shaka kaSenzangakhona was born into the much smaller neighbouring Zulu clan. Both Nyembezi (1958: 6) and Cope (1968: 123) give the following lines from Dingiswayo's praises:

Ungqwashi obomvu wawoHamuyana   Red [Rufous]-naped Lark of
those of Hamuyana

Omabal' azizinge sengathi bekiwe[5]	With circular spots as if placed there[6]

From the praises of the Qwabe Chief Phakathwayo kaKhondlo are the lines (Cope 1968: 143):

Ungqwash' obomvu wanguGodolozi	Red-naped lark like Godolozi[7]
Omabal' azizinge sengath' abekiwe	With circular spots as if placed there

It is clear that although these are extracts from the praises of different people, they are identical except for the personal names Hamuyana and Godolozi. In the following extract from the praises of Mpande we see the same basic idea, but with different phrasing, and without the Rufous-naped Lark (Nyembezi 1958: 67):

Inzingelezi kaNdaba	The circular spot of Ndaba
Sengath' abekwe ngabomu	Seemingly placed there deliberately

Also missing the lark, but expressing a similar core idea, are the lines from the praises of Shaka (Nyembezi 1958: 27):

UVemvane lukaPhunga	Mr Butterfly son of Phunga
Lumabal' azizinge sengath' abekiwe	It has circular spots as if placed there

These similar expressions, as illustrated here, are known variously in the literature as 'fixed phrases', 'core images', and 'formulas' or 'formulaic phrases'.[8] Today, we might call them 'memes': ideas, notions, phrases or visual images that are copied over and over again, and often twisted slightly while still retaining the essence of the original form. What happens is that an *imbongi*

---

5. Nyembezi explains these lines as: "*Izimbongi zazithanda ukuchaza amakhosi ngokuthi anamabala*" ['the bards liked to depict the kings as being marked with designs'].
6. Cope's more fanciful translation is "With colours in circles as if they had been painted on".
7. Cope's translation.
8. See Koopman (2001).

will create *ex novo* an unusual or striking phrase or image, so striking that it remains in the memory of all who hear it, including other *izimbongi*. When it comes to the time for these other *izimbongi* to compose phrases and images for their own chief or king, such memorable images 'pop up' in their minds and are incorporated into the praises they are composing.

In many of the examples from Zulu oral praises given below, we will find such copying of memes. It may seem, as in the section on swallows below, that certain species of birds are favoured more than others as symbols or images in Zulu oral poetry, but in fact this is not the case: they simply happen to be birds that first appeared in a strikingly memorable phrase, thus creating a successful meme to be borrowed over and over again.

8.1.4   Birds in Zulu praise poetry

The sections below concern loose groups of birds with different symbolic values in Zulu traditional poetry. Under each group of birds, examples are given from the three sub-genres of traditional poetry that I identified earlier: the praises of chiefs, the praises of commoners and the praises of clans. For all three types, the praises are in honour of specific people: named chiefs or kings, named commoners and named clans. The praises are, in other words, not the praises of the birds themselves. Birds do have their own praises, which are dealt with in the following chapter. In this chapter, the birds are used as images, mostly as metaphors and symbols.

I begin with the swallows that cross the oceans (see Plate 41).

8.2   *ULWANDLE KALUWELWA, LUWELWA YIZINKONJANE*: SWALLOWS OVER THE SEA

In section 8.1.1 above, in order to illustrate the poetic technique of 'cross-linking', we gave three lines from the praises of Sotobe kaMpangalala, one of two 'Zulu heroes' included by Cope in his 1968 collection of Zulu praises. Let us look at them again, this time from a memic point of view:

UMkhunjini wolwandle
  Great ship of the ocean
Ulwandle kaluwelwa
  The uncrossable sea

Luwelwa yizinkonjane nabelungu
>Which is crossed only by swallows and white people.

Cope's introduction to these praises has the following information about Sotobe kaMpangalala (1968: 177):

>Stuart records that he was a huge giant of a man, who wore his head-ring on the back of his head. His praise-poem gives an account of his mission as Shaka's representative to King George. The fact that he only reached Algoa Bay does not detract from his heroism in sailing out to sea in a ship . . . [Later] he rose to a position of political prominence under Dingane, and was made chief of a large area in Natal.

These details suggest that these lines may be the origin of the *lulwandle kaluwelwa* praise meme, a highly memorable one that has duplicated itself among many different praise-poems, including those of Shaka, and of the various clans mentioned below. Clearly Zulu people noted a long time ago that swallows disappeared at a certain time of the year, only to reappear later in the year. They were thus birds that left the country – perhaps over the seas – to visit elsewhere, and so were suitable as a metaphor for someone, like Sotobe, who left the country and travelled by ship to visit elsewhere.

The meme about swallows crossing the seas is seen in the *izibongo* of Shaka, in lines 435/6 (Cope 1968: 117):

Owalokoth' ulwandle engaluweli
>He who attempted the ocean without crossing it
Lwaluwelwa zinkonjane nabeLungu,
>It was crossed by swallows and the white people.[9]

How successful this meme is can be seen in the number of clans who have incorporated these lines into their *izithakazelo*:[10]

---

9. These two lines not only exhibit cross-linking, the verb *wela* appearing at the end of the first line and at the beginning of the second, but also exhibit the stylistic feature antithetic parallelism: X <u>doesn't</u> do something, contrasted with Y <u>does</u> do something. Here Shaka <u>attempts</u> to cross the seas, but doesn't do so, whereas the swallows <u>do</u> cross the sea.

10. The phrase *izithakazelo zikaDuma* means 'the praises of the Duma clan'. Rather than repeat this construction every time, I have abbreviated it to *ezikaDuma* ('those of the clan Duma') and the equivalent.

ezikaDuma: Sithole (1982: 24):

Duma lwandle!	Thunder[11] O sea!
Lulwandle aluwelwa	Sea which is not crossed
Luwelwa zinkonjane	It is [only] crossed by swallows
Zona zindiza phezulu	Those that fly above

ezikaKhumalo: Sithole (1982: 37):

Lulwandle kaluwelwa	Sea which is not crossed
Luwelwa zinkonjane	It is crossed by swallows
Zona ziphapha phezulu	Those that fly above

ezikaMlotshwa: Sithole (1982: 69):

Ntumbeza kaNtanzi	Ntumbeza son of Ntanzi
kaLwandle kaluwelwa	Son of Mr Sea-which-is not-crossed
Luwelwa zinkonjane	It is crossed by the swallows
Zona zindiza phezulu	Those that fly above

ezikaMnguni: Sithole (1982: 70):

Mzimela	Mzimela
Lwandle kaluwelwa	Sea which is not crossed
Luwelwa zinkonjane	It is crossed by the swallows
Zona zindiza phezulu	Those that fly above

ezikaMthombeni: Sithole (1982: 79):

Lulwandle aluwelwa	Sea that is not crossed
Lulwelwa yizinkonjane	It is crossed by the swallows
Zona zindiza phezulu	Those that fly above

ezikaNtuli: Sithole (1982: 96):

---

11. A play on words: the Zulu clan name Duma is based on the verb *duma* ('thunder' or 'be famous').

Nina bakwaLulwandle	You of the house of Lwandle-which-
Aluwelwa	is not-crossed
Luwelwa zinkonjane	It is crossed by swallows
Zona zindiza phezulu	Those that fly above

Swallows do not only occur in Zulu oral poetry in this meme about the sea that is not crossed. They can occur in other syntactic contexts as well. A very simple one is exhibited in the opening two lines from the *izithakazelo* of the Nkosi clan, again showing a play on words in the first line (Sithole 1982: 93):

Nkonjane yenkosi	Swallow of the chief
Mpangazitha!	Mpangazitha!

The swallow that crosses the seas is the swallow that leaves the country. It could be seen as a swallow that gets lost, as expressed in the praise of Mpande (Nyembezi 1958: 63):

Inkonjan' edukil' ezulwini	Swallow that got lost in the sky

Gcumisa and Ntombela (1993: 113) give exactly the same line for Mpande and explain as follows:

UMpande ubizwa ngenkonjane ngenxa yesivinini ebalekela esilungwini ebalekela umfowabo uDingane. Phela nguyena owawela kamuva waba ngowokugcina kulabo bantwana bakaSenzangakhona.

[Mpande is named as a swallow on account of the speed with which he ran away to the white people when fleeing from his brother Dingane. Indeed it was he who crossed over [the uThukela] and later became the last of the children of Senzangakhona.]

The image of the swallow becoming lost in the sky becomes a reduplicated meme when it is copied in the praises of the Nazareth Church leader Isaiah Shembe (Gunner and Gwala 1994: 75):

Inkonjan' edukel' emafini	Swallow that got lost in the clouds at
ekhaya kwaNhliziyo	the home of Nhliziyo (Mr Heart)

The translation above is mine. Gunner and Gwala themselves translate this line with a great deal of poetic licence: "Swallow that wandered among the clouds where its heart led it, to the heavenly vision."

The swallow as a symbol of movement from one place to another is seen in the following lines from Mangosuthu Buthelezi's praises (Gunner and Gwala 1994: 90–1). Note how Buthelezi, as the swallow, becomes 'lost' when he moves from KwaZulu-Natal to the University of Fort Hare in the Eastern Cape. The translation is that of Gunner and Gwala:

Inkonjane kaMaduka	The Straying Swallow,
edukele ngelaseKoloni	who strayed over somewhere in the Cape,
ith' isibuya yayisifak' iziqu	till he came back donned with the gown of honour
isifakwa ngabamhlophe	awarded by the whites.[12]

Rycroft and Ngcobo (1988: 92–3) show how the swallow is used in the *izibongo* of Dingane[13] kaSenzangakhona:

INkonjan' ewaba busephikweni	Swallow with black and white spots on the wing
Engenjengenkonjane zasendulo	Unlike the swallows of old
Zona zibuwaba buseqolo	Which had black and white spots on the back.

Interesting here is the use of the link between the markings of a bird and that of a certain bovine colour pattern, which we saw in Chapter 5. Doke and Vilakazi (1958: 847) give *iwaba* as 'black or red beast with a white patch on the flank' so Rycroft and Ngcobo's translation is not quite accurate. The extract from Dingane's *izibongo* shows the typical use of antithetic parallelism: 'olden-day' swallows were *iwaba*-coloured on the wings whereas today's swallows

---

12. The reference here is to Buthelezi's three years at Fort Hare University in the Eastern Cape from 1948–50. He did not graduate from there, however, as implied in his praises. According to Wikipedia, he was expelled from Fort Hare after student boycotts and completed his degree at the University of Natal. https://en.wikipedia.org/wiki/Mangosuthu_Buthelezi (accessed 26 September 2017).

13. Rycroft and Ngcobo prefer the spelling 'Dingana'.

are *iwaba*-coloured on the back. There is an ornithological conundrum here, which I must leave to the experts to solve.

* * *

Is it a bird? Is it a plane?

Harry Lugg (1975: 86–8) gives quite a different use of the swallow in the *izibongo* of the Royal Air Force, of which the following is an extract (translation by Lugg):

INkonjane endizela emafini	A swallow that flies in the clouds
Yehla yahululeka yangenyoni	Descending with the swooping flight of birds
Inyoni abathi isivuba	A bird they called the *isivuba* or churning bird
Ngekuvuba amadoda.	because it churned the bodies of men (1975: 86).

Lugg says that the praises were composed by a certain Nongejeni Zuma, who was a brother of a chief in the Nkandla district (1975: 86). In his youth he had worked as a 'stable boy' for Theophilus Shepstone. The swallow here is not symbolic of a journey into the unknown with subsequent return as it is for many of the praises quoted above. Rather, here, the swallow symbolises the flight pattern of the RAF aeroplanes. Lugg appears not to recognise the Zulu word '*isivuba*' as being that for the Giant Kingfisher, and calls it a 'churning bird' as a result of the cross-link with the word '*ngekuvuba*' in the next line.[14] One wonders where Nongejeni Zuma saw RAF aircraft in sufficient numbers as to stir up his poetic imagination.

8.3 THE 'RED BIRDS': *UNGQWASHI* AND *IGWALAGWALA*

The birds in this section are linked in imagery to the Zulu colour adverb –*bomvu*. Although this colour is what we would call in English 'russet', it is also the colour of blood, a theme that runs through most of the following praises.

---

14. The verb *vuba* means 'mix together', 'stir up', especially in reference to the preparation of food, but there is no reason why it should not also mean 'churn up bodies' when referring to the damage done by military aircraft.

In section 8.1.3 above, the following extract from the praises of Dingiswayo was given (see also pages 231–3 and Plate 38):

Ungqwashi obomvu wawoHamuyana

Cope translates *ungqwashi obomvu* as 'red-naped lark', which rather leaves out the word *obomvu*, as **umngqwashi** on its own is the name of the Rufous-naped Lark. A far stricter translation, although undeniably more clumsy, is 'Rufous-naped Lark which is red'. We find the same phrase in the praises of Phakathwayo kaKhondlo, viz. *Ungqwash'obomvu wanguGodolozi* (Cope 1968: 143), which Cope also translates as 'red-naped lark'. The same lark can also be found in Dingiswayo's praises as (Cope's translation):

Ungqwashi wawoZimangeye	Scarlet-throated lark of Zimangele[15]
Osuk' ebaleni wahlal' ezaleni	That flew from the courtyard and settled in the ash-heap

Note how Cope has translated *ungqwashi* as 'scarlet-throated lark' here. Whether it is a 'red-naped lark', or a 'Rufous-naped Lark which is red', or a 'scarlet-throated lark', it seems that the bird **ungqwashi** is always associated with *–bomvu*. This is the colour of certain cattle (or other animals), which are often given the name *uJamludi*, primarily an ox name, which is derived from the Dutch *Jan Bloed* (Jack-the-Blood), also a common ox name.[16]

As pointed out in Koopman (2005), the ox name *uJamludi* most often appears as the first element of the fixed phrase (the 'core-image', the 'meme'), *uJamludi obomvu njengentolwane* ('Jamludi who is red like the *intolwane* shrub', with its red roots used for making a red dye).[17]

The same lark appears again with the adjective *–bomvu* in the following extract given by Gunner and Gwala from the praises of Dingiswayo kaJobe (1994: 158):

---

15. In the translation, Cope gives this name as Zimangele, while in the Zulu original, the name is Zimageye. Neither one of these is an error, for the form Zimangele is in the standard Zulu of today, while Zimangeye is in the *thefuya*-style dialect of the Mthethwa clan in the late eighteenth century. In *thefuya*-speak, 'l' becomes 'y'.
16. See Koopman (2002: 215–16; 2005).
17. The lark, and this same red dye-producing shrub, reappear in the next chapter, where the praises of the **ungqwashi** lark are analysed. See pages 231–3 and 239.

Ugijimela emkhontweni
Agijimela emahaweni
Ungqengendlela; ubomvu; izingazi zamadoda.

He runs for his spear
He runs for his shields
Trail Blazer like the vulture along the path; he is red; with the blood
of men.

Gunner and Gwala translate *ungqengendlela* as "Trail Blazer like the vulture along the path", clearly seeing this word as **inqe** (vulture) + *ngendlela* (along the path). But given the reference to *gijimela* and to *bomvu*, I would say we have the word **ungqangendlela** here,[18] another name for the Rufous-naped Lark.

The adjective *–bomvu* also refers to the colour of blood, and that is part of an even deeper-seated meme that links birds with red plumage to blood. Take, for example, the following lines from the praises of Cetshwayo, for which three versions are given, one from Cope, one from Nyembezi and one from Gcumisa and Ntombela.

The Cope version is:

IGwalagwala likaMenzi elisuk' eNtumeni
    Red-winged Loury of Menzi that set out from Ntumeni[19]
Kwaye kwabhej' inDulinde kwabhej' uThukela
    And the Ndulinde[20] hills went red and the Thukela reddened (1968: 216).[21]

Nyembezi (1958: 87) has it like this:

IGgwalagwala likaMenzi
    The lourie of Menzi

---

18. This could easily have been a problem of transcribing from unclear handwritten notes.
19. Cope footnote 8: "A high plateau to the west of eShowe towards the Nkandla forest."
20. Cope footnote 9: "The heights to the west of Nyoni near the Thukela River."
21. Cope footnote 10: "Red with the blood of Mbuyazi's Zigqoza faction at Ndondakusuka on the north bank of the river. It is said that during the battle Cetshwayo was at nearby Ndulinde, kneeling on Mbuyazi's shield which had been medicated in order to bring about his ascendency over Mbuyazi. This praise indicates that the reddening of the Thukela originated at Ndulinde."

Elisuk' eNtumeni kwabhej' iShowe,
> That set off from Ntumeni and [the town of] eShowe turned red
Kwaze kwaya kwabhej' ulwandle noThukela.
> And eventually the sea turned red and the Thukela turned red.

Gcumisa and Ntombela give the following extract from the praises of "*Jinind'*
omnyama" (a praise name for Cetshwayo) (1993: 87–8):[22]

IGwalagwala likaMenzi
> Lourie son of Menzi
Elisuk' eNtumeni kwabhej' iShowe
> On his leaving eNtumeni eShowe turned red
Kwaye kwabhej' iNdulinde
> And then iNdulinde turned red as well
Kwabhej' ulwandle kwabhej' noThukela.
> And the sea turned red, and the Thukela river as well.

They go on to explain that it is indeed appropriate to *bonga* Cetshwayo in
terms of this bird, and that the 'turning red' is a reference to the spilling
of the blood of the iziGqoza faction during the battle of Ndondakusuka.
In this context, we should note also the Zulu expression *wamenza*
igwalagwala ekhanda: 'he did him a lourie on the head', [i.e. he hit him
and drew blood]. Bryant gives the expression in the form *uku-m-twesa*
igwalagwala, literally 'to make him carry a lourie-feather', and meaning
"to deal a person a blow on the head so as to draw blood" (1905: 212).
Moving away from praise poetry for a moment, but maintaining the link
between blood and the red plumage of birds, I would like to examine a name
that Adulphe Delegorgue records as giving to a bird in the first volume of his
travel narrative. I have left out all the descriptive details he gives:

---

22. Cope's lines 107–8 of Cetshwayo's praises (1968: 221) read:

UJinind' omnyama,	Black Jininda
Ongabubende bezingwe	Who is like the clotted blood of leopards
nobezingonyama.	and lions.

Cope explains Jininda in a footnote as follows: "A praise name meaning 'He who
turns the back on one'. After Ndondakusuka, Shepstone visited Mpande to offer his
condolences on the loss of his sons. He also conveyed his sympathy to Monase, the
mother of Mbuyazi. Cetshwayo was furious, and turned his back on Shepstone."

The 7th July 1841 was a memorable day in my career as a naturalist, for a beautiful bird, the most brightly coloured of the species, came into my possession: this was a lark which rose suddenly into the air fifteen feet ahead of me, displaying its pink abdomen. I fired and succeeded in bringing it down immediately . . . [Delegorgue decides it is a species new to science].

. . . Rather than name the bird for some friend, I have decided that I cannot do better than retain its native name, which will help other explorers to identify it: I shall therefore call it *Alauda hamgazy* (1990: 162–3).[23]

There are a number of interesting points in this brief extract that deserve further attention, such as the most memorable feature of Delegorgue's day being the shooting of a beautiful bird, and the importance of a name to help in the determining of the identity of a bird, but keeping within the theme of 'red plumage = blood', the focus should be on Delegorgue's new scientific name for this beautiful pink-stomached lark: *Alauda hamgazy*. Stephanie Alexander, in her scientific index to the first volume of Delegorgue's *Travels in Southern Africa* (1990: 309), says Delegorgue is wrong in calling the bird he has shot *Alauda* (lark). It is, she says, a Longclaw of the genus *Macronyx*, specifically the Pink-throated (now Rosy-throated) Longclaw (*M. ameliae*) (see pages 234–42 and Plate 46).

She usefully points out that in his "Zulu glossary"[24] in the second volume of his travels, Delagorgue identifies 'Hamgazy' as a Zulu word for blood (= Z. *igazi*) and that this is confirmed later when he refers to a request from 'Panda' (= Zulu King Mpande kaSenzangakhona) for certain beads, especially the ones that are "*om-bonvo hamgazy*", or blood-red (Delegorgue 1990: 190).

But to return to praise poetry: an intriguing example comes from Gunner and Gwala's version of the praises of Chief Albert Luthuli, in which they give the line (1994: 26):

Isigwe esithi singagweb' indoda yaze yafa.

23. A footnote at this point says "Compare the Zulu *Angazi*: I do not know. *Translator's note*".
24. The *Vocabulaire de la Langue Zoulouse* in his second volume. See also Koopman and Davey (2000).

Gunner and Gwala translate this line as "Red bird which can gore a man until he dies", which is fair enough, but does not identify the two types of birds in this single line, one quite transparent, the other more opaque. **Isigwe** is one of the names of the Southern Red Bishop, a well-known bird with its bright red back and head. The word *singagweb[a]*, quite reasonably translated as "which can gore/stab", suggests the bird **usibagwebe** (literally 'we stab them'), a Zulu name given generally to woodpeckers in reference to their 'stabbing' at trees when probing for food. A quick glance at illustrations of KwaZulu-Natal woodpeckers, such as in Chittenden, Davies and Weiersbye, shows all of them with red caps (2016: 331). Immediately the image of the 'lourie-feather which is the blood-drawing blow on the head' comes to mind. The Luthuli line quoted above sits between lines where Luthuli is shown as standing his ground against the English and the Afrikaners, and the line *UDlungwane kaNdaba, uSikhukhula sikaNdaba esimehl' amnyama*, which Gunner and Gwala translate as "Fierce Rager of Ndaba, Raging Flood of Ndaba with the fierce eyes".[25] In other words, Luthuli as the red-capped bishop bird and the red-capped woodpecker is seen in the context of fierce fighting, such fighting being graphically illustrated by the avian symbols of blood-drawing blows on the head.

The Southern Red Bishop is seen in one of its other identities in the following lines from the praises of Mpande (Nyembezi 1958: 68):

> *Intakansinsi kaNdaba*
>> Red bishop bird, son of Ndaba (see cover)
> *Okade kwasa besisithek' abakithi kwaZulu*
>> Who recently appeared when those of us from Zululand were in hiding.

Much simpler, perhaps, is the line that Gunner and Gwala give from praises of *babamkhulu* (grandfather) uNtungelezana (1994: 146–7):

> *Inyoni ebomvu ezahlala uPhunga noMageba.*
>> Red Birds that perched on Phunga and Mageba.

A footnote here explains that "Phunga and Mageba are Zulu royal ancestors and the red feathers of the lourie are a symbol of royalty".

---

25. Among the most well-known lines of the 'fighting king' Shaka.

And finally, as a closing comment on "Red Birds" and their link to blood, two names coined at the 2013–2017 Zulu bird name workshops were **ivuzikazi** (what oozes blood) for the Grey Waxbill and **ugazini** (in the blood) for the Red-headed Finch.

8.4   THE RAPTORS

The raptors are represented in Zulu oral poetry by eagles (particularly the Bateleur Eagle) and by falcons and hawks, which between them symbolise different aspects of the roles of warriors and chiefs: majesty, courage, fearsomeness, swiftness in attack, and so on.

We begin here with an unusually extended stanza where colonial administrator Sir Theophilus Shepstone (Somtsewu) is portrayed as an eagle when he sent Cetshwayo to England. The lines and the translation are from Cope (1968: 198):

Ukhozi lwakithi lumazipho
> Our own eagle with the sharp talons

Ebelubal' amadoda;
> That accounted for certain men;

Ngoba lubal' uCetshwayo kaMpande,
> For it accounted for Cetshwayo son of Mpande,

Lwamthatha ngamazipho,
> It took him in its claws,

Lwamphonsa phesheya eNgilandi,
> And threw him across the sea to England,

Lwamudla lwamyekelela,
> It destroyed him, then gave him a respite,

Lwabuya lwamkhafula.
> And eventually spat him out.

As can be seen, the imagery is graphic, with the talons of the eagle mentioned twice, and the eagle apparently playing with its prey: throwing it across the sea, 'destroying' it, but then letting it go.

Nyembezi (1958: 139) gives the following from the praises of Mbandzeni wasemaSwazini:

Lukhozi lumabhula ngezimpiko
> Eagle that beats with its wings

Laph' imihlamb' idla khona
> There where the flocks graze.

There is no indication of what sort of eagle this is, but the image of the beating wings suggests it is a bateleur, as this is the most common aspect of the Bateleur Eagle that is focused upon in Zulu oral poetry. See, for example, the following extract from the praises of early Natal settler Dick King, given by Lugg (1970: 52–4):

UNgqungqulu, udladla lwamafu
> Bateleur eagle, plundering talons of the clouds
Yashay' amaphiko kwaduma
> That thrashed its wings
Izulu ngokuthukuthela
> And the heavens thundered in rage.

The beating of wings, when done above a particular house, becomes a poetic meme, seen in the praises of Mangosuthu Buthelezi and of Isaiah Shembe. The following extract from Buthelezi's praises comes from Gunner and Gwala, with their translation (1994: 92):

Ingqungulu eshay' amaphiko phezu kwaseMona
> The Bateleur Eagle which hovered with its long wings above the river Mona
Izincele zamadoda zadudumela
> And men's chests thudded with fear

The extract from Shembe's praises is also from Gunner and Gwala (1994: 66), with their translation:

Ingqungqulu eshay' amaphiko phezu komuzi wakithi Ekuphakameni
> Bateleur eagle, hovering above our own place at Ekuphakameni
USambula-nkwezane kuvel' ukukhanya.
> Scatterer of the fog and there is light.

In the following longer extract from the praises of Dinuzulu, taken from Nyembezi (1958: 109), with my translation, two Bateleur Eagles beat their wings above a homestead. These represent Cetshwayo, Dinuzulu's father, and the long-running feud he had with Maphitha of the Mandlakazi faction of the Zulu clan:

> *Ingqungqul' egoq' amaphiko*
>> Bateleur that wards off blows from wings
> *Enhla komuz' ekuvukeni,*
>> Above the homestead at wake-up time
> *Ukhozi lukaNdaba,*
>> The eagle of Ndaba[26]
> *Olweqe luphindelela kwaMandlakazi*
>> Who repeatedly leapt across the house of Mandlakazi
> *Izingqungqulu zibethene phezulu,*
>> The bateleurs were striking one another above
> *NgekaMaphitha enye,*
>> One was that of Maphitha
> *NgekaJininindi enye,*
>> The other was that of Cetshwayo
> *Angiqondi nezobhabhalala*
>> I am not sure which one will fall flat
> *Kwathi ekaMaphitha yabhabhalala.*
>> It seems as if the Maphitha bateleur will fall flat.

In the praises of Senzangakhona, as given by Nyembezi, the Bateleur appears under another identity (1958: 15).[27] The name *ingqwayingqwayi*[28] does not appear anywhere else in the literature. Note the correct reference to the Bateleur's red feet (see Plate 2).

> *Ingqwayingqwayi ebomvu nezinyawo*
>> Bateleur red as to its feet

---

26. As noted in section 8.1.2 above (page 188), all members of the Zulu royal family are 'something of Ndaba', in oral poetry the 'founder' of the Zulu royal line. He was the grandfather of Senzangakhona, and therefore the great-great-great-great-grandfather of Dinuzulu.

27. The lines that follow are not in the version of Senzangakhona as given by Cope (1968).

28. Nyembezi's footnote here explains that this is another word for **ingqungqulu**. Doke and Vilakazi (1958: 564) include the word *ingqwayingqwayi*, explaining it as a "selected classification of persons, animals, or things". Nyembezi's comment on these lines is a rather obvious one: "*Ubuye futhi afanekiswe ngokhozi, ingqungqulu*" [Once again he is compared to an eagle, a bateleur].

Engiyibheke ngaze ngayejwayela
　　Which I watched until I became accustomed to it

In the following extract from the praises of Cetshwayo, given by Nyembezi, the Bateleur occurs as a personal name (1958: 86):

Washikezel' uMashikezel' omnyama
　　He was determined, Mr Black Determination
. . .
Eya ngoSikhonyana,
　　On his way to the people of Sikhonyana
Obezalwa nguNgqungqulu,
　　The son of Mr Bateleur
Uyawukhokh' imnyatheliso,
　　Going to make a presentation of cattle to the king
Iqabi lakwabo elixub' umbalo.
　　A small herd from their place, of mixed colours.

We move now to the smaller raptors: the hawks and falcons. The following extract from the praises of Shaka, as given by Cope, begins with the name **uhele**, used as a generic in Zulu for 'hawk' (1968: 99):

UHele engimbon' ukwehla kezikaMangcengeza
　　Hawk that I saw descending from the hills of Mangcengeza
Kwathi kwezikaPhungashe wanyamalala;
　　And from those of Phungashe he disappeared
Bathi 'Hele nangunangu',
　　They said 'Hawk, here he is, there he is',
Kant' uthul' emahlathini njengezingwe nezingonyama.
　　Whereas he was silent in the forest like the leopards and lions.

Nyembezi gives the same two first lines of the above extract from the praises of Shaka, replacing *uhele* with the name **uklebe**, a Zulu name for the Lizard Buzzard (1958: 24):

UKleb' engimbon' ukwehla kwezikaMangcengeza,
　　Buzzard which I saw descending from [the hills] of Mangcengeza
Kwathi kezikaPhungashe
　　And of those of Phungashe

The name **uklebe** is also used later in the version of the praises as given by Cope (1968: 113):

> *UKleb' owehle phezulu*
>> Hawk that descended from above
> *Waye wanqamula kuMadungela*
>> He went and passed through to Madungela

The image of the hawk descending, as created by Shaka's *imbongi* (unless he himself borrowed it from elsewhere), is borrowed by Dingane's *imbongi*, as seen in the following lines:

> *UKleb' engimbon' ukwehla kwezikaMagaye*
>> Hawk that I saw descending on those [cattle] of Magaye
> *Uth' eseMhodi wayamalala.*
>> On reaching Mhodi he disappeared (Rycroft and Ngcobo 1988: 82–3).

8.5 *UJOJO* AND OTHER FINCHES

The two Zulu bird names that occur here more than any other finch name are **intaka**, used as a generic for a number of female finches, and **ujojo**, used specifically of the male Red-collared Widowbird. Both of these names have, however, in historical terms, a much wider application.

The use of *intaka* as the word for 'bird' in Xhosa suggests that in proto-Nguni, before the Zulu-Xhosa split, this was the general word for 'bird'. For some reason, the word became semantically narrowed in Zulu to indicate 'finch' and the word *inyoni* came into the Zulu lexicon to replace *intaka* as the word for 'bird'. The word *inyoni* is used a great deal in Zulu oral praises, as the last sub-section below shows, but the earlier wider application of *intaka* in proto-Zulu might explain its common occurrence in Zulu oral poetry.

While **ujojo** today refers specifically to the male Red-collared Widowbird, in earlier times it was used as an equivalent of 'Everyman' or 'the man in the street'. Colenso gives the meaning of **ujojo** as "finch with long tail" (1884: 229), but in a number of entries for other words, *ujojo* simply means 'a certain person'. The examples that follow are but two of several similar examples in his dictionary.

For instance, on page 603 we find under Colenso's entry *vetula* (kick out) the example *ihashi lika'Jojo limtshaye izito ngesivetula* (Jojo's horse hit him with kicking yesterday); and on page 654, under the entry *zaleka* (be rich) is the example *wazaleka umfo ka'Jojo, uyise wamshiya nefa elikulu lezi-nkomo* (the son of Jojo is rich; his father left him a large inheritance in cattle).

It is extremely likely that this is how *izimbongi* have used the word in the praises that follow, although, in a few examples, it is clear that they are referring to the bird, as reference is made to carrying long tail feathers.

The word **intaka** occurs in the praises of Magemegeme of the Dube clan, as recorded by Gunner and Gwala, with their translation (1994: 149):

Intaka eyakha amadlangala phezu kweCush'.
Tiny Finch which built its rough nests above the Cush stream.

It is also seen in the following extract from the praises of Govan Mbeki,[29] as given by Gunner and Gwala, with their translation (1994: 107):

Unobhaca ngey'intaka[30] akhale ngelanga
He who shades under a flock of finches and moans about the sunheat
Kanti lomisile
Whereas the land is drought-stricken

Swazi King Sobhuza II is also praised as **intaka**, as seen in the following extract, recorded by Nyembezi (1958: 150):

Intak'emnyama
Black finch
Edlalel'kwezinde izintaba
Which plays among the high mountains

---

29. Stalwart of the African National Congress (ANC) and father of the one-time president of South Africa, Thabo Mbeki.
30. Note *ngey'intaka:* One might expect to see *ngezi[zi]ntaka* here. There are many occasions in Gunner & Gwala when a 'y' is used where a 'z' may be expected. Gunner recorded most of her praises in an area where the local dialect (a remnant of Lala?), uses a 'y' for a 'z'. Normally Lala replaces 'l' with 'y'. This must be a variant.

In the following extract, from the praises of the Zulu general Nozishada kaMaqhoboza of the Nzuza clan, as recorded by Cope, the word **intaka** occurs among other birds (1968: 188–9):

> *Umgwazi weziqanaqanazana*
>> Stabber of waterfowl
>
> *Unyawo lungangendlu yakwabo kwamaSizazana*
>> He whose foot is as large as a hut among the Sizanana people
>
> *Umanqe bathi adlani ngalaphaya kwaleziya ntaba na?*
>> Vultures of whom they asked what are they eating beyond yonder hills?
>
> . . .
>
> *Untak' eduze nahlambi*
>> Bird [finch] that is lost even in a flock

Cetshwayo is praised as an **intaka**, in the following lines recorded by Cope (1968: 225):

> *UNtaka yeVuna*
>> Finch of the Vuna River
>
> *Zal' umhlanga zikhothame*
>> They abandon the reeds and bow down

8.5.1   The *ujojo* finch

Coming now to the use of the more specific bird name **ujojo**, we can return to the praises of Swazi King Sobhuza II. The extract below is from Nyembezi (1958: 149):

> *Joj' onamagomb' ebusika*[31]
>> Mr Red-shouldered Widow who has long tail feathers in winter
>
> *Ojojo banamagomb' ehlobo*
>> [Whereas] other red-shouldered widows have long tail feathers in summer.

---

31. Doke and Vilakazi (1958: 255) give *umgomba (imi-)* as "long tail plume of a bird, often worn as a headdress".

As we saw in the introduction, where these same lines were given, this is an example of antithetic parallelism, where an idea is expressed twice – the second time being in contrast to the first.

Early Natal settler Henry Francis Fynn, named uMbuyazi (he who returns empty-handed) and Msifile (from 'Mr Fynn') by the Zulus, was also praised as **ujojo** in his *izibongo*, as recorded by Cope (1968: 192):

> *Ujoj' ovel' emaMpondweni*
> The long-tailed finch that came from Pondoland
>
> . . .
>
> *Ujojo kathekeli kanjengamakhafula*
> Finch that never begged, unlike the Kaffirs[32]

Cetshwayo is also praised as **ujojo**, in the following extract, taken from Cope (1968: 225):

> *UJoj' obethwe Zimpohlo*
> The finch that was beaten by the Zimpohlo regiment
> *Ziyohlobonga ngaye ngakubangoma*
> Going to 'hlobonga' on his account the admired ones.

Two types of finches can be found in the praises of a man called Mahlokohloko Mhlongo, one of them being in his personal name Mahlokohloko, derived from the plural of **ihlokohloko**, a Zulu generic for the yellow weavers. The following extract is from Gunner and Gwala, with their translation (1994: 170):

> *Umahlokohloko uyofa kusasa*
> The Yellow Weaverbird will die tomorrow
> *Izintaka ziyofa ntambama.*
> The finches will die this high noon.

---

32. Cope's footnote 1 here explains that for many years, Fynn always wore a bunch of the tail feathers of the *Sakabuli* (Long-tailed Widowbird) in his hat. He prized it because it had been presented to him by Shaka. Cope's translation here has been queried by Koopman (2017a). He suggests that the line should be interpreted as 'the person who speaks in a regular amaNtungwa fashion, and does not spit out his speech like the Lala and other southerners'.

Ujojomshololo
> Long-tailed Finch that lingers close to the ground.

The **ujojo** finch occurs in the praises of a certain Ngoqo of the Mbatha clan, recorded by Sithole (1982: 59). As with the praises of Mahlokohloko above, the personal name Ngoqo is also derived from a bird name: **ungoqo** is a name for the Button Quail:

UJojo odlala yedwa kwaMashawu
> Jojo plays on his own at Mashawu's place

Inyon' endaba isindwe yisisila sayo
> The bird of the affair is weighed down by its tail

Isijaka esithe ukusuka sathukuthela
> The ill-tempered, obstinate one who when about to leave became angry

In the praises above we see reference to the long tail of certain finches in the breeding season. Similar feathers are found in the praises of Tshanibezwe kaMnyamana (the grandfather of Mangosuthu Buthelezi), recorded by Gunner and Gwala, with their translation (1994: 119):

Intantane enjengekaBhuqwini
> Warrior's plumes like that of Bhuqwini

Usisaka singumyakanya.
> Black Finch Feathers that ripple and shake.

Two different finches appear in the same stanza in the praises of Dingane kaSenzangakhona as recorded by Rycroft and Ngcobo, with their translation (1988: 92–3):

UNtakansinsi zinqwamene phezulu;
> King-finches are in combat overhead;

Angiqedi nezokwaphuk' uphiko.
> I cannot tell which one will have a broken wing.

USomkhanda ngokub' akhandanisa.
> Great crusher for crushing people.

Inyoni kaMahube uMashubulezi!
     Bird of Mahube,[33] great swooper!

Rycroft and Ngcobo explain the use of the translation 'king-finches' for the Zulu **intakansinsi** as follows (1988: 197):

> The intakansinsi is listed [in Doke and Vilakazi] as either the 'Yellow-shouldered Wydah [*sic*] finch, Penthetria albonotata' or the 'Bishop bird, Pyromelana orix'. Our choice of 'King-finches' here as a translation is fanciful rather than factual, in order to convey the underlying implication that the image of birds is here being applied to kings, struggling for supremacy . . .

Their use of 'king-finch' for **intakansinsi** (today used exclusively for the Southern Red Bishop) is interesting, and calls to mind Woodward and Woodward's use of the popular name 'King of the Red-bills' for the Pin-Tailed Whydah (*Vidua macroura*). Woodward and Woodward (1899: 66) link their name to the Dutch 'roibek'.[34]

As to the line "*Inyoni kaMahube uMashubulezi!*", Rycroft and Ngcobo regard uMahube (in kaMahube) as a personal name, without recognising the finch within: **umahube** (or **umawube**) is the Zulu name of the male Fan-tailed Widowbird, formerly known as the Red-shouldered Widow. Samuelson glosses **uMa[w]ube** as "the large, black native finch, with red marks on its shoulders", and adds "N.B.– It is mentioned in Zulu songs of praise, for it swoops along, carrying its wings like a warrior his shield" (1923: 273).

8.6   MISCELLANEOUS OTHER BIRDS

In this sub-section, a miscellany of birds is found: the domestic hen, owls, the Hamerkop, wagtails, the egret, a honeysucker, a dove, waterfowl, vultures, a lark, glossy starlings (possibly), a partridge, a plover, and Southern Ground Hornbills. Where more than one reference is made to any of the above, these are linked, but no other attempt has been made to group any birds under themes, as was done above for 'red birds', raptors and swallows.

---

33. The **umahube** is the Fan-tailed Widowbird, formerly the Red-shouldered Widow.
34. Also *roodebek* and *roodebekje*, with Austin Roberts calling it "King-of-six . . . Koning Rooibek" (1940: 361) and Stark and Sclater in Vol. I calling it "Koning Roodebec of Dutch Colonists" (1900: 145).

8.6.1   Domestic hen

The domestic hen is found in the praises of Mnyamana of the Buthelezi clan, an early ancestor of Mangosuthu Buthelezi and a contemporary of Shaka. The extract comes from praises recorded by Sithole (1982: 18):

Isikhukhukazi esimaphiko
   Winged hen
Esifulele amazinyane eNgonyama
   Which covers over its young at Ngonyama

8.6.2   The owl

The owl (**isikhova**) is found in four different praises. Sithole records it in the *izithakazelo* of the Mthiyane clan, where it occurs as a personal name (1982: 78):

Mthiyane kaZikode kaBhambane
   Mthiyane son of Zikode son of Bhambane
KaSikhova sendel' amabele eMatikwe
   Son of Sikhova, going on a journey for the sorghum at Matikwe

Cope records the owl in the praises of Zwide, chief of the Ndwandwe (1968: 128):

Isikhova sikaMkhonto noLanga!
   Owl of Mkhonto and Langa!

Nyembezi's praises of Shaka have an owl (1958: 20):

Ushis' izikhova zaseDlebe
   He burns the owl from Dlebe[35]
Kwaye kwasha nezaseMabedlana
   Eventually those of Mabedlana burnt as well

---

35. Dlebe is given as a place name here, but as the noun *idlebe* refers to the ear of an animal or the protruding ear-tuft of a bird, this probably indicates one of the horned owls. As will be discussed further in Chapter 10, Godfrey (1941: 60) cautions against the imitating of the call of an owl among the Xhosa. Should one do so, all his blankets will be burned. This belief is apparently linked to the burning of an owl if it is caught, because of its link to witches and wizards.

Nyembezi has a simple two-word line from the praises of Dinuzulu (1958: 106):

Isikhova[36] *sikaMaphitha*
    Owl son of Maphitha

8.6.3   The Hamerkop

The Hamerkop has been found twice in Zulu oral praises, each time in the form of a personal name. The following lines from the *izithakazelo* of the Thwala clan are taken from Sithole (1982: 114):

Siwela!
    Siwela!
Mnyamanda!
    Mnyamanda!
Nina bakaThekwane
    You the children of Mr Hamerkop
Inyon' eyaqhamuk' eSwazini
    The bird that appeared among the Swazi.

The second appearance of Mr Hamerkop is from the praises of Mpande, recorded by Nyembezi (1958: 66):

Weza noMalambule kwabakaSobhuza,
    He came with Malambule among the people of Sobhuza
Weza no Sidubelo kwabakaSobhuza
    He came with Sidubelo among the people of Sobhuza
Weza noThekwane kwabakaSobhuza
    He came with Mr Hamerkop among the people of Sobhuza
Weza noMgidla kwabakaSobhuza
    He came with Mgidla among the people of Sobhuza

---

36. A footnote here reads "*Ngoba isikhova yinyoni yabathakathi*" (because the owl is a bird of witches).

8.6.4 The wagtail

Moving now to wagtails (**umvemve**), we take from Cope the following line from the praises of Cetshwayo (1968: 225):

Odl' uMvemve oncokazi kwabaMhlophe
> He who destroyed red-speckled Mvemve[37] among the white men.

Wagtails appear in the praises of Mbandzeni of the Swazi people, recorded by Nyembezi (1958: 140):

Mkhathazi wezimvemve[38] zabeSuthu
> Wearisome person of the wagtails of the Sotho people,
Bezishaya bezibethanisa lezo mvemve
> Hitting and striking one another, these wagtails
AbeSuthu bakaBhula nabakaSasawane
> The Sotho people of Bhula and of Sasawane

Nyembezi (1958: 144) gives the praises of a certain Bhunu, from which the following lines are extracted:

Umvemve[39] wakithi
> Wagtail/calf of our people
Ngeze ngawubek' emzaneni,
> Which I eventually placed under the ironwood tree
Ngiyawubeka kaZombede wamagugu.
> I went to put there the one of Zombede of the treasures.

---

37. Cope's footnote 1 here reads: "The identity of Mvemve ('Wagtail') is unknown. Malcolm suggests the red-coated commander of the British forces at Isandlwana, but if so, it is the only reference to the Zulu War in the poem."
38. A footnote here says "*Sengathi wuhlobo lomqangala* [musical bow instrument] *wabeSuthu osetshenziwa ekuzingeleni* ['This seems to be a type of musical instrument of the Suthu people playing during hunting']". The word *umvemve* is not recorded as such by Doke & Vilakazi who only give the meanings (1) wagtail, and (2) young, feeble calf (1958: 832).
39. A footnote here says "*Inkonyane esaqeda kuzalwa*" (newly born calf).

8.6.5   Mousebird and egret

We have already seen above how the bird name **ujojo** features in the praises of Henry Francis Fynn. These are not the only birds in his *izibongo* as we see in the two separate lines below extracted from his praises (Cope 1968: 193):

> *Ubuhle bangizindlazi zaseManteku*
> Beauty like the mouse-birds of Manteku[40]

and

> *Ilanda lakweth' elaphum' elwandle*
> Our egret that came out of the sea.[41]

8.6.6   Honeysucker

Above we gave an extract from the praises of Mahlokohloko (Mr Yellow Weaver) Mhlongo to illustrate the use of the bird name **ujojo**. There is yet another bird in Mahlokohloko's praises (Gunner and Gwala 1994: 171):

> *Incuncu ephuze kwezokude iziziba*
> The Honeybird [*sic*] that drank from the deep pools
> *Yaphuza kwemfushane yagunduk' umlomo.*
> Drinking from the shallow ones would have broken its beak.

This image of the honeysucker drinking from distant (not 'deep') pools is one of the most widespread memes in Zulu oral poetry. Koopman records these same lines from Mahlokohloko's praises (2001: 148), mentions how common the meme is in the praises of young Zulu men,[42] and cites the following lines from the *izibongo* of Dinuzulu as recorded by Samuelson (1923):

> *Incuncu ephuza kweside isiziba*
> Honeysucker that drinks from a deep pool

---

40. Cope's footnote 2 gives this as a river in Pondoland.
41. Cope does not comment on this, but a white bird, such as an egret, seems suitable as a metaphor for a white person.
42. See Koopman (1987).

Kuthi ingaphuza kwesifishane
   When it drinks from a shallow one
Iqundeke umlomo
   It blunts its beak.

8.6.7   The dove

Only one instance of a dove was found in the oral poetry searched. The line that follows comes from the praises of a certain MaHlalise, a woman of the Mkhwanazi clan, as recorded by Gunner and Gwala (1994: 205):

UMajub' avuk' adl' uthayela
   (I am) Doves-that-woke-and-pecked at the roof

8.6.8   Waterfowl and vulture

The stanza below, an extract from the praises of Nozishada kaMaqhoboza of the Nzuza clan, as recorded by Cope (1968: 188–9), was given on page 22 to illustrate the use of the word *intaka*. Here we look at two other birds:

Umgwazi weziqananazana[43]
   Stabber of waterfowl
Unyawo lungangendlu yakwabo kwamaSizazana
   He whose foot is as large as a hut among the Sizazana people
UManqe bathi adlani ngalaphaya kwaleziya ntaba na?
   Vultures of whom they asked what are they eating beyond yonder hills?

The vulture appears again in the praises of a certain Jeremiah kaMtekelezi, as recorded by Gunner and Gwala (1994: 127):

UNdaba yenqe, uSobhedlase.
   The Affair of the vulture, Aggressive Fighter.

---

43.   Doke and Vilakazi (1958: 687) gloss **isiqhananazana** as "species of water-bird". In its abbreviated form **isiqhanazana,** this name was assigned to the Black Crake in the 2013–2017 workshops.

A footnote here explains that Jeremiah came across a vulture in its nest and shot it, subsequently himself coining the phrase from that incident. The word *indaba*, meaning 'matter' or 'affair' is used here in its basic sense, but no doubt this son of Mtekelezi was well aware of the common use of the personal name uNdaba in Zulu oral poetry.

8.6.9   The lark *ucilo*

The lark **ucilo** is found in the praises of a certain Ntshidi son of Lindelihle, given by Gunner and Gwala (1994: 130). It is given in its *thefuya* form, where 'y' replaces 'l', and has the suffixes *–kazi* (female) and *–ana* (little) added to it:

> *Uciyokazana akafanga zidubulo zasekhweni lakhe kwaZibani ebulawa umkhwekazi*
>> How the Little Lark suffered blows at his in-law's place at the Zibani's, where he was worn out by his mother-in-law.

Doke and Vilakazi (1958: 121) give "**(u)cilo**: Species of lark, which feeds on grasshoppers." They add the sayings: "*Ucilo uyilahlile intethe kubani* (The lark has let go the grasshopper on So-and-so, i.e. it is all up with So-and-so)"; and "*Ucilo akafi izidubuli* (The lark does not die from his blows, i.e. don't mind hard knocks)". Clearly it is this second saying that is at the base of Ntshidi's praises.

8.6.10   Glossy starlings (perhaps)

In the next extracts, it is not absolutely clear whether the *imbongi* had glossy starlings in mind, or the morning star. Nyembezi gives the following lines in his recording of the praises of Dingiswayo (1958: 6):

> *Ilanga limdondoza*
>> The sun stunted him
> *Elaphum' amakhwezi abikelana,*
>> When it rose the *amakhwezi* were greeting each other
> *NakwaNtombazi nakwaLanga*
>> At the homes of Ntombazi and of Langa.

The word *ikhwezi* means both the Morning Star (Venus) as well as being a generic for glossy starlings. It is not clear in what sense Dingiswayo's *imbongi* was using the word, and as Nyembezi does not translate the praises in his 1958 anthology of praise poems, we do not have his guidance. The use of the word *ikhwezi* is far clearer in the following extract, taken from the praises of Mpande, and again as recorded by Nyembezi (1958: 64):

> *Inkwenkwez' ephum' izilwane zibikelana*
>> As the *inkwenkwezi* star came out the animals greeted each other
> *Kubikelan' iKhwezi neSilimela*
>> And the Morning Star and the Pleiades greeted each other

The word *inkwenkwezi* is unambiguously a star (of the Argo constellation) and not also a bird name, and Nyembezi explains the last two words of the second line as "*inkanyezi yokusa*" (morning star) and "*yizinkanyezi ezibonakala ngesikhathi sokuqalisa ukulima*" (stars which are visible when it is time to start ploughing). The meme is found again in the praises of Sobhuza II, also recorded by Nyembezi (1958: 130):

> *Amakhwezi kubikelana*
>> The morning stars[44] greet each other
> *Kubikelan' inkwenkwezi kanye nesilimela.*
>> The Argo star and the Pleiades greet each other.

Given the overall usage of *ikhwezi* and *amakhwezi* in these extracts, it seems likely that no glossy starlings are involved.

8.6.11   The *intendele* partridge

The bard who composed the praises for Isaiah Shembe chose the **intendele** partridge as an image in his praises. The lines below come from Gunner and Gwala, and the translation is theirs (1994: 71):

> *Intendele isibindwe emlonyeni*
>> The partridge is left breathless.
> *Unogobhoza enjeng' oThukela ongenakuvinjelwa.*
>> Violent flooder like the Thukela River, who cannot be restrained

---

44. A footnote here explains "*Inkanyezi yokusa, kodwa lapha ezibongweni lisho umnumzane noma inkosi*" (the morning star, but here in these praises it means the head of a homestead or a chief).

Isihlahla esihle somdlebe esingahlalwa i'nyoni.
> Deathly-Beautiful Euphorbia Tree on which no birds perch;

Siyasehlalwa 'zinyoni zeZulu.
> And then the birds of the Zulu perch on it.

There are two issues to enlarge on here: the 'breathless partridge' and the 'tree on which no birds perch'.

The partridge: Doke and Vilakazi (1958: 791) give the saying "*Intendele isibindwe yisidwa* ([lit.:] The partridge is being choked by a gladiolus bulb; i.e. Your exposure has made you speechless)".[45] Sobhuza's *imbongi* is clearly familiar with this saying.

The *umdlebe* tree: This highly poisonous tree (Deadman's Tree, *Synadenium cupulare*) has the reputation of being so poisonous that no birds will sit on it, and animals that draw near die right there. For details of this belief see Koopman (2015: 195).

8.6.12   The plover

Nyembezi (1958: 93), in a discussion of the role played by Cetshwayo's brother Mbuyazwe in the disputes between them, says, "*Kepha uMbuyazwe wayenomfowabo uMantantashiya*" (But Mbuyazwe also had [another] brother Mantantashiya) and he gives the following praises for this other brother:

UMatitihoy' akhal' exhaphozini,
> Titihoye plovers called from the marsh

UZululeka kuphum' ezakwaSikwata
> And Zululeka came out from Sikwata's place.

8.6.13   The Southern Ground Hornbill

We come now to praises that refer to the Southern Ground Hornbill (**insingizi**), one of the major birds of omen and portent in traditional Zulu bird lore. This hornbill is associated in traditional thinking with storms and rain, and it is interesting that where Cope translates **insingizi** as 'hornbill', Gunner and Gwala consistently translate the word as 'stormbird'. Our first example comes from

---

45. See also the 'partridge proverbs' in Chapter 9 (page 254).

the praises of Isaiah Shembe, and was recorded by Gunner and Gwala, with their translation (1994: 77):

> *I'nsingizi zakhal' esangweni kwaNduli waze wavuka*
> The Stormbirds cried out at the gates of Nduli's until he awoke.

In the praises of a certain Mqinisi, recorded by Gunner and Gwala, it is the feather of the Southern Ground Hornbill that is an omen for thunder (1994: 137):

> *Uphaphe lwensingizi olubik' izulu ukuduma*
> Feather of the stormbird that foresees approaching thunder[46]

It is again the feather of the hornbill that is significant in the following line from the praises of a certain Manqaba of the Mbonambi clan (Gunner and Gwala 1994: 154):

> *Uphaphe lwensingizi oluphezu kwendlu kaKhoto*
> Feather of the stormbird that is above the house of Khoto.

Two memes are combined in the single line above: the feather of the Southern Ground Hornbill and the notion of being above the house or homestead, which we saw earlier in connection with Bateleur Eagles clapping their wings. And as we will see in Chapter 10, on top of the hut or above the hut is a very significant place for omens and portents.

Cope records the following line in the praise of the Zulu general Ndlela kaSompisi (1968: 187):

> *Insingiz' edond' ukusuka*
> Hornbill that is reluctant to set out.

This line is an allusion to the battle fought between Cetshwayo and his brother Mbuyazi in 1852 on the Ndondakusuka Ridge above the uThukela River. Nyembezi has something similar in the praises of Cetshwayo, but it is a puff adder, not a Southern Ground Hornbill, which is reluctant to set out (1958: 89):

---

46. Interestingly, in a belief that I have not seen recorded in print, Nontuthuko Xaba, a young participant at the June 2017 Zulu bird name workshop, told us that in her community, a single **insingizi** feather is tied close to the surface of a river, and what happens to the feather will determine whether rain, heavy rains, thunderstorms, etc., will happen in the near future. See section 13.6.3 (page 425) for further details.

IBululu likaPhunga noMageba
>Puffadder of Phunga and Mageba

Elindond' ukusuka
>Reluctant to set out

Kwaze kwasuk' awezindlwana
>Until those of the young animals set out.

The Southern Ground Hornbill was found once only in clan praises, and, as with the occurrence of the Hamerkop (**uThekwane**) above, only as a personal name. The *izithakazelo* below of the Dlomo clan are from Sithole (1982: 22):

Mkhabela!
>Mkhabela!

Bhelezi!
>Bhelezi!

Nsingizi!
>Mr Hornbill!

Dinangwe
>Dinangwe

Mhlong' ungeyen' owaseLangeni
>Mhlongo who is not a member of the Langeni clan.[47]

8.7   *IZINYONI NJE* 'JUST BIRDS'

In this final section of birds captured in Zulu oral praises, we look at unidentified species of birds. There has been no attempt to group them in any way and the order of their appearance is purely arbitrary.

From Nyembezi and taken from the praises of Swazi King Mswati II are the lines (1958: 135):

Inyon' kaMabizw' asabele
>The bird of Mabizwa to be apportioned out

Ubizwe nguShila kaMlambo
>He is called by Shila son of Mlambo

---

47. The Mhlongo clan is an offshoot of the Langeni clan, thus they are closely related.

Nyembezi also gives us the following from the *izibongo* of Dinuzulu (1958: 108):

> *Umgwazi kaqaqi*
>> The stabber does not rip up [things]
> *Uqaqelwa zinyoni*
>> He himself is ripped up on behalf of birds

From the *izibongo* of Shaka, as recorded by Nyembezi (1958: 24), are the lines:

> *Inyon' edl' ezinye*
>> Bird that ate other [birds]
> *Yath' isadl' ezinye yadl' ezinye*
>> And while eating some, devoured yet more
> *Odl' imihlambi ingeyabahweba*
>> He who ate up the flocks which he was never going to trade.

We might well expect birds to appear in the praises of the clan Msomi (a clan name derived via the plural form **amasomi** from the singular **isomi** 'Red-winged Starling'). The lines that follow are from Sithole (1982: 74):

> *Izinyoni ezinhle*
>> Beautiful birds
> *Ezadla uvovo zadakwa*
>> Which ate the aloe flowers and became drunk
>> . . .
> *Izinyoni ezawusa umfula*
>> Birds which made the river come
> *Ukuba zewenyusa ngabe zafa zonke*
>> So they would go up it and all die

From Gunner and Gwala, and taken from the praises of Ngqengelele of the Buthelezi clan, are the lines (their translation) (1994: 38):

> *Inyoni kaMakhala eyakhalel' uZulu*
>> The Cry-Crying bird that cried for the Zulu people
> *Mhlamane uZulu engakulima kithi kwaBulawayo*
>> On the day that the Zulu nation couldn't plough at our place of Bulawayo

Eyolima ngensimbi edl' amadoda
> But had gone on to plough with the iron that devours men.

Gunner and Gwala explain in a note here that "Yet in Ngqengelele's praise there is also reference to his having chided Shaka for his destructive mourning after the death of his mother Nandi". Shaka's edict to the Zulu nation after the death of his mother is well known: they were not to plant any crops for a whole year, and the army was to go out and 'wash their spears' to mark his mother's death (ploughing with the iron that devours men). Ngqengelele's *imbongi* suggests that the very birds wept to see the position Shaka had put his people in.

Gunner and Gwala give the following extract from the praises of Mangosuthu Buthelezi (their translation) (1994: 89):

Unyikiz' uGodlankomo
> Shaker of rigid mountains
Kwayekwazamazama iNgome,
> Until Ngome Forest quivered

I'ntaba zodwa zobikelana,
> The mountains will report to each other in amazement
Izinyoni ziyowashaya amahlombe
> The birds will beat their wings
Zithi, "Zinkulu kwaPhindangene"
> Saying, "There is something stirring at Phindangene."

Mathole kaTshanibezwe was Mangosuthu Buthelezi's father. Gunner and Gwala give his praises, from which the following line has been extracted (1994: 125):

Makomane, inyoni engadliwa ikhanda
> Head-Dripper, bird whose head is not to be eaten.

It is Gunner and Gwala as well who record the praises of a certain Magemegeme, from which this line is taken (1994: 148–9):

Umahlal' emthini njengenyoni
> Percher on a twig just like a bird

And it is also Gunner and Gwala who record the praises of "a young Mjadu man", whence this line (1994: 198–9):

Unomasikisiki, iNyoni esindwa sisila
    Clatter-Clanger, Bird weighed down by its tail.

From the praises of Shaka, recorded by Cope is the line (1968: 115):

Inyon' ebizwe ngamakhwel' aseNgome
    Bird that is summoned by the whistles of the Ngome forest,

And then finally, to end off this section of birds found in Zulu traditional oral poetry, is the intriguing question found in the *izithakazelo* of the Malinga clan, recorded by Sithole (1982: 50):

Pho! kulula yini
    Well then! Is it easy
Ukubamb' inyon' izwa
    To catch a bird when it is fully aware?
Pho! ukhona yini
    Well then! Is there anyone
Owayengaphikis' izwi likaShaka?
    Who has not disputed the word of Shaka?

9 Praises, proverbs and riddles

9.1 THE PRAISES OF BIRDS

The previous chapter looked at the occurrence of various birds as symbols and metaphors in three different sub-genres of Zulu praise poetry: the *izibongo* of kings and chiefs, the *izihhasho* of the 'common man' and the *izithakazelo* of clans. We saw how swallows symbolised a departure and a return, how birds with red feathers were used as a metaphor for spilt blood, and we saw a variety of raptors, finches and other birds all playing their own literary roles. None of the poems, however, were composed in honour of birds, which is to say that birds were not the actual subjects of the praise poems. In this chapter, however, we look at the praises of certain birds themselves, i.e. praises composed where a particular bird is the subject of the praise poetry.

Later in the chapter we will visit other traditional Zulu oral literary genres: proverbs, then riddles, with a brief mention of the songs and children's games aligned to riddle play.

9.1.1 Introduction: praise names to praise poetry

Wainwright (1983), in an article on bird names and bird praises in Xhosa, proposes three stages of development, which run from the coining of a praise name to the development of a 'fully-fledged' praise poem. As examples of coined praise names for birds, in other words, names that represent the first stage of the development of bird praises, he gives the following:[1]

---

1. Wainwright has taken all of these from Godfrey (1941). Many of the names are also used by Zulu speakers.

indlanyoni 'eater of birds' – bateleur
udlezinye 'eater of other [birds]' – fiscal shrike
umdlampuku 'eater of mice' – [unidentified] kite
unoxwilimpuku [*unoxwil'impuku*] 'catcher of mice' – [unidentified] kite
ihlabankomo 'what stabs cattle' – swallow and/or swift
udlihashe 'what eats a horse' – swallow and/or swift
uhlalanyathi 'what sits on a buffalo' – oxpecker
intakembila 'bird of dassies' – puff bill [*sic*] shrike [Black-backed
    Puffback]
unomntan'ofayo 'the dying child' – black cuckoo
unozalizingwenya 'what gives birth to crocodiles' – goliath heron
umasengakhoth'idolo 'what licks the knee while milking' – African
    Jacana

Many of these names are straightforward, in that they are a simple statement of what the bird in question does: the Bateleur and the Southern Fiscal (formerly Fiscal Shrike) do eat other birds; kites do catch and eat mice; the oxpecker does sit on buffaloes. Other names employ recognisable literary devices such as personification in the case of *unomntan'ofayo*, and metaphor in *ihlabankomo*, *unozalizingwenya* and *umasengakhoth'idolo*. In these last three, the swallow, in flying close to cattle, only appears to stab them; the Goliath Heron only appears to have given birth to the young crocodiles that share the shallow waters of the pans; the African Jacana looks like someone lifting his knees right up while milking. For Wainwright, though, both the straightforward names and those containing allusions are similar to the praise names composed by ordinary people for themselves.[2]

What is a praise name for a bird, and what is a straightforward referential name for a bird is not always clear. Godfrey says (1941: 33):

From the Cis-kei to Flagstaff, the Bateleur has, as its distinctive Native name, *ingqanga*. It boasts, however, quite a number of nicknames:— *intaka yamadoda* (the bird of the warriors), *intaka yempi* (the bird of the army), *intaka yot∫haba* (the bird of the enemy), *intlaba mkhosi* (the raiser of the war-cry).

---

2. Wainwright's attitude towards these praise names is not always easy to determine. He says, for example, that they contain "some of the wildest imagery imaginable", while names like *ihlabankomo* and *udlihashe* are "bizarre attributes bestowed on the hapless swallow and … the swift" (1983: 296).

It would appear here that *ingqanga* is the Bateleur's 'name', while all the others are 'praise names'. If we accept that, then we must accept that in Zulu, the name **ingqungqulu** is the 'name' of the Bateleur, and *indlamadoda* (what eats men) its praise name. Yet *indlamadoda* is a straightforward description of what the bird does (it eats corpses on the battlefield), while **ingqungqulu** uses the poetic device of onomatopoeia to refer to the clapping sound of the Bateleur's wings. So we have a situation where the referential name is 'poetic', while the 'praise name' is not, yet 'praises' are regarded by all writers and scholars as examples of oral poetry. Take, too, the 'name' and the 'praise name' of the Puffback Shrike (now called the Black-backed Puffback). Wainwright has told us (see above) that he thinks the name *intakembila* (bird of dassies) is a praise name. The allusion, however, is not clear: surely the Black-backed Puffback does not eat dassies? Wainwright doesn't mention, however, the other name that Godfrey also gives for this bird (1941: 113): *unomaswana*, which Godfrey translates as "a little blob of calabash milk", a reference to the "snowy whiteness of the rump feathers". This seems to me a very successful example of a metaphor, making *unomaswana* a decidedly 'poetic' name.

The point I am trying to make here is that a distinction between what is a 'praise name' for a bird and what is not is often purely arbitrary. Nonetheless, such coined praise names constitute the base of Wainwrights' 'first stage' in the development of bird praises, and it is from these that he moves on to his second stage, explained as follows:

> Once the various *iinduna*, cattle, children and even birds have been given their praise names, a logical progression is the addition of a few, basic laudatory elements, and such rudimentary attempts at eulogy from the mouths of amateurs is excusable considering that even for gifted specialists *ukubonga* is an exacting discipline (1983: 297–8).

From the examples that follow this astonishingly patronising analysis, the addition of a few 'basic, rudimentary elements' consists entirely of verbalisations of bird calls. Thus, examples of the second stage of development of bird praises include the duck with its *Isifuba sam sithe: gaa gaa gaa!* (my chest goes *gaa gaa gaa!*), while the drake displays its own 'rudimentary elements' in *Uzithi, tshwe tshwe tshwe!* (you should say *tshwe thswe tshwe!*).

Another bird that exemplifies the "attempts at eulogy from the mouths of amateurs" is the Cape Parrot (**isikhwenene**):

Haha haha	Haha haha
Ndinabe nam abantwana	I also have children
Haha	Haha

One of his last examples of this second stage of development are the praises of the Black Cuckoo, and here we move away from pure verbalisation of the call to what I identified earlier (see Chapter 7, page 182) as an interpretation of the bird's call:

Ndina mntan' ufayo	I have a sick child
Ndiba ndiya mbiko	I think I am reporting him
Kanti akabikeki	But he is ignored

Godfrey, from whom these praises were taken, says that when he first heard the call:

The plaintive call was ascribed to *unomntan' ofayo*, and the bird was supposed to be continually wailing its sad condition (1941: 57).

Wainwright's third stage is the composition of a 'fully-fledged' praise poem. His first example of such is that of the Blue Crane (**indwa**), quoted below in its entirety:

UGaga kaMyeza	Gaga, progeny of Myeza
Intaka ehlonitshwa ngumthinjana	The bird revered by the maidens.

It is also his only example of a fully-fledged praise poem of a bird, as all his other examples are praises of people. However, with this single example, he has at least suggested an entry point to a discussion of our own on how bird names 'develop' into praise poems.

The first line of the Blue Crane praise above comprises two words: a name and a patronym: 'Gaga son of Myeza'. To my mind, it is this structure of NAME + PATRONYM that provides the stepping stone to praises. Let us consider the case of the Zulu bird name **uthekwane** (the Hamerkop, *Scopus umbretta*). Bryant tells us that in Zululand this bird was known as *ut[h]ekwane o'ziluba*, but in Natal as *uthekwane ka'ziluba* (1905: 619). Both these phrases mean 'Hamerkop of the plumes', in reference to the bird's crest, but there is a subtle difference. The grammatical structures here make the Zululand version the 'plumed hamerkop' and the Natal version 'Hamerkop, son of Mr Plumes'. This personification of both the bird and its plumes, linking them as son and father, immediately suggests praising. For example, such son-and-father links are typical of *izithakazelo* (clan praises) as seen, for example, in the praises of the Malinga clan:[3]

Mlotshwa kaGumakhulu kaZindlela	Mlotshwa son of Gumakhulu son of Zindlela
Malinga!	Malinga! ['the one who tries']
Owaling' ukuyihlomis'ebusuku	Who tried to arm himself at night.
Nzalo kaMagidela	Nzalo ['progeny'] son of Magidela
Omnyama ngentamo	He who is black on the neck
Owazil' inkatha bekuthezelile	He who respected the *inkatha*[4] while others talked wildly about it.

We see here how the first name – Mlotshwa – is extended by a double layer of patronyms. The second name – Malinga – is extended by a line of narrative praise. The third name – Nzalo – is extended by a patronym and two lines of descriptive and narrative praise. In this way, the praises are built up, in layer after layer of name, patronym and descriptive or narrative praise.

We see the basic start of this process when the Hamerkop is referred to as *Thekwane kaZiluba*. We see it again when Bryant gives **u-Ntloyile** as the name of the Yellow-billed Kite, but says that the phrase *untloyile ka'Gelegele* (or *ka'Mgubane*) is "a nickname, [i.e. praise name] for the bird" (1905: 452). And indeed, that is how the praises of the Yellow-billed Kite (see page 233), given below, start, followed by a descriptive phrase:

UNhloyile kaGelegele	Kite, son of Whirlwind
UMathumb' abek' ezulwini	The one who captured and headed upwards to the sky.

These two lines are the exact structural equivalent of the two lines of praise that I quoted Wainwright (above) as giving as his sole example of the 'third stage' in the development of avian praise poetry.

---

3. Sithole (1982: 47).
4. *Inkatha*: a large grass-based coil representing clan unity.

Table 9.1 Comparison of Xhosa and Zulu examples of bird name → praise phrase.

	Wainwright's Xhosa example of the Blue Crane, indwa	Zulu example of the Yellow-billed Kite, unhloyile	Syntactic structure
1st line	**UGaga kaMyeza** Gaga, son [progeny?] of Myeza	**UNhloyile kaGelegele** Nhloyile, son of Gelegele	Name followed by patronym
2nd line	**Intaka ehlonitshwa ngumthinjana** The bird revered by the maidens	**Umathumb' abek' ezulwini** Capturer that headed for the sky	Descriptive noun followed by verb-based qualifying clause

In the Zulu praises of selected birds immediately below, we will also see the pairs *Sobada wakwaZungu* (Sobada of the Zungu clan) starting off the praises of the Yellow-throated Longclaw, and in these praises the line *Qaza kaHlokohloko* (Seeker, [son] of Hlokohloko ('yellow weaver').

9.1.2 The praises of selected birds in Zulu

Note how all these birds in this section are involved in a cultural web and feature in oral poetry and traditional beliefs of all kinds (harbingers, tooth fairies, lucky charms, etc.). They have great cultural salience, and so do not only have names, but praise names, and praises. As poetry, the language is dense with imagery and allusion, so much so that we can talk of a 'web of allusions'. There are many examples, one of the best being just the first line of the praise of the **inqomfi**, given on pages 234–42 below. On the surface we would appear to have a personal name and a clan name: Sobada of the Zungu clan. But when these two words are taken in conjunction with Stuart's pencilled note in his original notebook, all sorts of allusions appear, linked to birds that manifest good luck ahead by running along in front of a traveller on the road.

I start with the shorter of the praises available, and go on to the longer ones from there. All the praises of individual birds given below were kindly forwarded to me by John Wright. Professor Wright and his late colleague Professor Colin Webb have worked since 1976 – over 40 years – editing and publishing the collected notebooks of late nineteenth-century and early twentieth-century magistrate James Stuart. However, in all six volumes published since 1976, Webb and Wright have consistently omitted the praises collected by James Stuart, but Wright, together with Mbongiseni Buthelezi, is currently putting these into a seventh volume. He was kind enough to allow me access to the praises of the birds that appear below, before they appear in print.

It must be emphasised that Stuart was the recorder, not the originator, of these bird praises. Wright has indicated that, according to the relevant notebooks, the praises were given to Stuart by Socwatsha kaPapu [modern spelling (ka)Phaphu], interviewed by Stuart on 5 October 1921. The original notebooks are held in *The James Stuart Archive* (*JSA*) of the Killie Campbell Africana Library in Durban. The bird praises that follow are in File 28, nbk 21, page 21.

The praises of inkanku: *the Jacobin Cuckoo* (Clamator jacobinus)

The praises recorded by Stuart for this bird are very brief, comprising only two lines, as follows:

> *Inkanku yadhl' amacimbi*
>> The inkanku cuckoo ate the caterpillars
> *Okwandulel' unyaka*
>> The harbinger of the year

Both the eating of caterpillars[5] and the announcing of the time of the year are well attested in the literature, as the following extract from Bryant (1905: 293) shows:

---

5. See Chittenden, Davies and Weiersbye (2016: 272): "Food: Spiny or hairy caterpillars".

i-n**Kanku** n. . . . Le Vaillant's Cuckoo (*Coccystes cafer*),[6] a bird whose appearance gives name to a month (see *u-Nkanku*), also from its conspicuous appearance about the end of July, supposed to announce the time for sowing.

Phr[ase]. *inkanku is 'iwatete amacimbi (*or *amacimbi okwandulela*),[7] the cuckoo has already taken the caterpillars – denoting that part of the season about early October time.

Woodward and Woodward (1899: 114) state:

Le Vaillant's Cuckoo (*Coccystes cafer*): "It has an extraordinary loud cry, and Dr Colenso says when its voice is first heard it announces to the natives that the time for planting has arrived, like the cuckoo in England".

Their reference to "Dr Colenso" is to Colenso's Zulu dictionary, where his entry for this bird reads:

inNkanku: black-and-white bush-bird which moults in winter, and whose piercing voice is heard in summer at night, announcing the time for sowing, corresponding to the cuckoo of England.

Ex. *sokupakati kwonyaka, innkanku seipelile is 'ipendule*, it's now midsummer, the *innkanku* is now full-throated, has changed its voice (1884: 379).

The praises of ungqwashi: the Rufous-naped lark (Mirafra africana)

Stuart's marginal comments in his original notebook are shown in the *Lucida Handwriting font*.

---

6. In the next chapter (pages 305–6) I give details about the confusion in both the scientific names and the English vernacular names of this bird: this section is headed "Jacobin Cuckoo" but I am quoting Bryant on "Le Vaillant's Cuckoo". Are they the same bird? Chittenden, Davies and Weiersbye (2016: 272) give them as separate (if strikingly similar species – to the lay observer) but earlier writers used **inkanku** of both and seemed to regard them as one and the same species.

7. Bryant (1905: 8): "**Andulela**: Begin first, or before another, as a woman beginning to hoe before the others of her locality".

1 *uNgqwatshi lo bomvu!*   *(or mangqwatshi)*

   The lark — this red one!

2 *uNgqengendlela:*

   The one that goes straight along the path

3 *Isitutukazane,*

   The little fool

4 *Esinamakizan'ekanda*   *akon' amakizan' ekanda*

                                       kuyena

   Who has ticks on his head

5 *uHuye!*

   Lark!

6 *Ugijima nge ndhlel'angaphumbuki*

   It runs along the path without deviating.

Comments:

This is the Rufous-naped Lark (*Mirafra africana*) (see Plate 38), also known as **unongqwashi, umangqwashi, usangqwashi, ungqangendlela** and **uhuye**. In the 2013–2017 Zulu bird name workshops, the names **untilontilo** and **uqaqashe** were assigned to this bird. Doke and Vilakazi's dictionary (1958: 684) give the name **iqabathule** specifically for this bird. Clearly this is a bird of many names. As we saw in section 8.3 above (pages 195–7: the 'red birds'), the Rufous-naped Lark occurs frequently in Zulu praise poetry, as well as having its own praises.

   The name **ungqengendlela** (also as **ungqangendlela**) for the lark is based on the ideophone *ngqe* meaning 'of rolling along straight' + *nga* + *indlela* (along the path). Doke and Vilakazi illustrate the ideophone with the example *ukuthi ngqe ngendlela* (to go straight along the path).

   Doke and Vilakazi define the word *isithuthakazana* as 'little fool' and illustrate the word with the phrase *isithuthakazana esinamakhizane ekhanda*, saying that this "is said of the lark, *ungqwashi*", and showing, again, that

Vilakazi was familiar with these praises of the lark. Why a fool should be characterised by having ticks on the head is unknown. Birds are usually regarded as 'fools' (and sometimes given names such as 'booby' and 'dodo') if they are unusually easy to catch. I have no idea if this is true of this lark.

The praises of unhloyile: the Yellow-billed Kite (Milvus aegyptius)

1 *uNhloyile kaGelegele*
   Kite, son of Whirlwind
2 *uMatumb' abek' ezulwini*
   Mr Snatch-up-and-away that headed towards the sky
3 *uTshungu la geq' elinye,*
   Mr Snuff-box that scraped out another
4 *Ngi nga bange ngize ngwiwabon' amatshung' ukugeqana.*
   I have never ever seen snuff-boxes scraping out one another.

Comments:
Line 1:
Doke and Vilakazi give *igelegele* as 'dust devil', 'whirlwind', and say '*uNhloyile kaGelegele*' is a praise name of the kite, meaning that Vilakazi was aware of this praise (see Plate 47).

Line 2:
The praise name uMathumba is based on the verb *t[h]umba* meaning 'take off captive, capture and take away' (Bryant, 1905: 661). Prefixing this with the name formative –*ma*– with its meaning 'characteristic of', gives a praise name that suits the kite perfectly: 'the one that snatches up [prey] and carries it away'.

Lines 3 and 4:

The word *ishungu* refers to a snuffbox, particularly if made from the dried-out shell of the fruit of a monkey orange tree (*umthongwane, Strychnos spinosa*). There is a clear link here to Bryant's (1905: 213) entry for *ugwayi-ka'kolo* and *ugwayi-ka'ntloyile*, both meaning the 'snuff of the Yellow-billed Kite', and referring to the puff-ball, a kind of powdery fungus growing on the veld.[8] Clearly the black spores of this fungus resemble snuff, and as the spores are not used as snuff by humans, they must 'therefore' be used by some animal or bird, perhaps the Yellow-billed Kite. From here it is a small cognitive step to seeing the kite as a snuffbox, taking away the snuff as it flies upwards. **Ukholo** is another name for this kite. Monkey orange fruits, as with calabashes, need to have the dried pulp scraped out (*uku-geqa*) before they can be used as containers.

Roger Porter makes the interesting suggestion that this image of the snuff and the snuffbox may relate to the fact that when there is a veld fire, the Yellow-billed Kites are the first birds to congregate in numbers, diving into the smoke and ash, (i.e. the 'snuff') and flying upwards with the insect they have caught, doing this again and again.[9]

9.1.3   The praises of the longclaw *inqomfi*

We are fortunate enough to have two praise poems for the **inqomfi**, a generic for longclaws, and used both for the Yellow-throated Longclaw (*Macronyx croceus*) and the Cape Longclaw (*Macronyx capensis*) (see Plate 46). The first is another Socwatsha kaPapu poem recorded by Stuart and part of the praises that Wright forwarded to me; the other is a more contemporary praise poem for the **inqomfi**, written by the late Zulu poet B.W. Vilakazi. Again, Stuart's marginal comments are shown in the *Lucida Handwriting font*.

The Socwatsha kaPapu praises

1 *USobada wakwaZungu*

Sobada of the Zungu clan

indoda ya ka Mfana wendhlela kaManzini ka Tshana. I do not know how this mans name

---

8. Also in Doke and Vilakazi (1958: 284). Doke and Vilakazi add that the fungus is also known as *ugwayi-kathekwane* (hammer-head's [hamerkop's] snuff).

9. Personal communication, 12 December 2017.

2 *Umboni wamagwala*   *came to be part of nqomfi*
   ukunyenyela   *sibongo*
   The one who sees the cowards slinking away

3 *Inganti uyena gwal' elikulu*
   Whereas it is he who is the great coward

4 *UQaza ka hlokohloko*
   Seeker of the yellow weaver bird [AK][10] / Qaza, son of Hlokohloko
   [W/B]

5 *Unxapanxapa, utebe zimanzi.*
   Clicker in annoyance; the meat platters are wet [AK] / The one
   whose speech is full of clicks, the one whose eating mats are wet
   [W/B]

6 *Unyama idhliwe ngamaklwa nayizolo*
   The meat was eaten by the warriors even yesterday.

7 *Ungubo itungwe osomtotwana*
   Clothes sewn by red-rooted shrubs [AK]   *izitungwana*
   The covering which was sewn by *usomthothwane* [W/B]

8 *Ongceda bona bepet' izixa zeminxeba*   *imzi yosinga,*
                                           etungayo
   As for the Neddicky larks [AK] / warblers [W/B],
   they were carrying bundles of fibres.

9 *Beyi gaba beyi bekela*   *hlaba ngo*
                                           sungulo betekeleza

   Piercing with awls, tying knots,
   and sewing on patches [AK] /   *sew on isiziba*
   They were making holes and sewing on patches [W/B]

10 *Umame onga kwazi ukubinc' isidwaba*
   Mother who cannot wear a leather skirt

11 *Uti ukubinca wenz' ikenke*   *expose, open out,*
                                          show nakedness

   When she tries to wear one, she exposes herself

12 *Wa bekisa nga ku mina*
   And shows herself to me [AK] / And she turns to me [W/B]

---

10. Here and there I have given both Wright and Buthelezi's translation, marked by 'W/B',
    as well as my own translation, marked as 'AK', when I feel there is a significant
    difference between the two.

13 *Uti ngi yo ba tshay' abafana*
    Saying "I will beat the boys

14 *Aba ku cupa nge ngange.*   *Small white ants*
    Who set traps for you with termites"   *for cupaing*
                                       found in ?izionti?

Comments:

At first glance, these praises appear to be a series of disjointed, arbitrary statements that have little to do with each other, and little to do with the bird **inqomfi**. As in any poetry, there is a great 'depth of allusion' here, and if one examines the statements carefully, a definite narrative appears. I do not say that every line is clear, but certainly a strong core discourse emerges, which is all to do with the longclaw lark, both its physical appearance as well as the layers of traditional beliefs associated with the bird. It is also important to note that the **inqomfi** is by no means the only bird in these praises: hidden in these lines are four other species of bird, and the praises as a whole are reminiscent of the traditional children's game "*Bhula 'Ntsentse Bo!*" This game, described by Letitia Samuelson is a game where a circle of children take turns to try to name as many birds as possible (1974: 88).[11]

I will try to analyse the poem line by line:

Line 1:
At the surface level, these two words are clearly a personal name and a clan name, translated as 'Sobada of the Zungu clan'. The depth of allusion become clearer if one looks at the note that Stuart has pencilled in next to this line: "*indoda ya ka Mfanawendhlela kaManzini ka Tshana.*[12] I do not know how this mans [*sic*] name came to be part of nqomfi sibongo".

Probing more deeply into this line and its accompanying marginal note, the first point is that the name Mfanawendhlela translates literally as 'the boy of the road'. The **inqomfi** lark is known for its habit of running along in front of a traveller on the road, and this is recognised traditionally as a sign of good luck. To give this lark the nickname '*umfanawendhlela*' would be entirely appropriate. Manzini is an *isithakazelo* (clan praise name) of the Zungu clan,[13] which identifies the 'man' in the first line as a Zungu.

The second point is that the name Sobada itself is constructed using the name-forming prefix –so– (father of) and the noun **ibhada,** which is the Zulu word for the Yellow-bellied Greenbul. As I said above, the Zulu word **inqomfi** refers to two species of the genus *Macronyx*, which occur in KwaZulu-Natal (KZN), and one of them is the Yellow-throated Longclaw. So if Sobada is, according to Stuart's pencilled note, the son of Mafanawendhlela, what we have in this line is not so much its surface meaning 'Sobada of the Zungu clan', but the underlying meaning 'yellow-throated (breasted) bird which is the [lucky] fellow of the road', a combination of a physical feature of the bird with a traditional belief <u>about</u> the bird.

Lines 2 and 3:
At first appearance, again, these lines are a puzzle: what have cowards to do with the Yellow-throated Longclaw? In answer, we can go back to line 1 where another species of bird, the **ibhada** (Yellow-bellied Greenbul) was linked to a

---

11. Details of the game are given in section 9.3 below, where I cover birds in riddles and children's games.
12. 'A man who is the son of Mfanawendhlela son of Manzini son of Tshana'.
13. The name is derived from *amanzi* 'water'. To a Zulu speaker the word 'Zungu' in line 1 would equate to 'Manzini' and so be linked semantically to line 5, where the eating mats (*izithebe*) are wet (*zimanzi*).

clan name. In lines 2 and 3 we have the same situation: the clan name Gwala (from the noun *igwala* 'coward') is apparently from the same root as the bird name **igwalagwala**, used for both the Knysna and the Purple-crested Louries (or Turacos, as they are now called). Repeating the stem of the noun *igwala* suggests augmentation, i.e. 'great coward' (*igwala elikhulu*). These two lines could then be rethought as 'The bird [the lourie] that thinks it sees the cowards (*amagwala*) slinking off, is in fact the greatest coward (*igwalagwala*) of all'. What has this to do with the longclaw? Nothing really, unless one is playing the game of how many birds, including their clan names, one can think of.

Line 4:
Wright and Buthelezi simply translate this line at face value: "Qaza, son of Hlokohloko". The verb *qaza* means 'to seek' and **ihlokohloko** is the generic name for the yellow-weaver-birds, so in a game where each player seeks a new name to add to all the other bird names that have gone before, the interpretation 'seeker of the yellow weaver' is not out of place.

Lines 5 and 6:
In a note below the praises of the **inqomfi**, Stuart has written "This bird is the greatest luck. You would never pass by two kraals without finding the most abundant food". Other sources, in earlier volumes of *JSA*, for example, have indicated that if the **inqomfi** runs along in front of you in the road when you are out courting, you will meet the most beautiful and receptive girls. If not, you will at least get much food at the next homestead. In line 5, however, someone is out of luck: when he (or she?) gets to the next homestead, he finds that the meat has already been eaten, and so the eating mats and/or meat platters have already been washed. Indeed, they are still wet. No wonder he clicks his tongue in annoyance.

Line 7:
This is a line with great depth of allusion. Wright and Buthelezi have kept the original '*osomtotwana*' as '*osomthothwane*' (using the modern spelling) but have indicated in a footnote that Bryant, in his dictionary, has given *umtoto* as 'small shrub . . . having very red roots and black edible berries", with a secondary meaning "any bright reddish thing, as some cows . . .". They have opted for this secondary meaning, suggesting an interpretation of 'the owners of the little red cows [*usomthothwane*]'.

Plate 1

The White-eyes (*Zosterops*) have names in many languages that refer to their striking eyes. In Zulu the two names for these birds are **umehlwane**, which means 'the little one with the striking eyes' and **umbicini**, a name derived from a word meaning 'discharge from the eyes'.

Plate 2

The Bateleur Eagle is known as **ingqungqulu** in Zulu. It is also known as **indlamadoda**, meaning '[the bird] that eats men', a reference to its eating of corpses on the battlefield.

Plate 3

Plate 4

Plate 5

Plate 6

Five different species of cisticola: Plate 3, the Lazy Cisticola (*Cisticola aberrans*); Plate 4, the Neddicky (*Cisticola fulvicapilla*); Plate 5, Levaillant's Cisticola (*Cisticola tinniens*); Plate 6, the Croaking Cisticola (*Cisticola natalensis*); and Plate 7, the Wailing Cisticola (*Cisticola lais*). In the Zulu folk taxonomy, none of these birds would have separate salience, and all would fall under the single name **uncede**.

Plate 7

Plate 8

The Hamerkop (**uthekwane**) is a bird of great omen in traditional Zulu thought-patterns. Its huge nest and its habit of apparently staring at itself in the water for long periods add to the bird's mystique. A Hamerkop landing on the roof of a hut and calling is a portent of great evil to come.

Plate 9

Plate 10

Plate 11

In the Zulu folk taxonomy, doves and pigeons have separate species-specific names: Plate 10, **unkombose** (Namaqua Dove); Plate 11, **ivukuthu** (Speckled Pigeon); Plate 12, **ukhonzane** (Laughing Dove); Plate 13, **isagqukwe** (Lemon Dove); and Plate 14, **ijubantondo** (African Green Pigeon). All come under the cluster name **ijuba**. This bird name is also used as a brand name for a commercially produced traditional Zulu beer (Plate 15).

© Ingrid Weiersbye

Plate 12

© Ingrid Weiersbye

Plate 13

© Hugh Chittenden

Plate 14

© Adrian Koopman

Plate 15

Plate 16

Black birds and their names: black plumage is reflected in a number of different ways in Zulu bird names. Here two such names are illustrated, one a coined name, the other a traditional Zulu bird name. The coined **unozila** for the African Oystercatcher (Plate 16) is derived from the verb *zila* (to mourn), while the traditional name **ummbesi** for the Southern Black Flycatcher (Plate 17) is derived from the verb *mbesa* ('cover over with a cloth or blanket'), suggesting that the bird is wearing a 'cloak of darkness'.

Plate 17

Plate 18

The Red-chested Cuckoo (**uphezukomkhono**) is the noisy announcer of the arrival of spring and the planting season. Its name has been interpreted in a number of ways, most commonly as 'on top of the shoulder', an indication that when this bird starts to call it is time for women to shoulder their hoes and set to work in the fields.

Plate 19

The Paradise Flycatcher has two Zulu names: **uve** (or **uluve**) and the onomatopoeic **inzwece**. The Zulu proverb *uluve ludl' isisilo salo* ('the *uluve* eats its own tail', i.e. some people do things that lead to their own destruction) is based on the erratic, swooping, darting flight of the bird.

Plate 20

Birds wearing colourful robes. The name **umambathingubo** for the African Hoopoe (Plate 20) means 'the one that wears a [colourful] blanket', while the coined name **umambathilanga** for the Yellow Bishop (Plate 21) means 'the one that wears the sun as a blanket'.

Plate 21

umathithibala

Plate 22

Birds with stripes may get names of different kinds that refer directly to their stripes. The Red-backed Shrike (Plate 22) has the Zulu name **umathithibala**, a metaphorical name using the same word *umathithibala* for the small succulent *Haworthia limifolia*. The female shrike is striped in exactly the same way. Plate 23 shows the Pririt Batis, with the Swahili name *ndege mpunda*, literally 'bird zebra'. For Zulu, the name **isadube** ('like a zebra') has been coined for Stierling's Wren-Warbler (Plate 24), and the decidedly striped African Cuckoo-Hawk (Plate 25) has the coined name **usomthende** ('father of stripes').

Plate 23

Plate 24

Plate 25

Plate 26

The Zulu name **unohemu** (Crowned Crane) is onomatopoeic, as is the case in various other South African languages. The Swahili name *ndege chai* (literally 'bird tea') has been recorded, supposedly because of the use of this bird as the logo of a popular brand of tea in Kenya. The actual name of the bird in Swahili is *korongo taji* (Mlinga 1993: 85).

Plate 27

This painting of swifts shows the birds flying high and low and swooping among cattle, as well as an approaching rainstorm, thus illustrating three commonly used Zulu names for swifts: **ihlabankomo** ('what stabs the cattle'), **ijiyankomo** ('what herds the cattle') and **ihlolamvula** ('what predicts rain').

Plate 28

The different species of Thick-knee (Dikkops) are known as **umbangaqhwa** in Zulu, the literal meaning 'what causes frost' referring to the white flecks in the bird's plumage so that when it lies flat in winter it looks just like a frost-covered rock.

Plate 29

The name **uzavolo** is used generally for nightjars, but the verbalisation of song as '*Zavolo, Zavolo, sengel'abanta bakho!*' ('Zavolo, Zavolo, go and milk for your children') refers specifically to the song of the Fiery-necked Nightjar.

Plate 30

Plate 31

© Hugh Chittenden

© Ingrid Weiersbye

Plate 32

Plate 33

© Hugh Chittenden

© Hugh Chittenden

In the Zulu bird names workshops, the Zulu prefix *−sa−* ('something like')
was used to create new names for birds when a similarity was perceived. On
this page we have **isaqola** (Fiscal Flycatcher, Plate 30) derived from **iqola**
(Southern Fiscal, Plate 31); and **usantiyane** (Green-winged Pytilia, Plate 32)
derived from **intiyane** (Common Waxbill, Plate 33).

Plate 34

The song of the Emerald Cuckoo is rendered as *Bantwana, ningendi!*
('Children, don't get married!'), but this call has been rendered in a variety
of other ways by various authors.

Plate 35

In Zulu, the word *undlunkulu* (literally 'big house') refers to the first wife in
a polygynous marriage: herself as a person, her status, her children, and
her part of the homestead. As a traditional Zulu name for the Cape Sparrow
(**undlunkulu**), the name really does mean 'big house' and refers to the huge
untidy nest this bird builds.

© Adrian Koopman

Plate 36

The name **umkholwane** has long been established as a Zulu name for the Southern Yellow-billed Hornbill. It was surprising, then, to discover that today a significant number of Zulu speakers call this bird **uzazu**, after the hornbill character in Disney's *The Lion King*.

Plate 37

The **inkwazi** (African Fish Eagle) is one of the most recognisable birds in South Africa, from its striking black-and-white plumage and its loud ringing call. Its arresting appearance has made it a favourite as a logo for a variety of companies, and the evocative nature of its call has made 'inkwazi' a popular name for a great number of South African holiday venues. In the plural *izinkwazi* gives the name to the iZinkwazi River on the KwaZulu-Natal North Coast and the name of the Zinkwazi holiday resort at the mouth of this river.

Plate 38

The Rufous-naped Lark carries more Zulu names than any other bird:
uqaqashe, **untilontilo**, **ingqangendlela**, **uhuye**, **iqabathule** and a number of
different forms of **ungqwashi**, such as **umangqwashi** and **usangqwashi**. It
is also a bird of omen, predicting either good luck or bad luck depending on its
behaviour. The bird appears in various genres of traditional oral poetry.

Plate 39

The **isiphungumangathi** (Long-crested Eagle) is asked for help when young herdboys have lost the cattle they are meant to look after.

© Adrian Koopman

Plate 40

This painting shows three birds that feature in Zulu proverbs. At top left is the **ithendele** (Shelley's Francolin, used also for other species of francolin). Middle right is the **inkwali** (Swainson's Spurfowl, but also, like **ithendele**, used for similar francolins). Bottom left is the **isagwaca** (Common Quail, used also for other quails).

Plate 41

The **inkonjane** (a name used generically for all swallows) appears in a number of different genres of Zulu traditional oral poetry, where it usually symbolises someone who has wandered afar and then returned.

Plate 42

At one of the Zulu bird name workshops, the name **impofana** was given to the Eurasian Golden Oriole because of the nickname 'Mpofana' of the Kaizer Chiefs Football Club. The name refers to the striking yellow and black club colours worn by the players.

Plate 43

Plate 43 shows a Pink-throated Twinspot. The name **umagumejana,** coined for this bird at one of the Zulu bird name workshops, is based on the name Gumede, the clan name of a young woman famous for her beadwork and finery. On the opposite page are two other birds whose coined names are linked to traditional body ornamentation: Plate 44 shows **unosongo** (the Common Ringed Plover), whose name is derived from the word *isongo* ('metal armlet'); and Plate 45 shows **unocu** (the Rose-ringed Parakeet), derived from *ucu*, the bead necklace a young woman gives to a man as a sign he can begin to court her openly.

Plate 44

Plate 45

Plate 46

In the traditional praises of the longclaw (*izibongo zenqomfi*) collected by magistrate/historian James Stuart, both the Yellow-throated Longclaw and the Cape Longclaw feature. Both birds are shown on this page.

Plate 47

The **unhloyile** (Yellow-billed Kite) is a greatly loved bird. Its arrival at the end of July after wintering to the north gives the alternative name uNhloyile to the month, and the name *idumbe likanhloyile* ('tuber of the Yellow-billed Kite') to the Snake Lily (*Scadoxus puniceus*), which pushes up from the ground at this time. The Yellow-billed Kite also plays the role of 'tooth fairy' in traditional Zulu culture.

Plate 48

Plate 48 shows the **ufukwe** (Burchell's Coucal), widely known as a 'rain-bird', i.e. a bird whose call presages rain. Opposite (Plate 49) is an illustration of the **inkanku** (Jacobin Cuckoo) whose call is an indication that summer has now fully arrived. These two birds are among several in Zulu culture whose call or arrival presages certain weather, or heralds a particular season or time of day.

Plate 49

© Ingrid Weiersbye

Plate 50

On these two pages we see birds with coined names that refer to their bills. Plate 50 shows the **idada elimlomophuzi**, an exact translation of its English name Yellow-billed Duck. Plate 51 is the Curlew Sandpiper, with the Zulu name **ungozwana**, from *ingozo* (the tiny Elephant Shrew, which has a similar long snout). Plate 52 shows the Pied Avocet, whose Zulu name **usipheshula** comes from *isipheshula* ('anything turned up'). Plate 53 is a photo of the Cape Shoveler, whose name **unofosholo** is derived from the Zulu word *ifosholo* ('shovel'). Plate 54 shows the Common Scimitarbill, whose name **unosungulo** is derived from *isungulo* ('awl').

© Ingrid Weiersbye

Plate 51

Plate 52

Plate 53

Plate 54

Plate 55

The painting on this page is of a Southern Fiscal (**iqola**). The word *iqola* also refers to any head of cattle with black and white markings on the back and sides like that of the fiscal. This is one of a number of such partnerships between bird names and cattle colour terms.

Plate 56

The red feathers of turacos (**igwalagwala**) have long been used as head ornamentation among Zulu and Swazi royalty.

Plate 57

The long tertial feathers of the Blue Crane (**indwa** or **indwe**, Plate 57) have long been associated with the Zulu king, especially since the popularisation of James Saunders King's nineteenth-century picture of Shaka wearing a single feather as sole ornamentation (Plate 58).

Chaka King of the Zoolus *(J. Saunders King)*

Plate 58

© Adrian Koopman

Plate 59

Plate 60

Plate 59 shows the Long-tailed Widowbird, well known by its Zulu name **isakabuli**. Bird feathers have long been used as ornamentation by Zulu youths and men, as shown in the head of uMbumbo (Isaacs 1970), a detail from a painting by G.F. Angas in the 1830s (Plate 60). Plate 61 is an illustration of current Zulu King Zwelithini kaBhekuzulu wearing red turaco feathers, a single blue crane feather and a quantity of **isakabuli** feathers, all above a cape of finely teased white ox-tails.

Plate 61

Plate 62

The cover of Jo Oliver's 1980 *A Beginner's Guide to Our Birds*.

Plate 63

Plate 63 shows the cover
of the 1989 *Izinyoni
Ezngamashumi Amane Nantathu
ZakwelaKwaZulu,* and Plate 64
the cover of the 2003 *Ibhuku
Lokucathulisa Abasaqala Ulwazi
Ngezinyoni Zethu.*

Plate 64

© Hugh Chittenden

Plate 65

On this page are the Pied Kingfisher, **ihlabahlabane** (Plate 65); the Giant Kingfisher, **isivuba** (Plate 66); the Malachite Kingfisher, **uzangozolo** (Plate 67); and the Pygmy Kingfisher, **isikilothi** (Plate 68). On the opposite page are the Woodland Kingfisher, **imbuyelelo** (Plate 69); the Mangrove Kingfisher, **unonkalankala** (Plate 70); the Brown-hooded Kingfisher, **usiphikeleli** (Plate 71); and the Striped Kingfisher, **unongozolwane** (Plate 72).

© Ingrid Weiersbye

Plate 66

© Ingrid Weiersbye

Plate 67

© Hugh Chittenden

Plate 68

Plate 69

Plate 70

Plate 71

Plate 72

Plate 73

Four Southern Ground Hornbills (**izinsingizi**) with thunderclouds behind.

I suggest a quite different interpretation of this line, which will refer directly to the physical appearance of the bird, especially if we refer to the Cape Longclaw, rather than the Yellow-throated Longclaw (both are found in KZN, and both are known as **inqomfi**). The first two words of this line will need to be interpreted slightly differently: *ingubo* can stay as 'clothing' (but should be seen here as a metaphor for 'plumage') but the verb *thunga* needs rethinking from its standard interpretation of 'to sew' into 'to prepare material for clothing'. Now we can look again at *usomthothwane*, noting that the prefix –*so*– is not just 'father of' but also suggests 'master of'. The word *umthothwane*, like its linguistic variant *intolwane*, refers to the shrub *Elephantorrhiza elephantina* with its red roots, which is the best-known and most frequently used source of a dye for staining something, such as rushes for weaving, a rich russet red (*bomvu*). Indeed, this is part of Zulu oral poetic imagery: a red ox or bull or often referred to as *uJamludi obomvu njengentolwane* (Blood-coloured one red like the *intolwane* roots). Look at the Cape Longclaw (top left of Plate 46) and think of it as the bird whose raiment has been stained red at the throat by using *umthothwane* roots).[14]

Lines 8 and 9:
The subject of these lines is '*ongceda*',[15] a word applied generically to the cisticolas. As '*ncede*', it is the root of the word 'Neddicky'[16]. These birds do not have any bright defining colours such as a bright yellow or a bright red throat. They could be seen, continuing with the metaphor of preparing clothing to wear, as the birds that are wearing scraps and patches of clothing sewn together. Stuart, in marginal notes, identifies *izixa* as *izithungwana* (bundles), *iminxeba* as *im*[*i*]*zi yosinga, etungayo* (fibres and sinews for sewing), the verb *gaba* as piercing with an awl and tying a knot, and the verb *bekela* as 'sew[ing] on *iziziba*' (patches). The Neddicky and other cisticolas are thus pictured as birds who stitch together their own patch-like plumage. The **uncede**, it must be noted, is a bird that features strongly in proverbs and other oral literature and lore (see below, section 9.2, page 256).

---

14. So perhaps Wright and Buthelezi's "little red cows" are not so much out of place.
15. Spelt variously as *uncede, unceda, ungcede* and *ungceda*, depending on the source.
16. Godfrey explains, in an early article, that the Xhosa word *ncede* as 'borrowed' by the Dutch at the Cape, who added the Dutch diminutive suffix –*tje* to the word to produce *ncedetje*. In Afrikaans this became *ncedetjie* (pronounced *n[c]edekie*) and this form was adopted into English as Neddicky.

Lines 10 to 12:
There is little problem in interpreting these lines literally (translating them at face value). Stuart explains in a marginal note that the verb *kenka* means 'expose, open out, show nakedness', so the reason the woman cannot wear a leather skirt (*isidwaba*) is that when she does she exposes her nakedness underneath. The problem here is associating these lines with the **inqomfi** or any other birds. Could the person composing the poem possibly have had the bird **inkenkane** (African Spoonbill) in mind, and this led him to thinking of the verb *kenka*?

Lines 13 to 14:
The **inqomfi**, like the wagtail (**umvemve**) is considered a lucky bird, and there is a taboo against killing it. Young boys out in the veld herding cattle are often both bored and hungry, and on occasion break the taboo by catching, roasting and eating birds such as the **inqomfi** and the wagtail. One of the favourite means of catching birds is by using a heavy stone propped on a stick, to which a string is attached. White ants or termites (*inganga*) are a common bait used to lure birds to this trap, where they are killed when the string is pulled and the stone falls on them. Should this eating of tabooed birds be discovered, however, the boys are chastised when they get home. Hence mother saying "I will beat the boys who set traps for you [= *inqomfi*] with termites".

Even deeper allusions

There may well be even deeper allusions than the ones I have assigned to the praises of the **inqomfi** above. It could well be that these praises are of some antiquity, going back to a proto-Nguni time before Zulu and the Xhosa dialects separated. Godfrey (1941: 103–4) has the following to say of the Cape Longclaw, a bird, as we have seen, well known to the Zulus and also named **inqomfi**. I quote him at length here because all he has to say about the names, praises and beliefs about the longclaws among the peoples of the Eastern Cape have direct reference to the Zulu **inqomfi** praises given above:

> Throughout the Native area [the Cape Longclaw] is known as *inqilo*, the only variation being *unqilo*, given in Clarkebury and Flagstaff lists.

In Bomvanaland it is nicknamed *igqwathiza*[17] . . . from the name
applied to it in one form of its *isibongo*:—

Ugqwathiza badi hloko-hloko
Nqabaz' igazi ngomlomo
Umabizwa yintlava esesigwini,

Repeated by a traveller in response to the bird's call, in full expectation
of his receiving a meal somewhere before sleeping time.

From the old men Mr. Luti has obtained for me the meaning of
the puzzling phrase *badi hloko-hloko*. The reference is to the habit of
the springbok, when walking, to follow one another. The long-claws
when they see the bait under the herd-boys' stone-trap are said to do
the same. The phrase therefore means "Follow-the-leader springbok".
Of this phrase there occurs a variation *gqwathi hloko-hloko*, whose
precise meaning remains undetermined (Godfrey 1941: 103–4).

Godfrey does not translate the three lines of praise given above, but we could
make an attempt here, assuming that these praises, like the Zulu **inqomfi** praises,
contain more birds than at first appear to be the case. We need to make the
following quite likely assumptions:

In line 1, '*gqwathiza*' is the Xhosa equivalent of Zulu *gqwayiza* 'bob up
and down', and '*badi*' is in fact *ibhada* (Yellow-bellied Greenbul, formerly
Yellow-breasted Bulbul) and '*hloko-hloko*' is *ihlokohloko* (a yellow weaver).

In line 2, '*nqabazi*' is unclear but '*igazi ngomlomo*' is immediately apparent
as 'blood at the mouth' or, more particularly, 'the blood-red splash of colour
at the throat'.

In line 3, '*umabizwa*' is 'the one who is called', '*yintlava*' is 'by the honeyguide' and '*esesigwini*' is 'at the place of the *isigwe* bird'. Both in Xhosa and
in Zulu, **isigwe** is the Southern Red Bishop, yet another bird with a glorious
splash of blood-red plumage. The honeyguide, I suggest, is brought in here
because it is associated with sharing a meal ahead, just as the longclaw is: the
lucky bird that promises a full meal at the end of the journey.

We have then, if our assumptions are correct, a brief three-line praise for
the **inqomfi**, which:

---

17. This is very likely related to Zulu *gqwayiza*, meaning 'bob up and down'.

- brings in at least four other birds: the Yellow-bellied Greenbul, the yellow weaver, the honeyguide and the Southern Red Bishop;
- refers to a head-bobbing behaviour, and a red splash on the throat; and
- refers to the belief that an **inqomfi** on the road is a sign of sharing food on the journey ahead.

If we then add in Mr Luti's interpretation of the words '*badi hlokohloko*' as being '*izinqomfi* coming in line to the stone-trap of boys', then this is another direct link with the Zulu **inqomfi** praises.

Another praise poem to the bird inqomfi

The Zulu poet Benedict Wallet Vilakazi (1906–47) was the first person to publish poetry in his mother tongue, Zulu. His first volume – *Inkondlo kaZulu* (Zulu Poems) – was published in 1935 and was followed by *Amal'eZulu* (The Dome of the Sky) in 1945. The first anthology contains many poems heavily influenced by the poetry of the English Romantic poets such as Wordsworth, Coleridge and Shelley. It is the latter's poem 'To a Skylark' that particularly influenced the young Zulu poet to write the poem 'Inqomfi' (Vilakazi 1935: 14–18). There is no space here to do a complete analysis and show how many of Shelley's thoughts and images were transferred in Vilakazi's poem, but the following will serve as an example. I give Vilakazi's stanza 15 first, then a translation of the Zulu, then Shelley's original, which is also stanza 15 in his poem:

Table 9.2 Comparative extracts from poems of Vilakazi and Shelley.

Lesisiphethu siqhunyiswa yin' ingom' emnandi?	This fountain is brought forth by what sweet song?	What objects are the fountains
Ngabe yimifula yona egobhoza kamnandi?	Could it be the rivers which flow so sweetly?	Of thy happy strain?
Ngabe yilo ulwandle lukanye namagagasi?	Or perhaps the sea together with its waves?	What fields or waves or mountains
Siqhunyisw' izintaba namagquma namafusi?	Is it brought forth by the mountains, the hills and the plains?	What shapes of sky and plain?
		What love of thy own kind? What ignorance of pain?

The influence of Shelley is evident here, and clearly it is no coincidence that Vilakazi has put these lines into an equally numbered stanza 15.

In some cases, Shelley's original thoughts are maintained, but the image by which he expresses these is replaced by one that relates more directly to the experiences of the Zulu people. A perfect example of this is when Shelley, in stanza 6, comes up with the curious image of the moon causing rain. The full stanza is:

> All the earth and air
>> With thy voice so loud
> As when night is bare,
>> From one lonely cloud
> The moon rains out her beams, and Heaven is overpowered.

Clearly Vilakazi was greatly taken by the image of the moon 'rain[ing] out her beams' and wanted to include it in his own poem about the **inqomfi** lark. His stanza 12 reads:

> *Njengoba ngibek' indlebe nawo wonk' umhlaba,*
>> And so I bend my ear with all the earth,
> *Nolwandl' olugubhayo, namaza, naw' umoya–*
>> And the sea, and waves, and the wind–
> *Sekuthe khemelele nakho kubek' indlebe,*
>> They too stock still in amazement,
> *Kudingiswa ongandile umlozi lowaya.*
>> Made to wonder at this never-ceasing sound.
> *Nom' inyang' inganethisa umhlaba kube*
>> Even the moon rains upon the earth to the extent
> *Izintingo zayo qwi, wen' uyibek' inxeba.* (stanza 12)
>> That the wattles stand naked, you having placed there your wound.[18]

We see here that in Vilakazi's description of the moon raining out its beams, the rain is heavy enough to wash out the mud between the wattles, an experience no doubt familiar to many people who have lived in a wattle-and-daub hut.

---

18. Presumably a reference to the blood-red slash of colour on the throat of the bird.

Another example of how Vilakazi has attempted to Africanise Shelley's thoughts can be seen when we compare Vilakazi's stanza 2 with Shelley's stanza 2, this time in connection with larks circling high and higher.

The 'lark poems' of the English Romantic poets make a big feature of the lark soaring higher and higher as it sings. George Meredith, for example, has the wonderful line, "As up he wings the spiral stair" in his 'The Lark Ascending'[19]. Shelley's lark ascends like this:

> Higher still and higher
>> From the earth thou springest
> Like a cloud of fire;
>> The blue deep thou wingest,
> And singing still dost soar, and soaring ever singest.

Vilakazi's **inqomfi**, alas, is not a bird that circles higher and higher in the sky as it sings, but he is determined not to lose this feature of English larks completely. To Africanise the notion, he turns to Zulu traditional praise poetry, specifically, to the *izibongo zikaSenzangakhona* (praises of Senzangakhona, son of Jama, and father of Shaka), where we find the lines:

> *Owaphoth' intamb' ende mntakaJama*
>> He who plaited a long rope, son of Jama
> *Owaphoth' intamb' ende waya phezulu*
>> Who plaited a long rope and climbed up (Cope 1968: 80).

Vilakazi incorporates these lines into his poem 'Inqomfi' as

> *Weluk' intambo ende uyobona amakhosi*
>> You plaited a long rope to go and see the chiefs
> *Iyilokhu isuke phansi*
>> Starting from down below

The image of plaiting a rope is the circling of Meredith's "spiral stair" and the climbing up the rope from below is what the English larks do, even if the African longclaws don't.

---

19. In Read and Dobrée (1952: 219).

One of Vilakazi's stanzas, however, refers directly to the Cape Longclaw, and has no equivalent in Shelley's poem. His stanza 4 contains the lines:

Usho ukuthini lowobala ongasuki?
> What means that everlasting colour?
Ngisho lelochaphaz' elibomvu engileni!
> I mean that splash of red on the throat!
Ubani okucibe ngemicibishel' emibi,
> Who has shot at you with hateful bows and arrows
Ley' eyabonwa kuBathwa behlel' edwaleni?
> Such as those seen among the San Bushmen dwelling on the rocks?

The notion of a splash of red in the plumage of a bird as being blood is not uncommon, and even for this bird the Cape Longclaw, Stark and Sclater have recorded that at the end of the nineteenth century, English-speaking colonists in South Africa referred to this bird as the "Cut-throat Lark" (1900: 238). And as we saw in the three-line praise of this bird supplied by Godfrey, the second line contained the phrase '*igazi ngomlomo*', literally 'blood at the mouth' but more specifically 'the blood-red splash of colour at the throat'. And in the Zulu version of the praises of the **inqomfi**, the allusion was to having used a red dye obtained from the roots of the *umthothwane* plant.

Another stanza that refers directly to the **inqomfi** lark and, again, which has no equivalent in Shelley's 'To a Skylark', is Vilakazi's stanza 5:

Ma uthi ntenene phambi kwalowo oyindlela,
> When you make little running steps in front of someone on the path,
Bathi abadala izindaba zimi kahle.
> The old people say that matters stand well.
Yeka kulowo enilandelene kuye nidweba,
> But oh! When you follow behind him and then veer off
Niqhekeze amaphiko nenz' umlozi kuhle
> Flapping your wings and making a whistle
Kwezinsingizi lizophendula. Lubi loludaba.
> Like ground hornbills [predicting that the weather] will change: That is a bad business.
Uyabhulalela nyondini, yeb' uyabhulela!
> You do prophesy, you amazing bird, yes, you do indeed prophesy!

Zulu traditional beliefs assign both good luck and bad luck to the behaviour of this and similar birds like the Rufous-naped Lark and the wagtails, and it is the good luck waiting ahead for the traveller that we saw recorded in the Zulu praises above together with Stuart's additional explanations.

To sum up, then, Vilakazi's praise poem to the **inqomfi** lark, although occasionally borrowing notions and images more suitable for the British skylark, at least attempts to record some aspects of the Cape and the Yellow-throated Longclaws, and to incorporate Zulu traditional beliefs about these birds.

9.2 BIRDS IN PROVERBS, IDIOMATIC EXPRESSIONS AND SAYINGS

9.2.1 Introduction to proverbs

Ruth Finnegan, in her seminal *Oral Literature in Africa*, says of the proverb:

> In many African cultures, a feeling for language, for imagery, and for the expression of abstract ideas through compressed and allusive phraseology comes out particularly clearly in proverbs (1976: 390).

For Zulu proverbs, I have relied on four main sources of Zulu proverbs: Colenso's 1884 *Dictionary*; Dunning's 1946 *Two Hundred and Sixty-four Zulu Proverbs*; Malcolm's 1949 *Zulu Proverbs and Popular Sayings*; and Nyembezi's 1974 *Zulu Proverbs*.

These four authors have not arranged their collected proverbs with birds in mind, nor are the 'bird proverbs' isolated from 'non-bird proverbs'. This is something I have done for the purposes of this chapter, and the categorisation and subsequent discussion is based on the relationship between a specific bird and the general thrust of the proverb. Two proverbs here illustrate this relationship:

> *Lapho kukhon' isidumbu, yilapho kukhona amanqe* (there where there is a carcass, there also will be vultures): Following a misfortune there will always be people, such as the relatives of the deceased, who are waiting for their time to get pickings. The link between the proverb and its central bird the vulture is obvious.

Inyon' ihluthuk' isisila (the bird has lost its tail feathers): This is said of someone who has fallen from prosperity to poverty. Again, the link is clear: birds need tail feathers as well as all their other feathers in order to be able to fly. A bird which has lost its tail feathers is distinctly 'in reduced circumstances'.[20]

In categorising these bird-inspired proverbs, I have chosen the characteristics of the bird in question that are the ones most central to the proverb. In the example with the vultures, it is the diet and feeding preferences of vultures that make them suitable for the proverb; in the second example with 'the bird', it is the fact that all birds need tail feathers as part of their plumage.

In some cases the link is not clear, but can be guessed at, as in the case of two proverbs involving the domestic fowl (**inkukhu**).

The first is *inkukh' inqunyw' umlomo* (the fowl has had its beak cut), a saying used of a talkative person who has been silenced. We could guess here that the domestic chicken is perceived as a bird continually making a noise, although there are certainly many other wild birds (some bulbuls, for example) that are continually heard.

The second is *Yimpi yakwamabonwabulawe, inkukhu nempaka* or often simply *Yinkukhu nempaka* ('it is the battle of be-killed-as-soon-as-looked-at: a fowl and a wild cat' or just 'it is a fowl and a wild cat'). This proverb is used of sworn enemies who always want to get at each other. There is no doubt that wild cats attack and kill a variety of wild birds in addition to domestic fowls, but perhaps it is because it is only the domestic fowls that are actually seen as being attacked and killed that has led to them playing this role in this proverb.

Again, we could consider the proverb *ithendele libulawe uqondo* (the partridge is destroyed by the grass), used of a person who is proud and obsessed with his own importance. Why the partridge **ithendele** and why the long river grass *uqondo* is not clear, but one could perhaps guess that the partridge believes it is well hidden in this long grass, but its hiding place is continually revealed when women came to pluck the grass for thatching? The link is tenuous, but possible.

---

20. Several proverbs are repeated in this section as they are used to illustrate various different notions.

Unclear relationships

For a number of proverbs, although the literal meaning is clear, there seems to be no valid or obvious reason why a particular bird should be chosen as the inspiration for the proverb. For example, take the well-known proverb *Hamba juba bayokucutha phambili* (Go dove, they will pull out your feathers ahead), said of someone with overweening pride, i.e. when you fall later, you will become a laughing stock. The proverb is also said of a stubborn person who will not listen to advice. There seems no reason why a dove should be chosen as the central bird of this proverb: any other bird would look ridiculous if stripped of its feathers while still alive. In the same vein, why should a guineafowl be the central bird of the proverb *Igeja liphamb' impangele?* This proverb, literally meaning 'the hoe has puzzled the guineafowl' is used of a very stupid person who considers himself as having accomplished something clever when it is others that have done the work. Then again, let us consider the situation where a commoner or a nonentity begets a child who rises to a position of importance or prominence. In such cases one can say *Yek'ungoq'ukuzal'isilomo* (to think that a quail should beget a favourite!) Why particularly a quail here? Are quails perhaps considered to be nonentities among birds?

9.2.2   Proverbs relating to the hunting and trapping of birds

A considerable proportion (22 per cent) of bird proverbs relate to the hunting or trapping of birds.[21] Young boys entrusted with cattle herding have always used two traditional methods of killing birds, which are then either roasted and eaten on the spot, or taken home to be shared. These two methods are the using of throwing sticks (*izagila*), or trapping. The most common method of trapping involves placing bait under a heavy flat stone supported by a stick, which can then be pulled away by a piece of string, as described in section 9.1.3 (page 240). Of seventeen bird proverbs recorded by Nyembezi that deal with the hunting and trapping of birds, seven relate to the use of throwing sticks, six relate to trapping with bait and stone, and the remaining four deal with other aspects of obtaining wild birds from the veld for the purposes of food (1974).

---

21. The hunting and trapping of birds in traditional Zulu culture is also dealt with in Chapter 11 (pages 329–32), where the emphasis is on the eating and sharing of the birds killed.

Using the throwing stick

The *isagila* is a relatively short-handled knobkerrie with the heavy head set off centre so that when the stick is thrown the handle whirls around in the air. This practice lies behind the following proverbs:

Noseyishayile akakayosi (even he who has struck it has not yet roasted it): Nyembezi equates this proverb with the English "A miss is as good as a mile" (1974: 102). It may equally be equated with 'Don't count your chickens until they've hatched'.

Ucishu kadliwa (almost is not eaten): The word *cishu* (or *cishe*) means '[I] almost [did it]'. When a boy throws a stone or a stick at a bird and misses it, he might say "*Cishu!*" He doesn't get to eat the bird, however. Another proverb that equates to 'A miss is as good as a mile'.

Ucil'uzishay'endukweni (the **ucilo** lark has struck itself against the stick): This is an expression of good luck. It is difficult to hit this little bird with a throwing stick, so if it hits itself against the stick, so much the better.

Ucil'kaf'izidubuli (the **ucilo** lark does not die of bruises): As seen in the previous proverb, this little lark is not easily struck with a stick, and it may need several attempts before it is killed. Try, try, and try again. Persevere against all odds.

Isigwaca silind'induku (the quail waits for the stick): This proverb is used of a stupid person who waits for trouble to overtake him.

Intendel'esuka muv'ikholwa yizagila (the partridge which rises last gets the worst of the sticks): When danger threatens it is best not to waste time. If you do not move away in time, you are likely to get into trouble.

Ungced'ukholwa yizagila (the warbler receives many sticks): The interpretation is similar to that of the previous proverb: a person who does not run away from danger is likely to get into trouble.

Trapping with bait and stone

Itshe limi ngothi, Nkombose kababa (the stone stands by means of a stick, Nkombose (the Namaqua Dove) son of my father): The Namaqua Dove is known to be one of the most suspicious of birds, not easily fooled by a stone trap. The proverb may be used in reference to a suspicious happening, or of someone who is not easily fooled.

Ayinyonki kabili (it [bird] does not steal (*nyonka*) twice): Once bitten, twice shy. If a bird escapes the falling stone of a baited trap, it is unlikely to try again soon.

Kubhajw'eshoshayo (it is the one [bird] which hops about that is trapped): Birds are safe when they fly in the air; it is when they hop around on the ground that they may be caught by various traps set for them. Those who travel around a lot are more likely to meet with accidents and mishaps.

Inkukh' iyawusol' ummbila (the chicken suspects the maize grain): To trap a fowl, one may take a mealie grain and tie it with a string. If the chicken eats the grain, it is helpless. The proverb is used when one suspects danger. A variant of this proverb is *Iyasol' impangele* (the guinea-fowl is suspicious), which refers to the same practice of tying string to grains of maize.

Zibhajwa ngezikudlayo (they [birds] are trapped by what they eat): Birds are trapped most successfully by using bait that is attractive to them. So also with humans: you often meet disaster in the things that interest you most.

Other proverbs that relate to accessing wild birds for food

The first two under this sub-heading refer to the practice of taking fledglings from the nest of birds, perhaps to rear them and either sell them or eat them at a later stage.

Zibanjwa zimaphuphu (they [birds] are caught while still fledglings): Nyembezi says that the proverb means that if a child is properly reared by its parents, it will behave properly as an adult (1974: 161). Yet the literal meaning of the proverb is of nestlings that do not have the chance of being properly reared by their parents. Perhaps the proverb can be interpreted as 'When a child is <u>not</u> properly reared by its parents, it will <u>not</u> behave responsibly as an adult'.

Inyoni kayikhulunyelw'eziko (the bird is not discussed next to the hearth): Nyembezi says of this proverb (1974: 170):

> When boys are out herding, they very often discover birds' nests. The discoverer marks the nest in a certain way, so that anyone coming subsequently may know that it has been claimed. A boy who has discovered a nest does not tell the others of his find as they sit around the fireplace. The belief is that if he does so, the bird will desert the nest.

Nyembezi interprets the proverb as "if a person wants to do something, he should not start by advertising to everybody his intentions, lest he be forestalled" (1974: 170).

Intendel' iw' enkundleni (a partridge has dropped into the yard): The **intendele** partridge, which is very good eating, is difficult to hunt with throwing sticks. It is fortunate indeed if one simply drops into the yard. The proverb is used as an expression of good luck.

Inyoni ishayelw' abakhulu (the bird is killed for the elders): Nyembezi says that it was not considered the proper thing for boys to kill a bird while out herding and eat it there and then (1974: 162). The correct thing was to bring it home and share it with their elders. Biyela shares her memories of how her brothers and her cousins used to bring their caught birds home to show their grandfather:

> Each of the boys who came with a bird would receive the head of his bird and my grandfather would take the rest and divide it amongst other small children who were not yet capable of hunting (2009: 39).

The seventeen proverbs examined above all look at one aspect of birds: their edibility. The quails, partridges and larks mentioned in the proverbs are all birds that are commonly killed by young boys for the purpose of food. We move on now to look at other aspects of birds that lead to them being selected as 'actors' in Zulu proverbs, for example, their plumage, their calls, their flight patterns, and so on.

9.2.3   Other aspects of birds leading to proverbs

One very specific characteristic of birds is that they are feathered. These feathers serve a number of different purposes: the wing and tail feathers are needed for flight; feathers keep the birds warm; and they carry the colours and patterns that make up the bird's appearance. All these feature in the following five proverbs:

Hamba juba, bayokucutha phambili (Go, dove, they will pluck out your feathers ahead). Nyembezi comments on this proverb as follows:

> A dove without its feathers is in [a] sad plight if still alive. Feathers protect it, and also enable it to fly. Then picture a human being with feathers, and others deliberately and ruthlessly pulling them out, so that one remains nothing more than a laughing stock (1974: 37).

The proverb is usually used of someone who refuses to take advice and goes ahead with an inadvisable project.

Akujoj' wamil' isisila wabonakala (it is not the case of the **ujojo** finch, which grew a tail that could be seen): The Long-tailed Paradise Whydah has a tail that is visible to everyone. This proverb is used of a person who feels that his efforts have not been seen and recognised.

Ukuz' ubon' inqe lihluthuk' intamo? (do you see for the first time a vulture with feathers plucked from the neck?): There is nothing to get excited about here, for all vultures have no feathers on the neck. The proverb is used when someone gets excited about a commonplace happening. According to Nyembezi, a variant is *Ukuz' ubon' inyoni ihluthuk' isisila* (is this the first time you have seen a bird with its tail dropped off?), but I cannot see this as an equivalent for it is certainly not normal to see a bird in this condition. This variant is more likely to go with Nyembezi's interpretation of "Is it then the first time you see someone in such misery as I am that you should deride me so?"

Continuing with the notion of lost tail feathers is *Inyon' ihluthuk' isisila* (the bird has lost its tail feathers): This proverb is used to refer to someone who has fallen from prosperity into poverty. Very similar in content and meaning is *Ith' ingangcothuk' isisil' ihlekwe* (when it [bird] loses its tail feathers it is laughed at): Someone may boast about his assets and his fortune. If he should later lose them, he is mocked and derided with this proverb.

More birds and laughter are found in the proverb *Uhlekwe zinyoni* (birds have laughed at him), a reference to someone who does something so strange that even the birds laugh at him. A variant of this is *Inyang' ihlekwe zinyoni* (the birds have laughed at the moon), a proverb used of the moon when it sets after daybreak.

Daybreak and calling birds (rather than laughing birds) feature in the next three proverbs:

Inyoni ikhala kusile (a bird sings at dawn): According to Nyembezi, this proverb is a warning to those who move about at night (1974: 215). The night is usually regarded as a time for witches (*abathakathi*) to be abroad, who need to be inside by daybreak lest they be discovered.

Nalá kungekho qhude liyasa (even where there is no cock, day dawns): This proverb is used when a person asserts himself as if he is indispensable and nothing would go on without him.

Zakhala kazehla (they [fowls] crowed but did not come down): it is unusual for cocks to keep crowing during the day but not come down from the perch. The proverb is used of a stupid person who never does anything properly.

Domestic chickens feature in other proverbs as well, as in the following five:

Usebenzel' emuva njengenkuku (he works backwards like a fowl): This refers to people whose work never seems to progress.

Kungaw' ilanga licoshwe zinkuku (the sun may fall and be picked up by fowls): This is said of someone who has been on a fruitless errand.

Inkuk' inqunyw' umlomo (the chicken has had its beak cut): This proverb is used of a talkative person who has been silenced.

Yinkuku nempaka (it is a fowl and a wild cat): This proverb refers to sworn enemies who always want to get at one another. A variant is *Yimpi yakwamabonwabulawe, inkuku nempaka* (it is the battle of be-killed-as-soon-as-seen, a fowl and a wild cat).

Udl' esulela phansi njengenkuku (he eats wiping his mouth like a fowl): Humans wipe their mouths only when they have finished eating, unlike fowls that wipe their beaks all the time while eating. The proverb is used of someone who wants to give the impression of being satisfied when he is not, or of someone always trying to hide his intentions.

From fowls wiping their mouths while eating we move to proverbs that refer to birds eating corn.

Sobona nyoni zowadla (we shall see which birds eat it [the corn]): Nyembezi interprets this proverb as meaning 'we shall see which one of us young men the girls will eventually choose as lovers' (1974: 77). Dunning has a completely different interpretation, seeing this proverb as being an ironic comment on a likely very poor harvest of corn (1946: 7). Another proverb involving birds eating corn is *Ziwadl' ebhekile* (they [birds] eat it [corn] in his presence): the proverb is used of someone who is easily fooled.

Many other proverbs, otherwise unlinked in content and interpretation, have an unidentified bird (*inyoni*) at their core.

Isisu somhamb' asingakanani; singangenso yenyoni (the stomach of a traveller is small, it is as big as a bird's kidney): A traveller asking for food at a strange homestead uses this proverb, claiming that it will not take much to satisfy his hunger. A shorter version of the proverb is *Isisu somhamb' asingakanani* (the stomach of a traveller is not big).

Wanyiwa yinyoni (he was defecated on by a bird): This is used of a person who is solitary and has been abandoned. A variant is *Washiywa yinyoni* (he was left behind by a bird).

Inyon' enkul' ingafa kubol' amaqanda (when the big bird dies, the eggs rot): In Nyembezi's words: "A family is kept together by its head and once he goes, everything seems to go to pieces" (1974: 110).

Inyoni yinhle ngezimpaphe zayo (a bird is pretty because of its feathers): It is because of one's good deeds that one is liked by people.

Inyoni kayiphumuli (the bird does not rest): This expression is used of someone always on the move.[22] With somewhat similar sentiments and content is *Akukh' nyoni endiz' ingahlali phansi* (there is no bird which flies and does not rest): This proverb suggests that everybody should take a break sometimes from work.

Although many of these proverbs, as seen above, refer to an unidentified bird, a greater number refer to specific birds, even if these are folk genera such as **inqe** (vulture) and **ukhozi** (eagle). For example:

Under the heading 'Unclear Relationaships' above, the Helmeted Guineafowl (**impangele**) appeared in the proverb *Igeja liphamb' impangele* (the hoe has puzzled the guineafowl). Another proverb featuring the guineafowl is *Impangel' enhle ngekhal' igijima* (a good guineafowl is one that calls while running): Guineafowl are clever enough to run away from danger, calling while they do so. The proverb warns people to avoid danger as soon as it threatens. A variant of this proverb is *Intendele' enhle ngekhal' igijima* (a good partridge is one that calls while running).

Partridges, francolins, and quails (see Plate 40) featured strongly in the section on proverbs relating to the hunting and trapping of birds above. They also feature in proverbs where there is no suggestion of hunting and trapping, as in the following:

Ithendele libulawa uqondo (the partridge is destroyed by the grass). Nyembezi says of this proverb "The origin of this expression is not clear, but it is used of a person who is proud and obsessed with his own importance" (1974: 57). Another is:

Intendel' ibindwe yisidwa (the partridge has been choked by the *isidwa* bulb). The *isidwa* bulb (*Gladiolus ludwigii*) is placed among the seed used by women when sowing to ensure a good harvest. Nyembezi suggests that "a partridge may find the seed-gourd and eat the bulb instead of the seed, thus choking on it" (1974: 119). The proverb is used of someone who is at a loss for a reply. Doke and Vilakazi gives the variant *Inkuku yamilwa yisidwa* (the fowl had a gladiolus bulb grown in her throat) with the same interpretation (1958: 176).

---

22. The name iNyonikayiphumuli was also given to the famous royal herd of white cattle of Zulu King Cetshwayo kaMpande. The implication was that the cattle were so numerous that the attendant cattle egrets never had a chance to rest.

Akukh' nkwal' ephandel' enye (there is no *inkwali* partridge that scratches for another):[23] The proverb means "Don't expect others to do for you what you should do for yourself".

Yek' ungoq' ukuzal' isilomo (to think that a quail could beget a favourite):[24] This proverb is used when a nonentity begets a child who becomes a powerful or prominent person.

In the proverb below, two birds are mentioned:

Negwababa lize liphath' umgodo nonhloyile' afise (even the crow may have excrement that the hawk envies):[25] The proverb is used of someone who refuses to help another in need. Even a poor person may have something that a rich person may desire someday. The proverb suggests that in Zulu thinking the crow is 'poor', (i.e. of a low status), while the Yellow-billed Kite is a high-status bird. The kite is of course the bird whose return marks the start of the Zulu calendar year, its name uNhloyile being an alternative name for uNcwaba.[26] Corvids, on the other hand, generally have a low reputation in societies all over the world (Cocker 2013: 384).

Unonele phakathi njengendlazi (he is fat internally like the mouse-bird):[27] According to Nyembezi, the mousebird appears lean when you look at it, but actually it is fat. The proverb describes a person who will not show his true colours (1974: 71).

Ubucub' obuhle buhamba ngabubili (good waxbills go in pairs): Nyembezi says that "waxbills will always be observed to go in pairs" (1974: 90). He interprets this proverb as "it is good to have friends who will help you in times of need". He gives the variant *Ubucubu bufa ngabubili* (waxbills die in pairs), saying that people who are always together are likely to be involved in the same accident or mishap. Bryant (1905: 81) has *Ub' uhamba wedwa nje; kawazi yini ukuti ubucubu buhamba nga'bubili na?*, translating this as "You were just going alone; don't you know that the waxbills (*i.e.* little children) go in pairs (*i.e.* never alone)?"

---

23. The word **inkwali** is used today for Swainson's Spurfowl (formerly Swainson's Francolin).
24. The word **ungoqo** refers generically to the buttonquails. It has been assigned specifically to the Black-rumped Buttonquail.
25. The word **igwababa** refers to the Pied Crow; **unhloyile** is the Yellow-billed Kite.
26. The month uNcwaba (or uNhloyile), starting about mid-August, is when the spring rains come and the ploughing season starts.
27. The word **indlazi** refers to the Speckled Mousebird.

Ucilo walahla' intethe (the lark has let go the grasshopper): The **ucilo** lark, which feeds on grasshoppers, is always flying about with a grasshopper in its beak. It doesn't let go of this until fatally struck by a thrown stick. The proverb describes a situation where irreparable harm has been done. During the 2013–2017 workshops, it was difficult to pin down exactly what species the **ucilo** is. Bryant glosses the word as "very small bird" in the entry:

> Very small bird, said to be difficult to hit with a stick, it generally managing to get through clear with [a] grasshopper still in its mouth. Hence the following proverb:–
>
> *ucilo walahla intete*, the *ucilo* has let go the grasshopper = it's done for this time, is dead – said *e.g.* when one breaks a pot to pieces (1905: 75).

Bryant also applies the proverb to an unidentified person in the form *Ucilo uyilahlile intete ku'Bani* (the *ucilo* has let go the grasshopper with So-and-so = it's all up with him; he's done for) (1905: 75).

The word **uncede** is used in Zulu generically for the cisticolas and for the Neddicky specifically. Various publications have different spellings of this bird's name, and Nyembezi (1974: 113) uses the spelling **ungcede** in *Kukwangced' omhlophe* (it is at the place of the white warbler), pointing out that as there is no white warbler, this applies to a place where there is nothing: no people or poverty of land. This bird was also mentioned in the section above on proverbs related to hunting and trapping: *Ungced'ukholwa yizagila* (the warbler receives many sticks).

Inkonjane yakhela ngodaka (the swallow builds with mud): The swallow builds slowly, bit by bit, but gets there in the end. This proverb is used as an expression of encouragement.

Uve ludl' isisila salo (the paradise flycatcher eats its own tail): This bird has such a wild darting movement that it often appears to be chasing its own tail. The proverb is used of a person who harms himself by his own actions (see Plate 19).

Umathebethebeni'usegoqile (the sparrowhawk has folded):[28] Sparrowhawks cruise in the sky and when they see a likely victim, they swoop down on it. The saying is an expression of danger.

---

28. That is, folded its wings for the dive.

Uyoz' ube nebala njengombalane (you will eventually have a mark like the wild canary):[29] Nyembezi says:

> The wild canary has a golden rump which may be regarded as a distinguishing feature. The saying is used as a warning to a person who constantly does things which are wrong. Such a person eventually becomes a marked man (1974: 173).

The next two proverbs feature vultures:

Lapho kukhon' isidumbu, yilapho kukhon' amanqe (where the carcass is, there the vultures will be): There are many people waiting like vultures, perhaps like the relatives of the deceased, for their time to get some pickings.

Amanq' akakakuboni, intuthan' isikubonile (The vultures have not seen you, the ant has): Vultures and ants both eat dead bodies. Generally humble people see more than the proud. Doke and Vilakazi, who give the same proverb under the entry for **inqe**, agree with this interpretation (1958: 591). Nyembezi goes further when he says:

> This saying is also a warning of worse and terrible things to come. The suffering that may be inflicted by an ant is nothing in comparison with what a vulture is capable of inflicting (1974: 174).

This second interpretation is puzzling. The first interpretation, i,e., humble people generally see more than the proud, can easily be linked to the fact that the tiny ant is much lower to the ground (and closer to the bone, so to speak), than the vulture, and so fits in with the literal meaning. Nyembezi's second interpretation seems, to my mind, to be divorced from the literal meaning.

The next two proverbs have owls as a core:

Ubuye nembande yesikhova (he came back with the shinbone of an owl): As Nyembezi notes, the owl is always associated with ill luck (1974: 109). So a person who goes away to fetch something and comes back with the shinbone of an owl is one who has met with bad luck.

Isikhova sidl'amehlw'aso (the owls eats its own eyes): Nyembezi explains this proverb as:

---

29. The word **umbalane** is used generically for a number of wild canaries, and specifically for the Yellow-fronted Canary.

The owl sleeps by day and hunts by night. The eyes of the owl are adapted to see at night, so they enable it to see its prey. Its eyes are therefore very valuable to it. This saying emphasises self-help (1974: 179).

To my mind, this is a puzzling interpretation. If the owl sees so well at night, this certainly "emphasises self-help". But the literal meaning goes completely against the grain of the usefulness of the owl's eyes, and the emphasised 'self-help'. Given the literal meaning, I would have thought a better interpretation would be something along the lines of, 'Some people are intent on destroying the talents they have been given'. Neither Bryant (1905) nor Doke and Vilakazi (1958) mention this proverb, nor does Dunning (1946).

Another proverb where there seems to be an inherent contradiction between the actual behaviour of a bird, the literal meaning of the proverb, and Nyembezi's interpretation is *Ihlokohloko lidla lilodwa* (the weaver-bird eats alone):[30] Nyembezi says "although only a tiny bird, the weaver may cause havoc in the fields" (1974: 181). I agree with Nyembezi's interpretation that although only a tiny bird, the yellow-weaver can cause great damage to grain crops, but that is precisely because it does not feed alone, but in great numbers. Given the inherent contradiction here between the behaviour of the bird and the wording of the proverb, it is difficult to see how it could be interpreted.

The role of the honeyguide is well known in traditional societies all over Africa, and everywhere there is the general belief that when the guide leads someone to a bees' nest, it is the right thing to do to share some of the honey with the honeyguide.[31] This notion leads easily to proverbs dealing with sharing, as in the following:

Inhlav' iyabekelwa (the honeyguide is given something): Nyembezi interprets this as "One should show gratitude towards a benefactor" (1974: 184); Doke and Vilakazi as "Provide for the morrow" (1958: 323).

Ungayishay' ingede ngoju (Do not strike the honeybird with honey):[32] Nyembezi explains this proverb as follows:

---

30. The word **ihlokohloko** is used generically to refer to the various species of yellow weavers.
31. In fact, it is important to leave the honeyguide with some pieces of honeycomb, so it can get at the bees and the larva inside.
32. Both **ingede** and **inhlava** are commonly used as generic terms for the different species of honeyguide found in KZN. The word 'honeybird' is Nyembezi's.

Although the bird also expects to get a share, it is an unkind act to strike it with honey. The proverb condemns acts of ingratitude to one's benefactors (1974:58).

Two proverbs, one about the eagle, another about the dove, complete this sub-section:

Ukhoz' olubambayo ngoluzulayo (the eagle that catches [prey] is the one that wanders): Eagles fly far and wide in search of their prey. In a like manner, a person should not expect to get anything unless he works for it.

Wahamb' okwejuba likaNoah (he has gone like Noah's dove): This proverb is said of someone who has gone on a long journey and is not expected to return for some time. This proverb differs from almost all the other proverbs in this section of the chapter in that it is not based on a salient feature of the dove itself but on the role assigned to the dove in Genesis 8:11.

9.2.4   Cognitive links between the three elements of a 'bird proverb'

From the point of view of birds as core images of proverbs, there are three elements to each proverb:
    (1) the bird that is at the core, together with its salient features (edibility, plumage, diet, behaviour, etc.);
    (2) the literal meaning of the proverb; and
    (3) the interpretation of the proverb, i.e. its application to humans.

In most of these proverbs, there is a cognitive link between the observed characteristics of a folk genus of birds (or a specific species) and some aspect of human frailty or the adverse circumstance in which humans find themselves. Proverbs are by definition pithy comments on human frailties or circumstances of adversity, and for a proverb to be 'successful' all three of the elements mentioned above must be in harmony, in other words linked cognitively. For example, take the three proverbs that were given on page 256 above:

Inkonjane yakhela ngodaka (the swallow builds with mud): It is a fact that the swallow builds its nest in mud, each added pellet of mud requiring a return flight. Transferring this avian activity to a situation where a human is slowly completing a difficult task is cognitively easy, and the link could not be clearer in the three words of the proverb.

Uve ludl' isisila salo (the paradise flycatcher eats its own tail): The observed activity of the flycatcher is its flight pattern, which suggests that it is chasing its own tail. Imagination is required to make the leap to the bird actually eating its own tail, and then another leap of imagination to link this to the human situation of a person harming himself through his own actions. The cognitive links are more tenuous here than in the proverb above.

Umathebethebeni' usegoqile (the sparrowhawk has folded): Sparrowhawks are observed as apparently appearing out of nowhere to stoop on a victim. Humans often find danger in unexpected areas. Making the link between these does not require much cognitive imagination.

Where proverbs do not 'work', to my mind, is where one of the three elements is at discord with the other two. For example, take the proverb *Isikhova sidl' amehlw' aso* (the owl eats its own eyes). Any observer of owls with their large front-facing eyes realises how important they are to the owl's successful hunting at night. Under what circumstances, then, can an owl be imagined to eat its own eyes? This proverb can only be interpreted, then, in human terms, as someone destroying something useful, even essential, to him. It is Nyembezi's interpretation of the proverb as expressing self-help that is discordant with the other two elements.

9.2.5   Proverbs and folk tales

Nyembezi gives the proverb *Insimba yesulela ngegqumusha* (the genet puts the blame on the bushshrike)[33] and says:

> It has been suggested that the origin of this proverb is based upon the fact that these two are companions of the thicket, and for that reason, when the hunters are out hunting, and they give chase to a genet, the bush-shrike may rise, and the hunters then follow it and the genet may give chase (1974: 81).

The proverb is used of someone who has been at fault but has shifted the blame onto someone else. There is, however, nothing about the factual behaviour of either the genet or the 'bushshrike' to suggest a relationship between them (despite Nyembezi's implausible suggestion that they are both "companions of the thicket"). Nor can any animal or bird be capable of placing the blame on

---

33. The word **igqumusha** has been used for the Southern Boubou. Doke and Vilakazi confusingly gloss the word as "Bush shrike, Butcher-bird, [i.e. Southern Fiscal]; name applied to several species of Oriole" (1958: 268). For the purposes of this proverb, it would be best to stick with 'bushshrike', vague though that might be.

any other, unless through extensive anthropomorphic imagination. But there is an *inganekwane* (folk tale) that lies behind this proverb, and which can be summarised approximately as follows:

The genet does something (anything, according to which teller is telling the tale) that is tabooed, or forbidden, during the darkness of night, such as stealing the honey or the meat of the king. He then smears honey, or fat, or mud on the feet of the sleeping bushshrike. The tabooed action is discovered by all the animals at dawn, and in order to trace the culprit the genet suggests that they look for anyone with honey, fat or mud on his feet. The traces of honey or fat would be obvious; the traces of mud would suggest that the guilty one would have walked in the dew during the night. The bushshrike is then punished for the crime, while the genet walks scot-free.

It is clear that the folk tale works no matter which two animal or bird characters are substituted (although clearly the proverb has helped fix these two as the central characters). The folk tale would have the same moral even if it was the tortoise who smeared honey or mud on the feet of the eagle, or the mongoose who did this to the weaverbird. It is the folk tale itself that carries the behavioural patterns of the actors involved, not the essential nature of the genet and the bushshrike. And it is this partnership between the proverb and the folk tale that makes this proverb different to the others that have been discussed in this section of the chapter.

This partnership between the proverb and the folk tale also makes this of interest to both avitourism and to what might call 'avi-education'.[34] The tale is an amusing one, and can (as with all folk tales) be expanded or otherwise adapted to suit circumstances. This folk tale, together with a range of bird-related proverbs, should be in the repertoire of any bird guide operating in KZN, as well as of any person working simultaneously with birds and children.

9.3 BIRDS IN RIDDLES AND CHILDREN'S GAMES

The two most important sources of material for this section were Cole-Beuchat's 1957 article on riddles in Bantu, and Khumalo's 1974 article on Zulu riddles specifically. Finnegan also has some interesting points to make (1976).

---

34. A phrase that is coined here to mean both educating schoolchildren about birds, as well as using birds to educate children about other issues, such as ecology and conservation.

Cole-Beuchat sees riddles as playing a "much more restricted role in the people's life than proverbs" (1957: 135), but nevertheless she sees them as "rich in colourful idioms, many of which reflect some aspect of the people's spiritual or material culture" (1957: 144). They differ, of course, from proverbs in that it takes a minimum of two people to articulate a riddle: one to pose the riddle, and one (or more) to propose an answer.

In the examples below, all the Zulu riddles are from Khumalo and all the proverbs from other Bantu languages are from Cole-Beuchat (1957). As with the proverbs in the previous section, I only look here at riddles that have birds in them. These riddles range from the simplest and clearest to those that are dense and impenetrable. As an example of the former, consider the Zulu riddle *Abantu abathanda ukucula* (people who like to sing) for which the answer is *Izinyoni* (birds). An example of the latter is perhaps the Lamba riddle *Ici tacikala kulutende?* (What does not sit on a grass stalk?) for which the answer is *Mbeŋguni* (A honeyguide). In between are all the riddles that in Cole-Beuchat's words:

> [D]escribe nature as a whole, particularly numerous being the riddles which refer to natural phenomena, and fauna and flora, from insects and grass to trees and elephants (1957: 145).

One of the more complex riddles, and one that uses a standardised 'template' in its answer, is the Northern Sotho riddle about the Hamerkop. The riddle is *Sehlaga sa Mmamalianoka, seôkama bodiba* (Nest of a Hamerkop hangs suspended over the deep pool) and the answer is *Yare gokwa kerialo: Ngwana-letswêlê lakgômo, kele bjang?* (On hearing I say: child of the udder of a cow, I being how?). Cole-Beuchat explains the riddle as follows:

> The essential part of the answer is 'the udder of a cow', while the deep pool over which it hangs is the bucket. The 'frame'[35] of the answer might be more freely translated as 'On hearing (your riddle), I say . . . how is that for an answer?' (1957: 137).

She notes that the essential part of the answer is almost always in some sort of possessive construction, thus, in the example above, the answer is not just simply 'the udder of a cow', but 'child of the udder of a cow'.

---

35. Which I have called the 'template'.

Different aspects of the same bird can be the subject of different riddles in different languages. Take (for example) the Secretarybird. In the following Southern Sotho riddle, the focus is on the head plumes: *Sejêre mahlaka, seaeba-eba* (It carries reeds on its head, it rocks, it rocks), to which the answer is *'Mamolangoane* (A Secretarybird). In the Zulu riddle, the focus is also on appearance but on a different aspect: *Insizwa yami egqoke imbilijisi* (My young man wearing riding breeches), with the answer **Intinginono** (Secretarybird). Khumalo explains this as "The legs of the bird get thinner below the knee. This thin-legged appearance is compared with that of a man wearing riding breeches" (1974: 224).

Even in the same language, two different riddles may look at the same bird from different angles. Take, for example, the Zulu riddle *Izinyoni ezikhipha umfazi ekhaya* (Birds that cause a married woman to leave her home), to which the answer is **Izinsingizi** (Southern Ground Hornbills), with Khumalo explaining that "It is an old Zulu belief that when a woman hears a hornbill call, it is a sure sign that her husband is going to expel her from their common home" (1974: 210).[36] 'Southern Ground Hornbills', in the form of their alternative name **izingududu**, is also the answer to the riddle *Abafundisi abagqoka okhololo ababomvu* (My priests wearing red collars), with Khumalo explaining that these black-clad birds have a red wattle around the neck (1974: 262).

In most of the examples above, birds feature as answers to certain riddles. There are, however, a number of riddles where the bird occurs in the question part of the riddle. A Zulu example is *Inyoni ezalela evungwini* (A bird that lays its eggs in an area of dry grass) and the answer is *Intwala* (A louse). More egg-laying is seen in the Swahili riddle that goes *Kuku wangu akazalia miiboni* (My hen has laid among thorns) and the answer is *Nanasi* (A pineapple).

The domestic fowl features again in the Tsonga riddle *Kuringe mbhaha, kukokole nkuku* (The hen has crowed and the cock has cackled). The answer to this riddle is *Namunthla kuni ntsuvi yobiha, ayikoti kuxa* (Today there is such a bad mist that the sun cannot appear). And we see the fowl again in the Ganda riddle, *Obunagenzere Buganda, bakamba enkoo yokuguru kumu* (I went

---

36. This is clearly tied up in some way with the interpretation of the call of the **insingizi**. In Chapter 7 above, on pages 175–6, we noted that it is the female calling first, saying "I am leaving home", while the male replies "Go now, you have been saying so for a long time." This does not give the impression, though, that the husband is actually expelling the wife from home, as Khumalo suggests in explaining the riddle.

to Buganda, and the people gave me a one-legged chicken to eat). The answer is the name of a type of edible mushroom.

A little more obscure, perhaps, is the Zulu riddle *Umuntu ongafuni ukuwenza umsebenzi ewubuka* (A person who doesn't want to look at the job he is doing) to which the answer is **Inkuku** (A fowl).

Two birds feature in the question part of the Mwera riddle *Nnamba tie, liwundi bata* (The glossy starling above, the owl beneath), with the answer given as *Liunde nalitaka* (Heaven and earth).[37] The owl (**isikhova**) is the answer to three Zulu riddles that all focus on the bird's nocturnal habits: *Umfan' olal' emini* (A boy who sleeps during the day), *Umuntu wam' olal' emini abheke ebusuku* (My person who sleeps during the day and is awake at night) and *Umuntu wami ozondana nelanga* (My person who hates the sun).

The word *ibhanoyi* (aeroplane) is the answer to three Zulu riddles as well. Two of them predictably use a bird as a metaphor for the aeroplane: *Inyoni ethwala abantu* (A bird that carries people) and *Inyoni yabelungu* (The white man's bird). Less predictable is the third version, *Inkunz' endiza emoyeni* (A bull that flies in the air). Perhaps an aeroplane 'roars' like a bull, and is strong and powerful.

Khumalo gives two riddles, to which the answer is **amanqe** (vultures). As both refer to the vultures' keen sense of smell, these could probably be regarded as two forms of the same riddle: *Umuntu odla okubolileyo; ubuzwa kanye* (One who eats what is rotten; he smells it at once) and *Abantu abezwa ngomoya* (People who smell in the air).

Migratory habits, diet, appearance and calling habits are the focus of the next four riddles:

Migration is seen in *Umuntu obaleka ebusika, othanda ukushisha* (A person who goes away in winter, who likes warmth) and the answer is **Inkonjane** (A swallow). As we saw in Chapter 8, the migration of swallows is one of the commonest reasons they appear in oral poetry.

Diet is the focus of *Umuntu odla izinyoka* (A person who eats snakes), to which the answer is **Unogolantethe** (White Stork).[38]

---

37. I am still puzzling over this one. Perhaps in Mwera, as in Zulu, the same word is used for both glossy starlings and for the Morning Star in the heavens. It could be just the colours, beautiful shiny, deep blue above and dusty, earthy brown below.

38. Chittenden, Davies and Weiersbye (2016: 104) give the diet of the White Stork as "insects, especially caterpillars, locusts and crickets; also mice, small reptiles, frogs, tadpoles and termite alates". No mention of snakes there.

The thick knees of the South African Thick-knee feature in *Umuntu wami onemicondo emikhulu* (My person with big [the adjective big refers to the knees] thin legs).[39] The answer is **Umbangaqhwa** (Spotted Thick-knee) (see Plate 28).

The riddle *Umuntu wami okhala egijima* (My person who cries while running), to which the answer is **Impangele** (Crested Guineafowl), is interesting in that the core notion is often found in personal praises. On more than one occasion I have found a happy, singing, person described as *UMpangele ekhal' egijima* (Mr Guineafowl who sings while running). And in the section on proverbs above (page 254), we noted the proverb *Impangel' enhle ngekhal' igijima* (a good guineafowl is one that calls while running).

Clothing can be a metaphor for plumage in riddles, for example:

Umuntu ogqoke isudi elimnyama (A person who has a black suit on):
Answer: **Igwababane** (black crow)
Abafundisi zami baseZiyoni (My priests of the Zionist Church):
Answer: **Amalanda** (white cattle egret)

Priests and their clerical garb feature strongly in these riddles. Besides the white-robed Zionist priests/white cattle egrets above, we have also seen Southern Ground Hornbills being riddled as priests with red collars. Priests, however, usually wear white collars rather than red, and the White-necked Raven is the bird most often associated with priests because of this. Khumalo links the raven to priests in two forms of the same riddle: *Umfundisi ontshontshayo* (A priest who steals) and *Umfundisi ontshontsha amaqanda* (A priest who steals eggs) both with the answer **Igwababa** (The White-necked Raven). **Igwababa** is also the answer to the riddle *Umfundisi wami ogqoka izingubo ezimnyama nokhololo omhlophe* (My priest who wears black clothes and a white collar) where the link between the White-necked Raven and the clerical dog collar could not be clearer.

Ruth Finnegan picks up the raven and the clerical collar when she talks about the 'bird-riddle':

In South Africa we hear of the 'bird riddle'.[40] This is a kind of competitive dialogue between two boys or young men in front of an

---

39. Khumalo explains (1974: 231) that 'big thin legs' means that the bird has thin legs, that are big (thick) at the knee.
40. A Finnegan footnote here refers to Jordan in *Africa South* 2.2. 1958, pp. 102–3.

audience. Each has to prove he 'knows the birds' by making an assertion about one, then an analogy likening it to a kind of person. Each in turn tries to show that he 'knows' more birds than his opponent. In this game 'freshness of idea, wit and humour count more than just the number of birds named'.[41] Thus a competitor with the following was declared the winner with a single attempt:

Challenger.	What bird do you know?
Proposer.	I know the white-necked raven.
Challenger.	What about him?
Proposer.	That he is a missionary.
Challenger.	Why so?
Proposer.	Because he wears a white collar and a black cassock, and is always looking for dead bodies to bury (1976: 432).

The question-and-answer format of the above, which moves away from riddling per se to more of a competitive game, is strongly reminiscent of the now forgotten Zulu children's game *Bhula 'Ntsentse Bo!* Below I give Letitia Samuelson's full explanation of the game:

ZULU EDUCATION IN THE KNOWLEDGE OF ANIMALS
The Zulus, during the Zulu kings' reigns, knew the names of all the animals, trees, herbs and grasses of their country. The knowledge of these was inculcated in them from their youth. One would often find a congress of Zulu children seated in a circle and rubbing into one another in a sing-song manner the names of these animals, which they had learnt from their elders. If any uncertainty arose, the doubtful matter would be referred to and settled by the elders, who were often present to witness the performance of the youths. The performance would be in the nature of a competition. Before the competition commenced a row or rows of maize (mealie) corns would be placed in order, and the competitors seated in a row or rows. The competition would be opened by a boy or girl seated at the end of the row, who would be allowed to name an animal and the next one would do the same thing, and so on, till all had had a turn, for each correct naming one corn would

41. A Finnegan footnote here refers to Jordan p. 103.

be taken out and placed to the credit of the youth who had made the correct naming.

The first man [*sic*] would start again and the rest followed until all had finished displaying his knowledge. The corns of each would be added, and the one who had the most would be declared the winner. This made each one keener to win the next time. The words that were uttered in a sing-song manner by each competitor while he was naming or trying to name some animal were "*Wena osemathafeni ungceda ijar' ekulu* [*sic*], – *bhula ntshentshe bo*" which means "you who art on the flats the tinkey bird (a diminutive species of the lark) is a great strong person – go on you clever fellow".

Bhula is addressed to the next competitor who is called "ntsentshe"[42] sarcastically, and bo is used to express and [*sic*] emphatic order. These words are repeated by all except that the place of ungceda is taken by the names of the animals named respectively by the other competitors (1974: 88).[43]

Samuelson goes on at this point to talk about 'Zulu superstition' about *ungceda*.

To perhaps clarify matters somewhat, here is Letitia's brother Robert Charles Azaria Samuelson on the same topic:

Bhula: thrash a mat, beat out fire; consult sangoma . . .
N.B. – There was a game played by native children called *bhula 'ntsentse bo*, which is a competition in the knowledge of the names of birds, but is has now almost died out since the Europeans conquered Zululand, and natives have got ignorant of the names of most birds as well as of grasses, trees, animals, etc. The game consisted of each boy or girl naming in a pleasant sing-song way a bird, then another would name the next, and so on, until all were exhausted, and the one

---

42. Letitia Samuelson uses both *ntshentshe* and *ntsentshe* in this description of the game.
43. Presumably what Samuelson is saying here is that the chant *Wena osemathafeni **ungceda** ijar'ekulu , – bhula ntshentshe bo* is repeated each time a player is challenged to name a new bird, with the word *ungceda* replaced by all the names (in the order of their utterance?) by all the previous competitors. If this is so, then there is a double challenge: a competitor must not only name a bird, but in challenging the next player, must remember every other name previously given in the game. This is a tall order.

who has named the largest number would be the winner. The naming would be performed in the following manner:

E.G. Wena osemathafeni isigwaca ijara elikhulu, bhula'ntsentse bo!, you who are on the flats the quail (*isigwaca*) is a strong big fellow, divine you clever fellow! (R.C.A. Samuelson 1923: 40).

In connection with this competitive game to see who could name the most birds, and keep their wits about them while doing so, the following may be of interest:

In June 2017, at one of the Zulu bird name workshops, which are the subject of Chapter 12, a discussion was going on about the use of birds in children's education, more specifically how to keep the attention of the younger children by playing games involving birds. One of the younger participants, Nontuthuko Xaba, was enthusing about an application ('App') she had on her mobile phone that had bird quizzes suitable for younger children, and how everyone in her school wanted a mobile phone so that they could play this 'App'. She lamented the fact that these bird quizzes were not in Zulu. I took advantage of a lull in the discussion to bring up the game *'Bhula'Ntsentse Bo!'* and offer it as an alternative to an English-only bird quiz available only on mobile phones, which most of the children did not possess. There was a distinct silence as the dozen or so Zulu-speaking participants of the workshop stared at me blankly. One of them then kindly explained that no, they had never heard of this game,[44] and that, moreover, no self-respecting Zulu child today would be interested in any game that could not be accessed on a mobile phone.

But to return to ravens and how they featured in the oral activities of children before the time when mobile phones were de rigeur among young, black rural children, let us look at Godfrey's 'Raven's Song', which he says is "a fragment only of a Native nursery-rhyme heard throughout Kafraria" (1941: 83).

Ye! Thuthu! Aph' amathole?	Hello Tutu! Where are the calves?
As' esapha!	Over the hill.
Wenza ni apho?	What are you doing there?
Nditya amatye.	Eating stones.
Yint' ikh' ityiwe?	Are they ever used as food?
Akuyazi na wena?	Don't you know that?
Kha utye sibone!	Come, eat (some stones) and let us see!

---

44. See section 13.6.1, 'The dying out of a meme', in Chapter 13 of this book.

Ndakutya ngomso.	I'll eat (them) tomorrow.
Khona kunani?	Why then?
Khona ndovuba.	I'll be mixing up food.
Nganto ni?	With what?
Ngomqothonga.	With—[45]
Uya kuwuthatha phi?	Where will you get it?
Endlwini kaGantsa.[46]	In Gantsa's house.
Athi ni uGantsa?	What is Gantsa saying?
Athi:—	Gantsa says:—
Ndakubodloza ngenduku yam le	I'll stab you with this stick of mine
EMabodloza	at Mabodloza's,
Yabodloz' inja yaphaphatheka	That beat the dog, and the dog fled
Yabeka eluSuthu	To Basutoland
Yadibana namakwababa emabini,	And fell in with two ravens
Lathi elinye:– vuk' uvuthele	One raven said:—Get up and light the fire!
Lathi elinye:–	The other replied:—
Ndakuvuthela njani na	How can I light it
Ndixholiwe nje ngamakhwenkwe	Damaged as I am by Sabe-sabe's
Akwa Sabe-sabe?	young fellows?
Sabe-sabe, xhel' inkabi le,	Sabe-sabe! Kill this ox,
Sinqunquthe.	That we may gnaw it!
Msila wenja, ulukhuni	Dog's tail, you are hard
Nje ngesonka.	As bread.
Se kuza kusa.	Already dawn is breaking.

I have to say that the first part of this song reminds me of the song 'There's a hole in the bucket, dear Liza, dear Liza'. Godfrey says that the song, which seems to combine elements of song, riddle and competitive game, was in use at Somerville among the Pondomise, and was taught to his eldest child (whose Xhosa name was Nontutuzelo), by the Pondomise girl Lena Botya (1941: 84).

---

45. A Godfrey footnote here reads: "The meaning of the word *umqothonga* seems to be lost. Though the word itself is known to all through this rhyme, its meaning cannot be obtained from my helpers."

46. A Godfrey footnote here reads: "*Endlwini kaGantsa* (in Gantsa's house) would appear literally to be: 'from the lanner's eyrie,' — Gantsa being the form which, through long repetition, has evolved from *khetʃhe*."

10 Bird beliefs
Portents and heralds

10.1 INTRODUCTION

This chapter and the following one are effectively about how birds are integrated into the Zulu world view. They look at two different but interlocked aspects: traditional beliefs about birds and traditional uses of birds.

The traditional beliefs concern birds as omens and portents, namely that certain avian behaviour – patterns of flying, patterns of calling, settling on a hut – are portents of evil to come, of death, sickness and of war. The beliefs include bird-related taboos: taboos against killing certain birds and taboos about eating certain birds. Traditional beliefs also include the belief of the effectiveness of certain birds as ingredients in charms such as love charms and protective charms, particularly those charms that protect against bad weather.

The traditional uses include a variety of different ways in which birds have been used in Zulu society. The largest section is on the use of feathers, primarily as a form of regimental insignia in the days of active Zulu regiments, but also as pure ornamentation as part of festive dress. The section on birds as food is comparatively short, reflecting the minor role birds have played as food resources in Africa generally. Also included are such practical aspects as birds as aids in hunting, rodent eradication, cattle herding (and finding lost cattle) and even as a 'tooth fairy'.

Although the two aspects of traditional beliefs and traditional uses are treated as discrete entities in different chapters, there is naturally considerable overlap between the two: a bird that heralds the spring (covered under the section on beliefs) is a very useful reminder of the start of the ploughing season; using a bird to help find strayed cattle rests on a belief that this will be effective.

In all the sections of these chapters, while the emphasis is naturally on Zulu traditional beliefs and practices, much supporting material will be given from Xhosa beliefs and practices. These are much the same as the Zulu beliefs and practices, probably because these beliefs and practices go back to an earlier proto-Nguni stage before the various dialectal varieties settled into their separate languages of Zulu, Xhosa, Swazi and Ndebele of Zimbabwe. Indeed, these language divisions are modern, artificial constructs, superimposed on a much earlier set of bird-related traditions.

Further contextual material is given from a variety of other sub-Saharan African beliefs and traditions, ranging from southern African cultures and language groups such as Southern Sotho and Shona, to those further away in the Congo, Tanzania and Madagascar. Where it has seemed relevant, further context is provided in the way of reference to similarities in Euro-Western avian beliefs and usage. Many groups of birds such as the larger raptors and the corvids seem to inspire similar beliefs around the world.

10.1.1   Explanation of terminology

This chapter is based on research from material ranging from the late nineteenth century to today. The earliest work consulted is Layard's 1867 *Birds of South Africa*, the latest works published in 2016 and 2017. The earlier works use terms that were acceptable in their times, but which are unacceptable today. For example, the word 'Kafir' is now decidedly offensive, but was used regularly to refer to the Bantu-speaking inhabitants of the Eastern Cape specifically, and to Bantu-speaking inhabitants of South African generally.[1] The term appears in the titles of books, such as Kidd's 1904 *The Essential Kafir*, which has provided data for this chapter. Most of the earlier writers use the term, and when they have been quoted in this chapter, the offensive term has been removed and replaced by 'Xhosa' or 'Zulu' in square brackets. Obviously, the term 'native' with its pejorative overtones has not been used in this chapter, unless in a direct quote. Nor has the outdated word 'tribe' been used. Where such identification is required, links to languages have been preferred, as in 'Zulu-speaking inhabitants' or 'Xhosa speakers'. The neutral term 'people' is also used, as in 'the Mbuti people of the Ituri Forest' or 'the indigenous people of Madagascar'.

---

1. 'Bantu' as a synonym for a black South Africa was used offensively in the apartheid era, but is and has never been an offensive term when used of a related group of languages.

Many of the earlier books use the term 'superstition' as well. This is another loaded word, such as 'tribe' and 'native', and carries overtones of superiority and patronage. The phrase 'traditional belief' (or just simply 'belief') is used instead. The word 'magic' has been avoided for similar reasons.

And then there is the ubiquitous word 'witchdoctor', surprisingly still used even today. Not only does this still carry the colonial weight of 'tribe', 'native' and similar words, it is confusingly ambiguous. The term is used in the literature of four quite distinct groups of people:

1. the *isangoma*: the shaman or diviner, who is neither a doctor nor a witch, but who may seek out and block the power of witches among her several other duties;
2. the *inyanga*, or traditional healer, who <u>is</u> a doctor, but not a witch;
3. the *inyanga yezulu*, sometimes referred to as a 'sky-doctor' or 'heaven-herd', who 'manages' weather, but is not a healer; and
4. the *umthakathi*, the evil witch who seeks to destabilise society.

Where the term 'witchdoctor' occurs in a supporting quote, a footnote will identify which of the four above it is meant to indicate. When it comes to birds, it is mainly the *inyanga yezulu*, the sky-doctor, who is involved.

10.1.2 Sources of material

By far the two most important sources of material on Nguni bird lore have been Robert Godfrey's 1941 *Bird Lore of the Eastern Cape Province* and A.T. Bryant's 1905 *Zulu-English Dictionary.* Jacobs says of the Reverend Godfrey's book:

> A particularly rich source of vernacular avian environmental knowledge, this book is filled with wonderful detail on avian human relationship: it tells about birds' behaviours, abilities, powers and diets. It shares proverbs and stories about bird's characters (2015).

Jacobs' quote is not only useful for giving an enthusiastic review of Godfrey's research, it also explains exactly what this chapter is attempting to do for Zulu bird lore in a wider African context.

Father A.T. Bryant is known for a number of important ethnographical and historical publications about the Zulu people, but it is his dictionary that provides most of the material about bird lore. Fortunately for us, Bryant was not just content to gloss a bird's name by simply identifying the bird,[2] he frequently offered much more in the way of cultural information. The following is a typical extract from the dictionary:

> **i-nDhlazi**, *n*. Mouse-bird (*Colius Capensis*) whose long tail-feathers are used as a head ornament.
>
> P. *nginonele pakati njengendhlazi*, I am fat inside like a mousebird, *i.e.* my feelings, thoughts, anger or revenge, is not seen by you, but you may come to feel it – may be used as a threat, or of a person with a brooding ill-feeling.
>
> *N.B.* The *amafuta* of this bird is used as an *isi-betelelo* (q.v.) 'because it is always sticking at home in its nest!' (1905: 100).

This extract illustrates exactly what Jacobs means when she talks about "avian human relationship[s]": the bird '**indhlazi**' [**indlazi**] is identified by an English vernacular name and a scientific name, the use of its tail feathers as ornamentation is mentioned, the fat body of the mousebird becomes the core of an idiomatic phrase that refers to human feelings, and the fat of the bird is used in a charm for 'fixing' the love of a woman for a particular man. These are all examples of bird 'lore' – the beliefs that articulate the relationships between birds and humans. Bryant's details in his dictionary illustrate how birds are integrated into the Zulu world view, and this extract shows how useful Bryant has been as a source of information.

While Godfrey and Bryant have, without doubt, been the most useful sources of data about bird lore, some other sources from the range of sources consulted can be specifically mentioned as well. R.C.A. Samuelson published his *The King Cetywayo Zulu Dictionary* in 1923, and his combination of reminiscences and ethnography *Long, Long Ago* in 1929. Although there is much repeated material across both books (and indeed much repetition within the dictionary itself), Samuelson often comes up with something unique. It is he alone who comes up with the intriguing story of the partridge in the thumb (see Chapter 11, page 338). Useful for a wider African context has been Ichikawa's 1998 article on the bird-related beliefs of the Mbuti hunter-gatherers of the Ituri Forest in the Congo. Moreau (1940, 1942) and Sibree (1881a, 1881b, 1882)

---

2. Which he consistently took from the 1899 book *Natal Birds* by the brothers R.B. Woodward and J.D.S. Woodward.

have made useful comments about bird beliefs in Tanzania and Madagascar respectively.

For contextual material on Euro-Western bird beliefs, use has primarily been made of Armstrong's 1958 *The Folklore of Birds*, Waring's 1978 *A Dictionary of Omens and Superstitions* and Cocker's 2013 *Birds and People*. All other sources are fully referenced and can be found in the bibliography.

10.2   OMENS, PORTENTS, TABOOS AND CHARMS

In this sub-section, I have divided the material into what appear to be discrete categories such as 'manifestation of omens and portents', 'omens of war', 'bad and good luck', etc., but there is considerable overlap of information between these categories.

10.2.1   Manifestation of omens and portents

Birds that are portents of evil, or omens of death, or are otherwise 'telling' humans of good or bad luck to follow, do not convey this information simply by appearing and being the species that they are. They have to call in a certain way or at a certain time, or fly in a certain way or direction, or settle on a hut or a fence. In other words, bird-related omens and portents depend on bird-related behaviour. In this sub-section, the behavioural patterns that manifest the omens or portents are examined.

Waring, in a general statement about birds as omens and portents, emphasises their behaviour and positions: any bird flying in and out of an open window signals the death of someone in the house; tapping on a windowsill also portends a death; and the direction of a flight of birds crossing in front of you when you leave for a journey is important, as if they fly to the right all will be well for your journey, but if they fly to the left, you should rather stay at home (1978: 32). She concludes this section by saying that if bird droppings fall on you this is a sign of misfortune. As we will see in the case of the Bateleur (**ingqungqulu**) and the Hamerkop (**uthekwane**) below, this is more than just a misfortune in Zulu thinking.

In Zulu thought patterns (and in sub-Saharan African thinking generally) the worst thing any bird can do is sit on top of a person's hut. It is even worse if the bird calls while doing so, and worst of all is a bird calling at night while sitting on a hut. Samuelson makes this point twice about the Southern Ground Hornbill (**insingizi**), (which he calls a 'turkey buzzard'). In his dictionary, he says:

This bird is a great snake-hunter and destroyer; it is held sacred and not killed by the Zulus; if it sits on a hut it is taken to be a bad omen by the owner (1923: 471).

Then, in his book *Long, Long Ago*, he says of the bird (now referring to it as the Southern Ground Hornbill) that "if this bird happens to settle on the hut of a kraal, the native considers that some evil will befall the kraal" (1929: 407). Samuelson illustrates the general point with a specific example while glossing the term *isihlabamhlola*:

> **isiHlabamhlola**: any extraordinary or unusual occurrence . . . a bad omen.
> *E.G.* A turkey-buzzard came and settled on one of the huts of the Ulundi kraal and a duiker ran into the same kraal before the Zulu war and it was taken as a bad omen (1923: 165).

Bryant adds his voice here, saying, "Whosoever strikes a [ground] hornbill will as surely die! And should one ever alight on a hut, it is an omen so evil, that the hut-owner would immediately consult a witch-doctor" (1905: 654).[3]

Josiah Tyler adds an extra detail to the dangers of killing a turkey buzzard (Southern Ground Hornbill), when he says "No-one dares kill a turkey buzzard, lest the arm with which it be done be paralyzed" ([1891] 1971: 111).

To the Southern Ground Hornbill (alias the Turkey Buzzard), Godfrey adds the rook, Hamerkop and 'the owl'. Note that his 'rook' is today's Black Crow:

> The rook is classified with a number of other species including the hammerhead, the turkey buzzard and the owl, as being in league with the *magqwira*.[4] Should it perch on a hut, it would be regarded as a messenger from the departed spirits or the *magqwira* to foretell approaching evil to one of the inmates (1941: 80).

---

3. Here Bryant is presumably referring to an *isangoma* (diviner).
4. The Xhosa word *igwira* (plur. *amagqwira*) is the Zulu *isangoma*. Godfrey is saying here that the diviner actually sends the birds that portend evil (in addition to the diviner's normal role of interpreting the omens).

There is obviously something about perching on a hut that intensifies anything ominous about birds of omen.[5] There does not seem to be anything obvious connecting these four bird types that would lead to them falling into the same traditional thought patterns. The Southern Ground Hornbill and the crow are both black birds, and the owl is a creature of the black night.[6] The Hamerkop, however, is neither black nor a nightbird. 'Nightness' (if one may coin this term) is a feature of nightjars, but there is no record in the literature of them being birds of omen. There may be something about the Southern Ground Hornbill and the owl being 'liminal' creatures, that is, sitting on the boundary between bird and human: the hornbill walks about on its two legs rather than flying as other birds do; the owl has two large eyes facing forwards more like humans do, instead of having them on either side of the face like other birds. Moreover, both utter cries that could be said to be human. Liminal creatures, such as the half-human, half-mongoose character uHlakanyana of Zulu folktales; like snakes that are animals but have no legs; like *imikhovu* (zombies), which are humans but not living; and like the mischievous *tokoloshe*, also human-but-not-human – all these function in the shadowy world of spirits and witches. In some ways, the Hamerkop is a liminal creature, too: it builds a nest large enough to be called a hut (with two 'doors' according to traditional belief), and it sits admiring itself in the water of a pool, as humans do before mirrors. The crow, however, is surely just a bird. But perhaps there is something about the *Corvidae* that marks them in this way – the raven, for example, is universally regarded as a bird of omen. Waring has the following to say about crows, making numerous points about portentous behaviour relevant to a number of sections in this chapter, particularly as regards placement, position and direction:

> Throughout the world and from earliest times the crow has been looked upon as an omen of misfortune and death. The bird is widely associated with witchcraft and is said to have the gift of prophecy, hence it has proved a major ingredient in many potions of fortune telling. A crow flying about a house and cawing signifies a death to come, while to

---

5. In fact, any time a bird 'intrudes' into human space, this can be seen as an omen. Waring says that in Western thought patterns, a goose is said to be giving a warning of death when simply flying around a house (1978: 104).

6. Then again, there are a considerable number of purely black birds (from black oystercatchers to black flycatchers to black drongos), which have no record of being birds of omen.

see one of the birds perched alone is a sign of hard times ahead, while
their activities at dawn and evening are weather portents. If they are
seen flying towards the sun in the morning the weather will be fine
and dry, but to be spotted stalking around water at nightfall is the sign
of a storm in the offing (1978: 69).

10.2.2   Omens of death and illness, witchcraft and evil

Quite a number of birds can be portents of death, illness, evil and witchcraft,
but the main birds are the Hamerkop (*uthekwane*), the Bateleur (*ingqungqulu*)
and the owls (*izikhova*). I start with the Hamerkop.

Quite why this bird has these connotations of evil and bad omen is difficult
to say. Cocker suggests that the bird is linked to witches (*abathakathi*), in that,
like witches, the Hamerkop collects together a weird assortment of human
detritus (2013: 138).[7] The witch does this in order to have power over an
intended victim; the Hamerkop to decorate its nest, but the collecting of this
material (*insila*) has led to traditional beliefs of the Hamerkop as a familiar
of witches.

Godfrey states that when the Hamerkop is not predicting rain, or causing
thunder and lightning, it is seen as a bird of evil omen, especially if it flies over
a hut or settles on it. A Xhosa respondent told Godfrey, "It is a bird that brings
bad luck, for, should it hover above a village, something evil is about to befall
you" (1941: 17). And woe betide anyone found destroying a Hamerkop's nest:
Godfrey quotes a local informant as saying:

> When it has found a person destroying its nest, it speedily (seeks
> vengeance); sometimes it hovers over the spoiler and lets its dropping
> fall on his head, thus ensuring his instant death (1941: 13).

Socwatsha kaPapu, in Vol. VI of *The James Stuart Archive* (*JSA*), also tells of
the 'deadly droppings' in referring to the Bateleur as an *umhlola* (evil omen),
saying that "if it excretes on a person, he becomes sick or dies" (Webb and
Wright 2014: 78). Regarding the Hamerkop, he says that this bird is sent by
an *umthakathi* and must be warded off by sprinkling preventative medicines at

---

7. Cocker has an astonishing list of the detritus found in the construction of a single nest
   in Zimbabwe (2013: 138).

the doors of huts, and beneath the fence (Webb and Wright 2014: 75). Bryant echoes this in saying that the flesh and nest of the Hamerkop are used for the purpose of *ubuthakathi* (witchcraft) (1905: 619).

As for the Bateleur eagle, Godfrey sees this as "one of the outstanding birds of omen". Just its cry, he says, indicates trouble somewhere, and the bird's call is rendered *lof' ilizwe* (the country will die), which suggests that war is imminent. Like so many other birds, its potency is intensified if the bird flies over a hut or village – a sure sign that some evil is about to happen in that village. And like the Hamerkop, its droppings are potent: "If its droppings fall on a man's head, he dies forthwith" (1941: 34). Godfrey quotes one Nimon [*sic*] Ndingi of the eMfundisweni Mission, who has much to say about the Bateleur:

> The Bateleur has an awe-inspiring spell. When it appears high overhead, some evil fortune is sure to happen. When any misfortune is about to happen in a person's village, such as the death of one of his family or perhaps of his cattle, the bird proclaims it by flying straight over the huts, uttering a cry that makes one's blood run cold, flapping its wings and calling *ha-a-a-a!* It maintains this sorrowful cry, as it circles round that person's village. Off he goes that very day to summon the witch-doctors[8] to his home. The bird usually appears when war is imminent, appearing in numbers when anything of that kind is about to happen. The birds wail over the contending armies. When one of the warriors is killed, down comes a Bateleur maintaining this sad wail, plucks out one eye and then leaves him. For these reasons, this bird is held in high repute among the men; it is not killed by any Tom, Dick or Harry,[9] but by witch-doctors only,[10] and even witch-doctors must take precautions by rendering their family immune before killing it, for great damage might be done to their villages in the event of their killing it (1941: 35).

Krige contributes to the discussion:

---

8. Godfrey translates from the original word *amagqwira* here, so the reference is to diviners.
9. The original reads "*ayibulawa nangubani nje*" (it is not killed by just anyone).
10. Here the original term is *amaxhwele*, equivalent to the Zulu *izinyanga*. Other sources refer to the specialised killing of Ground Hornbills and Bateleurs for the purposes of rainmaking, so possibly the *inyanga yezulu* is meant here.

The *iNgqungqulu* (the largest of all birds [??]), is a great omen-bearer; if it smites its wings together as it goes, it reports the arrival of an enemy; if it cries while flying, it foretells rain; while if it drops its dung on a man, evil will befall him (1950: 288).

These beliefs among the Nguni are echoed in Western thought patterns. Waring says of 'the eagle' in Western thought that when it is seen hovering over the plains for any length of time, it is presaging disease and death for those below. The sound of its screeching, she adds, is also a death portent (1978: 83).

Owls are creatures of the night, and night is the time of the *abathakathi* (witches) in African thought patterns. For this reason, owls are often thought of as witches' familiars. Samuelson says of the **isikhova** (a generic term for owls):

[T]his bird is hated and detested by the native as a bird of bad omen; it is said to be one of the companions of the Umthakathi, "Witch", travels about with him, at night, while he is performing his nefarious deeds and acts as his winged messenger (1929: 406).

Krige has in all likelihood picked up the idea of the winged messenger from Samuelson when she says, "When an owl hoots near a hut, it is believed to forebode evil or death, because it is the messenger of some *umthakathi*" (1950: 325).

The name **isikhova** is used as a generic name for all the owls found in KwaZulu-Natal.
© Adrian Koopman

The link between owls and witches is found elsewhere in Africa. Ichikawa says of the Mbuti people of the Ituri Forest in the Congo that "Owls are avoided because they are thought to be the 'watchmen' of a witch or a sorcerer" (1998: 106).

Bryant's entry for 'isiKova' refers to the verbalisation of its call as "*Vuk' ungibhule!*", which he interprets as "Get up and whack me!" (1905: 321). A more relevant interpretation of the call uses the meaning of *bhula* as 'divine', 'smell out', or 'consult a diviner'. If the owl is saying 'wake up and consult a diviner' then it is playing the same role as *umswelele* in Colenso's entry:

> **umswelele**: the name of a bird, like an owl, which cries at night when it sees a person, buck, dog, &c., the sound being like a whistle, *swi! swi!*
> Ex. *lalelani ngizwe umsweleswele;*[11] *uy'eza umtakati!*, listen to the cry of the *umsweleswele!* the *umtakati* is coming! (1884: 531)

To emphasise the point, Colenso enters the bird again under the spelling *umtshwelele*, glossing the word as a "sort of owl that is said to warn against the presence of a witch" (1884: 582). On the other hand, an interesting notion comes from Themba Mthembu, a participant of the June 2017 Zulu bird name workshop. While discussing owls and witchcraft, he suggested that when an owl 'spoke' the words "*Vuk' ungibhule!*" (Consult a diviner!) it was taunting the listeners in the nearby homestead, effectively saying "Yes, I have been sent by an *umthakathi*. You will need to consult a diviner to find out which one."

Given these various entries on owls, it seems that they play an ambivalent role, simultaneously being the familiars of witches and sorcerers, and being birds that warn against witches and sorcerers.

Unlikely birds of evil omen include the inoffensive Cape Robin and the familiar and friendly wagtail. Of the robin, Godfrey says, "In former days, when an army was on the warpath, the cry of this bird was held to portend bad luck" (1941: 89) and of the wagtail he says that this bird is a messenger of death, although this is not a general belief but one held only by a few men and boys (1941: 102).

Given that the dove is universally held to be a symbol of peace, it is surprising to hear from Godfrey that "in days of old, when doves settled on your cattle-kraal fence and called, it was said they foretold some evil" (1941: 51).

---

11. Colenso adds the extra syllable /*swe*/ to this example.

However, "Nowadays it is not so! Even the doves don't go to our homes lest they be shot." Waring says that in Western thought the dove has been considered as sacred since earliest times as it is the one bird into which the devil cannot turn himself. However, even in Western thought, doves may have an evil aspect:

> If one is seen flying near the window of a sick room or actually knocks upon the pane, then this is a sign of death, while no miner would dream of going down a mine if a dove was seen near the pit shaft (Waring 1978: 79).

Another perhaps unlikely bird is the Neddicky (**ungceda**). Letitia Samuelson says that the Zulu people believe that "if this little bird flies along and drops on or near anyone a grasshopper carried by it, that person is doomed to meet death" (1974: 89). Krige says the same of the **ucilo** (a small species of lark), but as the **ungceda** is also a small species of lark, this may be the same bird under a different name (1950: 288).

The Knysna Lourie,[12] according to Godfrey, is specifically an omen for hunters (1941: 54). When hunters enter a forest, the bird sings at the edge of the forest where the hunt begins, and then it goes deeper into the forest. When the hunters hear its "sad and sorrowful tones" they will know that danger threatens them.

Zwayi kaMbombo (recorded by Stuart in *JSA* Vol. VI)[13] and Krige both link evil omens to bad weather: Krige, after saying that a 'doctor' must be consulted at once if a Southern Ground Hornbill (**insingizi**) alights on a hut, says that the **insingizi** is "the bird that does not fear thunder", and, furthermore, it is the Bateleur (**ingqungqulu**) that is responsible for inducing this thunder (1950: 288). Zwayi kaMbombo says that with all evil omens, for example, the coming of a Hamerkop, a Southern Ground Hornbill or a monitor lizard, a doctor was quickly resorted to or a storm would come. We must assume that the 'doctor' referred to by both Krige and Zwayi kaMbombo was an *inyanga yezulu*, the 'doctor' who controlled the weather.

Finally, the vulture, a bird that one might expect to be a bird of many omens and portents simply appears as a bird with connotations of presentiments of danger. It is Colenso who gives the meaning of the word **inqe** as "vulture; dread;

---

12. Now the Knysna Turaco.
13. See Webb and Wright (2014: 405).

presentiment of danger" and offers as an example of usage "*lo 'muntu unenqe lokuwel'amanzi*" (that man has a dread of crossing the water) (1884: 286).

10.2.3   Omens of war

Previously it was mentioned that the cry of the Cape Robin was held to portend bad luck for an army on the warpath. Charles Pacalt Brownlee, in his 1896 *Reminiscences of Kaffir Life and History*, tells of a similar belief involving a 'crested hawk eagle' (the Long-crested Eagle, **isiphungumangathi**, whose role in finding lost cattle is described in Chapter 11 on pages 334–5). Brownlee's narrative concerns a crested eagle that flew over a Xhosa army on the warpath "uttering its shrill piercing shrieks" (1896: 244). The elders said that this was an omen of ill luck, and said the warriors should all go home. The young chief in charge of the warriors ignored this and pressed on ahead, but when his second in charge was shot,

> the sight of the dying councillor, and his horse covered in blood, struck a panic in the hesitating [warriors]. The omen was being fulfilled; they turned and fled (1896: 244).

Two other sources specify that it is the Bateleur that is an omen of coming warfare. One is Singcofela kaTshungu of the emaBomvini clan, who told James Stuart:

> The *ingqungqulu* bird, should it, when seeing a body of men armed with shields, flaps its wings loudly and makes the noise 'Who!', is believed to be observing an *impi*, i.e. war is imminent (Webb and Wright 2001: 405).

The other is R.C.A. Samuelson, who makes three separate but overlapping statements about the Bateleur:

> **Ingqungqulu**, the Bateleur Eagle, which is a very rare bird with blood-red eyes and thus named because when any two of them have declared war against each other, they cause the sounds "ngqu, ngqu" as they collide, in mid-air;
>
> . . .

This eagle lives and moves about with his female companion only and very rarely shows himself; the Zulus believed that if they came soaring about, more than usual war would follow (1929: 408).

inGqungqulu: a very rare, large eagle, brown-coloured, reddish underneath, with red eyes flaring like balls of fire, when he and another of his kind fight, as they do, high in the air, they cause the booming sound of 'ngqungqu'; these eagles are said to foretell a war when they soar about so much (1923: 142).

This (the Bateleur – *ingqungqulu* in Zulu) is a very rare bird, an inhabitant of Zululand, with fiery red eyes, very powerful, and with a loud ringing call; when two fight in the air their collision sounds like a booming gun in the distance; the Zulu connects the movements of this bird with warfare (1923: 315).

Samuelson seems to have been much taken by the blood-red fiery eyes that 'flare like balls of fire'; he is likewise convinced that their name is an onomatopoeic one taken from the 'booming' sound their wings make when they fight. There is a subtle difference in the ways Samuelson links the bird's names to the clashing of its wings and to warfare: in the first of the three quotes, the birds have declared war on each other, and their collision causes the sound *ngqu ngqu*, which leads to their name. The omen of warfare is in their circling about. The second quote reinforces this: it is the fighting eagles colliding that produce the *ngqu ngqu* sound, and it is their circling that predicts war. In the third quote, however, it is their collision in mid-air that is the omen of war, through the 'booming' noise sounding like a gun in the distance.

10.2.4   Omens of luck, bad and good

With these birds linked to messages and portents of evil and death, it is a relief to know that some birds are 'good luck birds'. The "Hadadah" Ibis (**inkankane**) is one of these, with Godfrey simply saying that it is sometime regarded as a bird of good omen (1941: 20). Swallows also bring good luck, especially if they build their nests under the eaves of a house (Godfrey, 1941: 73). This is also a common Western belief: Waring confirms the swallow as a bird of good omen, but also shows how a bird can be both good luck and bad luck when she says, "Throughout much of Europe it is said fortune favours the household where the bird builds its nest, although it is an ill omen if it deserts the nest for any reason" (1978: 223).

Godfrey says of the longclaw that it is universally accepted as a bird whose call portends good luck. He quotes a local informant as saying, "It is a bird that tells of good fortune, as it flutters in front of a traveller" (1941: 104). We saw references to this in the previous chapter, in the lines from Vilakazi's poem 'Inqomfi' (see page 245 above).

Further details are given by Samuelson (1929: 406) who says

> **Inqomfi**, a lark with red under the neck and which is credited by the native, who is out on a girl-courting tour, with giving luck to the venture of the said tourist.

A source in Vol. VI of *JSA* also notes that the Lark (**inqomfi**) brings luck if it repeatedly crosses the road or path, specifically as to food and beer that can be expected at the next homestead (Webb and Wright 2014).

The notion of both positive and negative portents in one bird is seen in various comments on the Hamerkop (**uthekwane**) reported in Vols V and VI of *JSA*:

Ntazini,[14] recorded in *JSA* V (Webb and Wright 2001: 189) says "If a hammerhead gets up in front of one and defecates, it is an omen that one's journey will be prosperous and fortunate". Earlier we noted that if a Hamerkop's droppings fall <u>on</u> a person, death is instantaneous. Here we see that if similar droppings fall in front of a person, the "journey will be prosperous". As it is often observed, it's all in the details . . .

This notion of prosperous Hamerkop droppings is repeated by Zwayi kaMbombo in *JSA* Vol. VI, when he says that if the **uthekwane** gets up and defecates ahead, then the journey will be one of good fortune (Webb and Wright 2014: 405).

Socwatsha kaPapu in *JSA* Vol. VI reports:

> *Hammerhead with the tuft, bird which turns the doors, the umtakati of important matters, who reports good and then reports evil*[15]

---

14. Stuart does not give a patronymic for this informant.
15. This dual nature of the bird is reported elsewhere. It is an interesting and salient feature of the bird.

[presumably praises of the bird]; if goes over kraal, shouting 'Ke, Ke!' a girl will marry at the kraal or a boy will be chosen as a husband, [i.e. good luck]. If [it] sits on hut or on post of fence of the cattle enclosure, that is a bad omen (*umhlola*); means some person will die at kraal. They, knowing this, go to doctor and then break off intelezi medicines, and sprinkle them about (Webb and Wright 2014: 75).

The evil omen of the Hamerkop sitting on the roof of the hut has been noted several times in this chapter; the detail about it sitting on the fence of the cattle kraal is new, as is the notion of good luck being expressed as a forthcoming marriage. Again, it is all about position: sitting on hut or kraal fence is a bad omen; flying overhead, however, is a good omen. Zwayi kaMbombo, recorded on the topic of the Hamerkop by Stuart on a different occasion, confirms what Socwatsha has to say and adds some interesting and new details about Hamerkop beliefs. The extract is in Stuarts 'telegraphic' notebook style:

[H]ammerhead bird (*utekwane*) not killed, builds nice nest with headrest (*umcamelo*), possibly for its headfeathers. If killed, person killing dies. Boys nowadays tend to avoid custom. If flies past kraal, flies straight on for some days and simply goes 'keke-keke' etc., that means that a girl will marry at kraal at which this is done. This always comes true. Some boys have medicines to cause *tekwane* to go round and round a kraal and sit on a hut, this means that a storm will come and burn the kraal. If many *tekwane* fly past or many fly round, it is a very good or a very bad sign (Webb and Wright 2014: 405).

The detail about the "nice nest" with a headrest for the bird's crest is entertaining. The other details have been noted above, especially about sitting on a hut, but it is curious that "some boys have medicines" to actually cause this to happen. Which boys, one wonders, and why boys specifically, and why should boys or anyone else seek to anticipate in this way a storm with lightning that will strike the hut and burn it down? The potential in a single bird for either good or bad luck is also expressed here, but this time the position or action of the bird seems irrelevant. Presumably a diviner would be able to work out whether this is bad luck or good luck.

Godfrey, in referring to owls, does not record any difference between their sitting on a hut and flying overhead calling:

The owl is believed to be in league with the killing witch-doctors (*amagqwira*) and is ranked along with them as *amagqwira*, Should one settle on a hut, it is regarded as a messenger of death. Even if it merely screams in flying over a hut, it is believed to be predicting some misfortune to the inmates (1941: 59).

It seems that you just can't win with owls.

Nor can you win with Black-headed Paradise Flycatchers if you are one of the Mbuti people living in the Congo. Ichikawa says of this bird:

Black-headed Paradise Flycatchers are omen to a very bad luck [*sic*],[16] to say nothing of mentioning their name, just encountering them by chance is an ill omen (1998: 110).

Moving now to other birds, Zwayi kaMbombo (*JSA* Vol. VI) says of the **umvemve** (wagtail) that if it comes frequently and keeps on whistling, this means a friend is coming (Webb and Wright 2014: 407). The bird was not killed in former times because it was looked on as a 'letter' (*incwadi*) or messenger carrying news. When it called '*Tiyoh! Tiyoh!*' this was seen as a greeting, and a sign that the inhabitants of the homestead would soon be greeting friends in the same way. Ichikawa provides a similar example from the Ituri Forest in the Congo, but here the portent of an imminent guest is actually captured in the bird's name:

The bird *manbuekendu* (meaning a bird of guest) heralds, by its call, that there will be a guest in the near future, as its name suggests (1998: 112).

Zulu belief in the good luck inherent in the **intendele** partridge is expressed in a proverb, given here by Colenso[17] as *intendele iwe enkundhleni* ('the partridge has dropped in the yard', expressing gladness at a person's luck) (1884: 456). As with so many other birds described in this chapter, the partridge can portend bad luck as well. At least, that seems to be the case with another **intendele** proverb also given by Colenso: *intendele ibindwa isidwa* (the partridge is

---

16. Curiously, the Paradise Flycatcher familiar to South Africans, which also occurs in the Congo, has no portents, bad or good, attached to it.

17. But found in several other sources listing Zulu proverbs.

choked by an *isidwa* root), used of a man whose lies have been so exposed that he remains speechless.

And to round off our description of birds that predict either good luck or bad luck, here is a bird from a distant part of the world. Waring tells us that in New Zealand the emu is believed to be a harbinger of good luck and its flesh a cure for many illnesses (1978: 86). To kill one of these birds, on the other hand, is to bring down on oneself misfortune.[18]

10.2.5   Taboos against killing, eating and imitating birds

Taboos against killing certain birds

Linked to the notion of birds of omen and connections to witchcraft are taboos against killing certain species of birds, and the dire consequences of breaking these taboos. From one of Godfrey's Xhosa informants comes the statement:

> The [yellow-billed] kite[19] is not greatly afraid of the Native people, for it is not usually killed thoughtlessly. When a person kills a kite, all the hairs of his head fall out, as the feathers of the kite do in winter-time (Godfrey 1941: 30).

Godfrey also tells us that the Bateleur Eagle is "immune from harm at the hands of the ordinary Native; but on account of its awe-inspiring qualities, it is greatly sought after by witch-doctors" (1941: 34).[20] Again from Godfrey comes the information that even though the rook, [i.e. the Black Crow] is very destructive in the maize fields, it is "practically immune from danger" since everyone knows that if anyone kills a single rook, the other rooks will immediately take revenge (1941: 77).[21] For Bryant, though, it is the Southern Ground Hornbill that must not be killed: "Whosoever strikes a hornbill **[insingizi]** will as surely die!" (1905: 654).

---

18. There seems to be a sort of catch-22 situation here: should one be so unfortunate as to suffer from one of the many illnesses mentioned, surely one would have to kill an emu in order to access its flesh to be cured. But to do so brings about misfortune.
19. Godfrey calls this bird the "Cape Kite".
20. Once again, this term causes confusion. Does Godfrey mean *izinyanga*, *izangoma* or *abathakathi*? The context suggests that he is talking about *izinyanga*.
21. What form this revenge might take, Godfrey does not say, but given that both the Hamerkop and the Bateleur have droppings that are lethal, who know what a flock of crows might do.

There is much detail in the literature about the taboo against killing a wagtail (**umvemve**). Godfrey has a whole section in his book entitled 'The Wagtail as a Sacred Bird' (1941: 102ff.). He points out that the boys are warned by their elders that, if they kill the wagtail, the cattle in the kraal will die, and that anyone who is venturesome enough to eat a wagtail will be a poor man, i.e. he will never have cattle. He goes on to say that if a boy accidentally kills a wagtail, "he carefully buries it and puts two white beads along with it in the grave." Then he offers a prayer asking for forgiveness. Godfrey adds:

> This method of propitiating the wagtail is somewhat analogous to the action of the women when they turn up *nocebeyi* (the donder-padde[22] or Jan Blom) while hoeing. They return it to its little hole in the ground and put a few maize grains beside it to conciliate it, so that it may not bring down rain upon the hoers (1941: 102).

The notion of atonement here also applies to the killing of a Crowned Crane or a Southern Ground Hornbill: Godfrey records Rev. John Brownlee as saying, "If a person kill by accident a mahem (crowned crane) or a bromvogel (turkey buzzard) he is obliged to sacrifice a calf or young ox in atonement" (1941: 66).

Returning to the wagtail, Socwatsha kaPapu in *JSA* Vol. VI has some interesting (if somewhat cryptic and telegraphic) comments to make about wagtails and hair:

> Hair – not hidden but preserved (*londoloza*'d). Hidden for fear *abathakathi* take his body-dirt (*insila*). Others say wagtail (*umvemve*) will take and build house with and if this does so hair will fall off and he will accuse the *abathakathi* whereas the *umvemve* did this. When the *umvemve* builds the hair falls off, i.e. until it hatches young; the hair will then begin to grow (Webb and Wright 2014: 79).

Ntazini in *JSA* Vol. V makes the brief statement, "The wagtail (*umvemve*) [is] not killed by [a] great majority of people" and adds that the "hammerhead (*tekwane*) – 'feather in head' is not killed because if it is killed a hut will burn down because it is an evil bird (*inyon'embi*)" (Webb and Wright 2001: 188).

In Western thought patterns related to the killing of various birds, Waring notes that albatrosses were particularly revered by sailors because each was said

---

22. This is presumably the *donder-padda*, literally 'thunder-frog', of the genus *Breviceps*, known as a 'rain-frog' in English.

to contain the soul of a dead sailor, and for any sailor to kill one was to bring bad luck upon himself for the rest of his life. As Waring points out, this was a central theme in Samuel Taylor Coleridge's poem *Rime of the Ancient Mariner*.

Taboos against eating certain birds

Closely linked to taboos against the killing of certain birds are taboos against the eating of the flesh of certain birds. Gcumisa and Ntombela in their 1993 *Isilulu Solwazi Lwemvelo, Umqulu I*,[23] say of a number of birds that "*inyama ayidliwa*" (the flesh is not eaten), listing for example the Blue Crane, the vulture, the Bateleur and the Southern Ground Hornbill.

Godfrey says the flesh of an owl is not eaten by boys (1941: 60). Whether or not this means the flesh is eaten by girls, men and women is not clear. Probably what Godfrey has in mind is that it is normally boys who hunt and eat wild birds while herding cattle. In any event, the reason they do not eat owls, according to an informant, is that "[the owl] is a slut; its body is full of scurf, which causes it to smell as if it were dead". On the other hand, it is only the boys who eat the very tough flesh of the Steppe Buzzard, as "big people do not eat the bird, because it eats lizards and mice which are disregarded by Native people" (Godfrey 1941: 36).

Bryant tells us that the flesh of the Grey-backed Bush Warbler (**ibhoyi**) is not eaten by girls, as if they did so, they would bear children with scraggly legs (1905: 50). Then again, Bryant says that Burchell's Coucal[24] (**ufukwe**) is only eaten by old women and young boys (1905: 155). Bryant does not explain the reason for this food taboo, nor the selection of old women and young boys out of all the possible members of a community. Krige also mentions the very young and the very old when she says, "Eaten by very young and very old persons only are duck, paauw, birds' eggs, porcupines and fowls" (1950: 388).

Moving further afield to the baNyabonga people of the Lake Kivu area in the Congo, we find Hendrickx saying of the name *nyange* ("*Bubulcus ibis*" [cattle egret]) that it is derived from the verb *kuyanga*[25] (non-edible bird or thing). He explains:

---

23. 'Grain-basket of Knowledge of Nature, Volume 1'.
24. Bryant calls this bird the "Lark-heeled Cuckoo".
25. Probably a class 15 gerund, similar to Zulu *ukudla*, which means 'to eat', 'eating', and 'food'. The Nyabonga *kuyanga* would surely be in the negative, comparable to Zulu *ukungadli* 'not to eat', implying also 'something not eaten'.

For pastoral tribes this bird, which always lives near droves of cattle, was considered as something taboo. Killing and eating it involved very heavy penalties (1944: 210).

Not too far from Lake Kivu is the Ituri Forest, also in the Congo, and Ichikawa gives a detailed picture of the complex system of bird-eating taboos among the Mbuti hunter-gatherers of this forest, worthy of a sub-section of its own (1998):

A system of bird-eating taboos
Ichikawa remarks:

> The Mbuti consider most of the birds as their food, including the birds of prey. There are only five kinds of birds which they would not eat even when available. These are pied wagtails, owls, crows, nightjars and various species of swallows and swifts (1998: 106).

Explanations are given for the above tabooed species: owls are associated with witchcraft; swallows and swifts are used for hunting rituals; crows are "polluted" because they feed on human waste; nightjars ("probably", says Ichikawa) because of their nocturnal habits. No reason is given for not eating wagtails, but as in so many parts of Africa wagtails are considered 'sacred' birds and must not be killed.

Ichikiwa's listing of the five different groups of birds suggests that all other species are eaten freely by everyone, but this is absolutely not the case. His article details a different set of birds, this time not defined by family or bird type, but by taboo type, of which there are four:

1) the *nginiso* birds: these are tabooed as a food resource because they are clan totems;[26]
2) the *kuweri* birds: these are tabooed for suckling infants and both their parents because they cause various sicknesses such as diarrhoea, fever and skin rashes;

---

26. Zulu society does not have a system of clan totems, but it would be interesting to find out whether the Msomi clan eat the Red-winged Starling (**isomi**). Subsequent investigation has shown that there is no totemic relationship (or any kind of cultural relationship) between the **isomi** and the Msomi clan.

3) the *ekoni* birds: these are tabooed for both parents when a woman is pregnant, because they will cause a difficult birth and there will be problems with the baby when it is born; and

4) the *ekuse* birds: these are tabooed for both parents of very young children as eating them will cause the children to die.

A considerable number of birds are involved in this complex system of dietary taboos, and we can only give a few examples here.

Quails are *ekoni* birds: "If a pregnant woman or her husband eat this bird, a baby with an extraordinarily large head will be born", while the forest francolin is an *ekuse* bird: "Young couples should avoid eating this bird, as their offspring may be lost" (1998: 115).

Birds that are clan totems, (i.e. *nginiso* birds in this taboo system) include different types of hornbill, the white-browed alethe, the great blue turaco and the Senegal coucal (1998: 109). The great blue turaco is also an *ekoni* bird, because it "has a very loud voice, which may cause deafness in a newborn baby, if its parents eat the bird during the pregnancy" (1998: 116).

The African Nicator (now called the Eastern Nicator) is also subject to a dietary taboo, although it is not clear whether this bird belongs to one of the four taboo types listed above, or whether this is a different type of taboo. Ichikawa explains:

> As African nicators (*amapopo*) [are] said to be the bird of circumcision, from the way they shake the tail like the dancing initiates in circumcision rite[s], they are eaten only by a circumcisor (1998: 112).

The implication here is that the youths taking part in the circumcision rites are not allowed to eat the bird because they 'shake their tails' like the bird does. The person doing the circumcising, who presumably does not 'shake his tail', is exempt from the prohibition. Ichikawa adds that the Green Pigeon is also denied to the initiates of a circumcision rite, but no explanatory details are given (1998: 116).

To round off this section about birds and food, we can quote Waring here as telling us that in certain parts of Europe it was believed that if you ate three lark's eggs on a Sunday morning before the church bells rang you would have a sweet voice (1978: 138). And on the topic of bird's eggs, we can note what she says about ducks laying dun-coloured eggs: this is an omen of such misfortune that the duck responsible must be killed immediately and hung with its head down afterwards so that the evil spirits it harbours can leak out (1978: 82).

She does not say so, but one assumes that after the evil spirits have all gone, the duck can be eaten.

Taboos against imitating the call of the bird

Samuelson says of the Hadeda Ibis:

> **Inkankane**, the Adeda, so named on account of its call being from, as it were, the nasal passages — the natives believe that if one imitates its call he will break out in boils on his seat (1929: 406).

Samuelson is referring to the word *amankanka* (nasal passages) here, and saying that the bird is named **inkankane** because of the nasal quality of its call. The derivational information does not appear to be linked to someone imitating its call breaking out in boils on his 'seat'.

Bryant says of the word *ingqangamathumba* (lit. 'breaking out in boils') that it is a name usually associated with a large, brownish grasshopper, but is sometimes applied to the **inkankane** (hadeda) and the **ungceda** (Neddicky) "because a person who mocks them will break out in abscesses" (1905: 197). Unlike Samuelson, Bryant does not specify the site of the boils or abscesses.

Godfrey cautions against the imitating of the call of an owl among the Xhosa. Should one do so, all his blankets will be burned. This belief is apparently linked to the burning of an owl if it is caught, because of its link to witches and wizards (1941: 59).

We quoted Ichikawa above as saying about the Black-headed Paradise Flycatcher that its name is not mentioned. As is well known, the taboo against mentioning names is widespread among different societies around the world. Names of different kin types are not spoken aloud, and in many societies, such as Zulu society, even words that are similar to the names of different kin are not used.[27] This leads to what is known, at least among the Nguni, as '*hlonipha*' language: the respectful substitution of non-similar words for words that resemble personal names. Ichikawa gives no further details about the non-mentioning of the name of the Black-headed Paradise Flycatcher, but one wonders if this is part of a *hlonipha* system among the Mbuti people as

---

27. See Koopman for further details of '*hlonipha*' name avoidance (2002: 18–20).

well. There is no record of Zulu people avoiding the names of certain birds, unless of course a person (usually a woman) has a senior male relative whose name is derived from a bird name, which is quite possible.[28]

Sibree illustrates this kind of name taboo when he says:

> [T]he Cape Dove has the strange name of Tsiázotonònina (i.e., "Unspeakable") among the Tanàla or forest tribes, probably because its more common name had become tabooed or sacred through having formed part of the name of one of their chiefs (1891b: 559).

10.2.6   Birds used as charms

No examples have been found in the literature of bird or bird parts used for purely medicinal purposes. The quote below from Bryant could possibly come under this heading, although the medical condition is said to be a result of witchcraft. It concerns the Southern Ground Hornbill, the bird that has featured so strongly in this chapter as a bird of omen and a bird of rain. According to Bryant:

> A person suffering from unusual prominence of the eyeball – which is said to be due to an *umtakati* – may have the defect removed by the application of a little *inTsingizi* eyeball, whereupon the offending organ will return to its normal size! (1905: 654)

Birds as ingredients in love charms

Bryant gives examples of a number of birds whose fat is used as an ingredient in the concocting of love charms (1905):

- The mousebird (**indlazi**): The fat (*amafutha*) of the mousebird is used as an *isibethelo*, a generic term for any charm that is used to attract a girl to a young man and make her 'stick' with him. The reason given by Bryant is that "this bird is always sticking at home in its nest" (1905: 100).

---

28. 2013–2017 workshop participants had no problem in assigning the name **inkayishana** (or its abbreviated form **inkashana**) to the Red-billed Firefinch. Given in Doke & Vilakazi (1958: 575) as "species of bird", it is also the name of uNkayishana Solomon kaDinuzulu, grandfather of the current king of the Zulu people.

- The Bateleur (**ingqungqulu**): Similar ideas link the fat of the Bateleur to young men's desires. Bryant says that "parts of this bird are in great request among young men as love charms" (1905: 202). He gives the following recipe: "Take the fat of this bird's eye,[29] mix it with some *umkhando*,[30] for example of the *ulangazine*,[31] and you will have a powerful *ihabiya*."[32]
- The owl (**isikhova**): The fat of the owl is used also used in this way for the same purpose. Bryant refers to Woodford's Owl [now called the African Wood Owl] (**umabhengwane**) when he says the fat of this bird is mixed with *isokalakwazulu* (common washing soda) to make an *ihabiya* (1905: 371). The word *isokolakwazulu* is itself a powerful ingredient in such a love charm, as it literally means the lover boy (*isoka*) from Zululand (*laKwaZulu*).
- The Rufous-naped Lark (**umangqwashi**): Bryant says the fat of this bird is used by young men in concocting a love charm, but does not give any details (1905: 376). There is no reference to this use of the bird in its praises, discussed in great detail on pages 231–3 above.

Woodward and Woodward add to Bryant's list of birds used for love charms: their captive African Fish Eagle (**inkwazi**) was killed, they believe, "by a native for the sake of its heart, which is used as a love charm" (1899: 149).

Not a love charm, but a charm nonetheless, brings us back to the subject of eyes, and concerns the reputed sharp sight of eagles. The reference is from Samuelson and it is not clear what species of eagle he is referring to:

> **Ukhozi** was the name of the largest eagle in Zululand, white and black coloured, [and it] was noted for its farsightedness. It was specially required for the concoctions which were used by the Zulu king, at the great Feast of First Fruits (1929: 409).

---

29. Which, as we recall from Samuelson are "fiery red" and so may make the targeted girl 'burn' with desire.
30. Any medicinal charm used for doctoring or magic.
31. Bryant mentions this word in this description of the charm made with Bateleur fat but does not include the word as a separate entry in his dictionary.
32. Bryant explains *ihabiya* as "Medicine or love-charm of any kind used by young men to cause a girl to *hayiza*, i.e. to throw her into fits of shouting hysteria in which she repeatedly cries *hayi! hayi!* or *hiya! hiya!*" (1905: 220).

Charms as a protection against lightning

Most anti-lightning charms use plant material and information on birds used in this way is limited.

Koopman quotes Berglund (1976) on the use of the fat of the 'lightning bird' (*impundulu*, or *inyoni yezulu*, i.e. 'bird of the sky')[33] as one of ten ingredients that must be used by an *inyanga yezulu* to make anti-lightning charms (Koopman 2011a: 44). The 'sky-doctor' will smear the concoction onto pegs that are driven into the ground and placed in the thatch of the huts to ward off lightning. Krige tells us:

> The commonest fat used as an ingredient in this [anti-lightning] peg-medicine is that of the *Ngqungqulu* bird (*Helotarsus ecaudatus*) [Bateleur Eagle] which, when flying quickly, makes a noise like thunder, and to this is sometime added fat of a "peacock" which is said to cry and ruffle its feathers before thunder (1950: 315).

10.3   HARBINGERS AND HERALDS: BIRDS OF WEATHER, SEASONS AND TIMES OF DAY

In this section we look at birds whose calls announce approaching weather, or herald spring and summer and times for planting,[34] or mark the start and the end of the day. Under the 'weather birds' section we will also look at birds used to cause certain weather, especially those that figure in rainmaking rituals.

10.3.1   Weather forecasters: rain birds, storm birds and lightning birds

Birds predicting rain, storms and lightning

Godfrey says that if the Hamerkop's nest is destroyed, the sky becomes overcast on the spot as the bird keeps calling. If you run into the hut for safety the bird will "sit on the roof and call until you are struck by lightning and then it will

---

33. See Koopman on 'lightning birds' for a full description of this mythical bird in Nguni culture (2011a).
34. No birds in any culture have been found in the research for this book that announce autumn or winter.

go away". Another source told Godfrey that "if a person destroys its nest, great thunderstorms take place that very day" (1941: 15). Similar comments from other Godfrey sources suggest that this is in a way a 'thunder bird' or 'lightning bird', but not in the same way as discussed in my article 'Lightning birds and thunder trees', where the emphasis was on the *impundulu* or *inyoni yezulu* (bird of the sky, bird of the weather), a bird that is the avian incarnation of a lightning bolt (Koopman 2011a). Godfrey continues describing the Hamerkop as a 'rain bird', saying "The hammerhead is classed with the ground hornbill as a rain bird (*intaka yemvula*) and like the latter species it is believed to foretell rain by its cry" (1941: 16). Bryant confirms this, saying that the crying of the **uthekwane** (Hamerkop) is said to foretell rain (1905: 619).

The Southern Ground Hornbill, normally named **insingizi** (or *intsingisi* and other close variants) is referred to as **ingududu** when it utters its booming cry. This is normally taken to presage rain. The Woodward brothers combine the predicting of rain, the bringing of rain, and evil omens in their narrative of the Southern Ground Hornbill:

> It is generally heard crying before rain, from which the Natives think it has the power of bringing rain; they are very superstitious regarding this bird, and believe that if one is killed near their kraals some misfortune will be sure to happen (1899: 97).

Nyazini, recorded by Stuart in *JSA* Vol. V makes a similar statement about this bird, saying that it causes consternation if it comes to a home; when it does so a doctor[35] must be consulted "for this is the bird that brings *izulu* [weather]" (Webb and Wright 2001: 189). Zwayi kaMbombo in *JSA* Vol. VI has a lot more detail about this (Webb and Wright 2014: 405).[36] He says that the **insingizi** was not killed. If it was killed accidentally, this would be reported to the chief – possibly by the person who killed it. A report is made because it is feared that a storm would come. The chief would then direct the dead hornbill to be carried off to 'some precipice' where there was seldom much rain, and there it would be carefully concealed behind a stone, just as if it was the corpse of

---

35. Of the different types of doctors this might mean, the *inyanga yezulu* (sky-doctor) is the most likely.
36. Stuart's telegraphic notes of Zwayi kaMbombo's testimony have been paraphrased here.

some great chief. The rain would then come down 'properly'.[37] If the bird was left in the open, very bad rains would come (but not necessarily lightning and storms). The death of an **insingizi** would normally be reported to the chief, but if a man of high standing in the community were to see the dead **insingizi** there would be nothing amiss in his going and burying it and hiding it himself, so long as this was in a place where there were no precipices.[38] He would then report to the chief what he had done.

Nyazini in *JSA* Vol. V notes that the Bateleur (**ingqungqulu**) brings a lot of heavy rain (Webb and Wright 2001: 189). When this happens a 'doctor', presumably an *inyanga yezulu*, will pluck out a handful of grass from the top of the hut. Stuart's cryptic notes at this point do not clarify at what stage of the heavy rain the doctor plucks out the grass, or what the purpose is.

Both these birds are connected to lightning as well as heavy rain. Hammond-Tooke tells us:

[I]f the *uthekwane* (hammerkop) or *indlazanyoni* (red-faced coly)
flies over a kraal or alights on it, it is said that lightning will strike
the homestead, but if the bird is killed or driven away the evil will be
averted (1960: 288).

Hammond-Tooke has mistakenly interpreted the name *indlazanyoni* as being a combination of **indlazi** (mousebird or coly) and *inyoni* (bird). The name is surely a variant of **indlazinyoni**, a well-attested name for the Bateleur, and based on the verb *dla* (eat) and the noun *izinyoni* (birds), giving the name the meaning 'what eats birds'. The Bateleur is also widely recognised as a bird associated with thunder and rain, whereas the Red-faced Mousebird has no such reputation.[39]

The Grey Crowned Crane, which Godfrey calls the Mahem (Xhosa *ihem*, Zulu **unohhemu**) is also a weather bird, presaging a heavy fall of rain when it calls daily. Godfrey likens it to the Scottish 'Teuchits' (peeweeps)[40] whose

---

37. Presumably 'properly' here, in the context of the following sentence, means rain falling in a non-violent way: persistent, soaking rain.
38. It is not clear why the chief chooses the hornbill to be carried off to "some precipice" while any other man must choose a place where there are no precipices.
39. Besides which, the Red-faced Mousebird is **umtshivovo**. The name **indlazi** only refers to the Speckled Mousebird.
40. Greenoak gives both 'teuchit' and 'peeweep' as Scottish folk names of the Northern Lapwing (*Vanellus vanellus*) (1997: 87).

annual return to their Perthshire breeding grounds also predicts heavy rainfalls, "the Teuchits' storm" (1941: 45).

While the peeweeps and teuchits of Perthshire are singing of the coming rain, in King William's Town in the Eastern Cape, the 'Dutch farmers' believe that the Rufous-naped Lark (Zulu **unongqwashi** and **ingqangendlela**) does the same. Godfrey cites Dr J. Brownlee of this town as saying that "among the Dutch, the sight of one of these larks sitting on a tree and singing was considered a sign of a storm" (1941: 45).

Biyela quotes Mapopa Ntonga (1994: 335) as saying that when boys go hunting birds, they are warned not to kill "the Diderick [Diederik] Cuckoo or rain bird" (2009: 41), and Nyazini in *JSA* Vol. V says that the '*udoye*' is also a rain bird.[41] If it comes to a homestead, lightning will surely come, as is the case with the **insingizi** (Webb and Wright 2001: 189).

As opposed to all these 'rain birds', the coming of the White Stork, according to Godfrey is regarded as presaging drought (1941: 19).

Rain in bird names

Predicting rain also features in some bird names. The generic name for swifts in Zulu is **ihlolamvula**, which means exactly that – 'what predicts rain'. At the 2013–2017 Zulu bird name workshops, the participants allocated this name specifically to the African Black Swift (*Apus barbatus*), which meant that other swifts had to acquire species-specific names. It was decided at the time to keep within the semantic field of weather prediction, and the name **umvuliyeza** (the rain is coming) was coined for the Little Swift (*Apus affinis*) and **inhlolazulu** (what predicts the weather) for the Alpine Swift (*Tachmarptis melba*). Both name choices clearly reflect the deeply held cultural belief that swifts are predictors of rainy weather.

Moving further afield, to Tanzania, we find Moreau recording the name *langavura* for the Jacobin Cuckoo (1940: 55). The name is in the Kami dialect and literally means 'call-rain', with Moreau explaining that this bird is "especially noisy at the break of the rains". The Red-chested Cuckoo is also a 'rain bird' for in the Bindei dialect it has the name *semchocho*, which Moreau translates as 'deluge fellow', saying that the bird is thought to call especially

---

41. This name is unknown elsewhere as a bird name. Stuart adds in a side note that it is a crane, and also gives the name **intungunono**, which is the Secretarybird.

in very heavy rain (*chocho*) (1940: 56). In Madagascar, the Rev. Sibree tells us of *Rallus gularis* [*caerulescens*] ("one of the ten birds in *Rallidae*"):[42]

> According to M. Pollens account, this rail is regarded with great respect by the north-western Sàkalàva, as they believe it brings them rain in very dry weather, so they will not kill it (1892: 104).

This belief is surely the basis of one of the bird's names: *akòholàhundràno*, meaning 'water-cock'. And from Zambia, Torrend records the names *kowa* and *kilwa*, both meaning 'let it [rain] fall' for the "Rain-bird or black and grey Cuckoo" (1931: 62). The bird is also known as *mu-kwe wa Leza* (God's son-in-law), which may or may not have to do with rainfall.[43]

The notion of 'rain birds' is a long-established colonial belief as well. Burchell's Coucal (**ufukwe**) is frequently called the Rainbird among English-speakers of KwaZulu-Natal (KZN),[44] and a number of writers have associated this notion with the liquid, bubbling sound of the bird's call, like water pouring out of a bottle.

The famous South African artist Barbara Tyrell, known for her depictions of the traditional attire of various South African ethnic groups, captured this 'liquid call' in verse:

> I have heard the fukwe in the cane
> First, their beseeching call then the
>   pattering rain
> And stood; then that liquid call again
> Quietly now, appeared, in the dripping rain (Tyrell 1945: 34, cited in Jenkins 2012: 5).

Perhaps less well known as a 'colonial rain bird' is the "Rufous-headed Lark *Mirafra rufipiles*",[45] of which Stark and Sclater (1900: 218) write:

---

42. *Rallidae*: the family of rails.
43. Cf. comments by Berglund quoted in the following section on the relationship between the Southern Ground Hornbill, rainmaking, and the 'Lord of the Sky'.
44. And, indeed, often called 'ufookwe'.
45. It has not been possible to track down this lark species, but surely it is the Rufous-naped Lark (*Mirafra africana* formerly *M. rufipilea*).

Mr. T. Ayres writes in the *Ibis* for 1880 "This bird is called amongst farmers the 'Rain-Bird', as they consider it a sign of rain that it rises during the breeding season for some yards in the air with a fluttering flight, descending with a loud '*whew*' when this action is repeated, but it is very certain that the same habit prevails during a succession of dry weather; in fact, it is one way in which the cock bird pays its addresses to the hen, and weather has very little to do with it."

Quite a put-down of the farmers' belief by Mr Ayres, there.

Edward's Armstrong's classic 1958 *The Folklore of Birds* has much to say about 'rain birds'. In Britain, birds such as the cuckoo and the green woodpecker are known by folk names such as 'rain-crow', 'rain bird', 'rain-pie'[46] and 'rain-fowl'. The green woodpecker, according to Armstrong is known throughout Europe as a rain bird and has names in French such as *pic de la pluie* (woodpecker of the rain) and *pleu-pleu* (rain-rain); in German such as *Giessvogel* (rain bird) and *regen vogel* (rain bird), and in Danish – *regnkrake* (rain-crow) (1958: 94). Hand in hand with these rain birds is the snipe, considered to be a 'thunder bird' because of the drumming sound made by the air rushing through the bird's feathers. Translations of its name in a variety of European languages include 'weather bird', 'storm bird' and 'thunder-goat' (or 'thunder's she-goat').

Birds used for rainmaking

Godfrey says of the Bokmakierie (**inkovu**) that "[t]he Natives have a belief that if a man kills [the bokmakierie] and puts it in water, rain will fall" (1941: 111). Godfrey is the only writer to record the Bokmakierie in this role, but a number of writers give details of the Hamerkop and the Southern Ground Hornbill used in this way. Layard is the earliest writer to record the Southern Ground Hornbill as used in rainmaking, and he quotes a certain Mr T.C. Rickard of East London and a Mr H. Bowker of an unspecified location as giving him the following information:

Mr Rickard writes:– The [Xhosas] have a superstition that if one of these birds is killed, it will rain for a long time. I am told that in times of drought, it is their custom to take one alive, tie a stone to it, and then

---

46. 'Pie' here is a variant of 'pied' and is also seen in the name 'magpie'.

throw it into a 'vley'; after this rain is supposed to follow. They avoid using the water in which this ceremony has been performed. When I had one in my yard we were getting a great deal of rain, and I often heard the [people] blame me for keeping the bird a prisoner. [Xhosa] name 'Insigees' (Layard 1867: 122).

Mr Bowker adds a unique (if not necessarily entirely believable) interpretation of the practice:

There are many superstitions connected with the 'Brom-vogel', the bird is held sacred by the [Xhosas], and is only killed in time of severe drought, when one is killed by order of the 'rain doctor', and its body thrown into a pool in a river. The idea is, that the bird has so offensive a smell that it will 'make the water sick', and that the only way of getting rid of this, is to wash it away to the sea, which can only be done by heavy rains.

Mr Bowker's rather unlikely interpretation of these traditional beliefs and practices is taken even further by Kidd (1904: 114):

The Pondos kill certain birds with bright red feathers on their breasts [i.e. the Southern Ground Hornbills]; they hunt for these in times of drought, and then when they have caught the birds they throw them into the river after they have killed them, a practice which is probably a remnant of some old custom of offering a sacrifice to an ancestral spirit living in the water.

Berglund has a far more reliable interpretation of the activities and beliefs associated with both Southern Ground Hornbills and Bateleurs in relation to rain, in that he quotes the diviners and *izinyanga zezulu* he has interviewed (1976: 57ff.). Moreover, in writing more than a hundred years after Layard, he shows how lasting these beliefs are. Rather than quoting his extensive and detailed description of the rainmaking practices, what he has to say is summarised here.

When there is drought, rainmakers catch a Southern Ground Hornbill. It must be a multi-coloured bird so that the sky sees that they are taking away "the friend of the rainbow". Once the bird has been killed and lowered into a deep pool, according to the diviners and sky-doctor quoted by Berglund, the sky will take note and:

The sky will weep because of its [bird] with the beautiful colours. It (the sky) says then, 'Are there no colours (i.e. is there no rainbow) in that land?' Then it remembers that there are no colours in that land which has lacked colours, there being a great drought in it. So it sends rain to the land which has lacked colours, so that the killing of its [bird] be not repeated. To the sky the death of this bird is painful (1976: 57).

The reasons given to Berglund for the use of the Southern Ground Hornbill as a rainmaker are twofold and subtly different. Firstly, the bird, with its beautiful colours,[47] acts as an aide-memoire to the sky, reminding it that it has not sent a rainbow to the drought-stricken for a long time; secondly, the death of the bird makes the sky feel guilty for not remembering. Berglund adds, again obviously translating from his sources, that the bird and the rainbow are the "beautiful ladies" of the Lord-of-the-Sky and "his children, whom he loves".

Of the Bateleur, Berglund begins with information that we have already noted in previous sections of this chapter, such as the sound of its wings sounding like thunder, and its droppings falling on a person being an evil omen (if not actually death, as recorded by others) (1976: 58). It is interesting to note, in parenthesis, that the droppings of the Southern Ground Hornbill, even when falling on a person or a homestead, are a sign of good luck. Berglund continues that a Bateleur may be used in the place of a Southern Ground Hornbill in rainmaking if a hornbill cannot be found, but the subsequent rains will not be the persistent soaking rains wanted after a drought but heavy downpours. To counteract this, the body of the Bateleur should not be lowered into a deep pool as is the case with the hornbill, but into a shallow pool ("If the pool is shallow, then the rain will not be so violent"). Nor is the Bateleur killed before being placed in the water. Instead it is drowned in the pool, because then (in the voice of one of the sources):

[T]he breath (*umoya*) goes upwards, bubbling through the water, and reaches the sky. There it reports that it has suffered at a certain place. Then the sky weeps, mourning this strong bird known for fearlessness (1976: 58).

---

47. The Southern Ground Hornbill is essentially a black bird, with white on its wings, which are more visible in flight, and naked red skin on neck and around the eyes. It is certainly not 'rainbow-coloured'.

10.3.2   The heralds: birds of spring and summer

C.E. Hare, writing on bird lore generally, says:

> Movement of birds played a large part in the Greek calendar. Ships were beached for the winter when the cranes migrated to Africa, and the autumn sowing was then started. Sheep-shearing began when the kite returned in the spring, and the swallow's arrival was the signal to leave off winter clothes (1952: 20).

As in Greece in earlier days, so also in traditional Zulu society, the movement and behaviour of certain species of birds marked certain parts of the year. Hare specifically mentions 'the kite', and the Yellow-billed Kite (**unhloyile, ukholo**) was a major actor in the Zulu calendar, as shown in the section that follows.

Harbingers of spring

The Zulu lunar year begins in late July when the Pleiades first appear over the horizon.[48] The first lunar month of the year is uNcwaba (the glossy, fresh, clean month). It has the alternative name unNhloyile, which is also the name of the Yellow-billed Kite, a clear reference to the fact that this bird reappears from its winter migration at just this time of the year. I have previously noted that this is also the time when the Snake Lily (*Scadoxus puniceus*), dormant and hidden during winter, first appears above the ground.[49] The link between season of the year, bird and plant is seen in the plant's Zulu name *idumbe likanhloyile* (tuber of the Yellow-billed Kite). It is this kite that Ndukwana kaMbengwana in *JSA* Vol. III is referring to when James Stuart records him saying that the appearance of a hawk (**ukolo**)[50] marks the start of the new year (Webb and Wright 1982: 167). Ornithologists may not necessarily agree with the further details that Stuart records of Ndukwana's testimony:

---

48. See Koopman (2002: 250ff.), where there is a great deal more detail about the Zulu lunar calendar and the names of the lunar months.
49. Koopman (2015: Plate 1, facing p. 142).
50. The Yellow-billed Kite has a number of Zulu names. **Unhloyile** and **ukholo** are just two of them.

As regards the *ukolo hawk* which appeared when the *Great uluTuli was drawing to a close*,[51] this bird after appearing, would vanish again, go to the trees, live on food previously accumulated[52] and moult its feathers, and come out again once more when the grass fields were being burnt, ready to *catch the grasshoppers* which flew up off the ground in the smoke, and this would be in the month of *uNcwaba*. As soon as the *ukolo hawk* had actually been seen there would no longer be any dispute (Webb and Wright 1982: 167).

Among the Xhosa, Godfrey, quoting the early Xhosa lexicographer John Bennie, says that it is the arrival of the African Hoopoe in spring that "informs the Native that winter is past" (1941: 64).[53] On the other hand, Samuelson (1929: 406) refers to "**Inongqwatshi**, the South African lark that ushers in the spring by its sweet notes . . .". In his dictionary, an earlier work, the bird was **uMangqwatshi**,[54] and the entry reads:

> [T]he South African skylark, which appears as the harbinger of spring; it sits on anthills and tops of bushes and sings beautiful notes; it also soars towards heaven, warbling sweetly and beautifully; its first notes are a sure sign that spring has arrived; it is brown in colour and about as large as an English lark; it is also called **unongqwatshi** or **uhuye**; these two words are much more generally used than *umangqwatshi* (1923: 271).[55]

In Western bird traditions, the cuckoo is the quintessential bird of spring, so much so that in England a tradition arose of who could be the first to write

---

51. Webb and Wright have chosen as a matter of editorial policy to put in italics all of Stuart's notes that he had originally written in Zulu. The month *uluThuli oluKhulu* (the Great uluThuli) is the last month of the lunar year, and accordingly marks the end of winter.
52. See the section 'Where birds winter' on page 336.
53. Chittenden, Davies and Weiersbye describe this bird as having "residential, nomadic and migratory populations" (2016: 314). Perhaps in earlier times, they were all migratory, otherwise they would not have been very reliable seasonal markers.
54. Doke and Vilakazi recognise the forms **ungqwashi**, **umangqwashi**, and **usangqwashi** (1958: 564). The form **unongqwashi** is just as likely. These are all used of the Rufous-naped Lark.
55. Which rather begs the question of why he chose to enter the gloss under the lesser-used term.

to the venerable newspaper *The Times* to announce the earliest call heard. In Britain there is only 'the cuckoo'; in South Africa there are a number of cuckoos, and it is **uphezukomkhono**, the Red-chested Cuckoo, which plays the major role as a herald of spring and the start of the ploughing season. The name **uphezukomkhono** (on top of the arm) has been variously explained, with two of Godfrey's explanations marking time of day and time of year respectively (1941: 51). The first explanation (which smacks heavily of folk etymology) is that Xhosas customarily sleep with their head on their arm, and so when the cuckoo calls at dawn, is saying "You that are sleeping on your arm, get up! It's time to be in your garden!" The other explanation is the better-known one, and is the one most often given by Zulu informants, namely that the cuckoo is saying, "Put your hoe on your arm *(=* shoulder) – it is time to cultivate the fields".

An interesting extension of the role of the Red-chested Cuckoo and the explanation of its name is given in Koopman when he mentions Zulu schools with the name uPhezukomkhono (2002: 181). This is explained as the name is linked to an exhortation to hard work. However, further (unpublished) research by Koopman into school names has shown a consistent use of imagery that sees school and formal education as 'planting the seeds for future success', with names such as iMbewenhle (good seed) being found. A name like uPhezukomkhono for a school can then be seen as representing a time for planting the seeds of education.

Links between bird calls and cultivating and planting seeds can be found elsewhere: Wilson (2012a: 49) notes that in Malawi, the Chewa name *chinsoso*, given to both the Red-chested Cuckoo and the Emerald Cuckoo, means effectively 'get your garden cleaned'. Wilson notes that these birds "start calling in October, shortly before the planting season".

Harbingers of summer

The quintessential bird of summer is the **inkanku** cuckoo (see Plate 49), although there is a lack of clarity as to which species of cuckoo this actually is. Colenso is one of the earliest writers to refer to this bird by its Zulu name **inkanku** (1884: 379):

> **iNkanku**: black-and-white bush-bird which moults in winter, and whose piercing voice is heard in summer at night, announcing the time for sowing, corresponding to the cuckoo of England.
>   Ex. *sokupakati kwonyaka, inkanku seipelile is'ipendule*, it's now midsummer, the inkanku is now full-throated, has changed its voice.

Note that Colenso does not identify the bird as a species, simply referring to it as a 'black-and-white bush-bird". Doke and Vilakazi say "species of cuckoo, *Coccystes cafer* and *C. Jacobinus*" (1958: 381). They illustrate the use of the word with "*Inkanku isiwathathe amacimbi* (The cuckoo has already taken the caterpillars, i.e. it is early October)". Now, the two cuckoos Doke and Vilakazi refer to are the Levaillant's Cuckoo (*Clamator levaillantii*)[56] and the Jacobin Cuckoo (*Clamator jacobinus*), but the question is: which of these two species is the 'real' **inkanku**? Woodward and Woodward are sure it is "LeVaillant's Cuckoo *Coccystes caffer*" and say:

> Dr. Colenso says when its voice is first heard it announces to the natives that the time for planting has arrived, like the cuckoo in England (1899: 114).

On the other hand, in the 2013–2017 Zulu bird name workshops, the name **inkanku** was confirmed as a long-standing and well-known name for the Jacobin Cuckoo. Levaillant's Cuckoo did not even come up for discussion, having only a very limited distribution in the far north-eastern corner of Zululand. Chittenden, Davies and Weiersbye describe both species of cuckoo as breeding migrants arriving in South Africa in October (2016: 272). This ties in well with Doke and Vilakazi's suggestion above that the 'inkanku cuckoo' returns in early October. The Zulu month that runs from the new moon in September to the new moon in October is known as uMfumfu, but it has the alternative name uNkanku, as Bryant has pointed out (1905: 293): "**inkanku**: bird whose appearance gives a name to a month (*uNkanku*) also . . . announces the time for sowing." Bryant incidentally refers to this bird as Le Vaillant's Cuckooo, but as a careful comparison between Woodward and Woodward's 1899 *Natal Birds* and Bryant's 1905 dictionary shows that he took all his English vernacular and scientific names of birds from Woodward and Woodward, and this cannot therefore be regarded as an independent confirmation that the name **inkanku** refers to Le Vaillant's Cuckoo.

Whatever the identity of the summer-heralding **inkanku**, it is clear that Samuelson greatly favoured this bird, for he makes no less than four references to it, twice in his dictionary, entering it first under 'K' and then again under 'N' (1923), and twice in *Long, Long Ago* (1929):

---

56. Formerly known as the Striped Crested Cuckoo.

inKanku: a kind of jay; a bird marked brownish-black and white, with crest, which announces that summer has set in; it lives on caterpillars and such-like, which become plentiful at this period; it has a loud, cheerful, ringing call (1923: 211).

iNkanku: a white and black bird which utters a piercing call at commencement of summer, thus announcing it (1923: 321).

Inkanku, the black and white coloured Jay bird, which ushers in early Summer by a loud ringing call, while looking for caterpillars in trees (1929: 406).

At set intervals the **Inkanku**, the black and white Jaybird . . . would suddenly be heard calling out in a high and gleeful manner, "Kle-kle-kle," as if announcing his joy over the Winter having passed and the early Summer having set in bringing with it its favourite morsel . . . The **Inkanku** was the certain harbinger of early Summer, while the **Uhuye**, alias **Unongqwatshi**, was the certain harbinger of Spring (1929: 419).

We note that Samuelson, like Colenso writing more than 40 years before him, does not help us identify which of the two species of cuckoo this is. We have instead "a kind of jay", a "white and black bird", "the black and white coloured Jay bird" and the "black and white Jaybird". At least he is quite clear that this bird announces the summer, and that it does so in a loud voice.

The call of the **inkanku**, as we have seen above, is supposed to announce the time for sowing. Godfrey, giving the Xhosa name *ilunga legwaba*,[57] says more specifically that the call of this bird announces the time for planting millet (1941: 56). He identifies his summer herald as the Black Crested-cuckoo [*sic*], which rather deepens the mystery of the summer cuckoo. McLachlan and Liversidge, however, in the index to *Roberts* second edition, say that this is an earlier name for the Jacobin Cuckoo. Godfrey, incidentally, says that the name *ilunga legwaba* is not used universally in the Eastern Cape. In Pondoland, he

---

57. A curious name, when seen from the viewpoint of Zulu bird naming: **ilunga** is one of the names of the Southern Fiscal (formerly Fiscal Shrike), and /gwaba/ is the main core of names like **igwababa** and **igwababane**, which are names for crows and ravens. The Xhosa name makes this cuckoo a 'corvid fiscal'.

tells us, "[t]he name in use there . . . is *inkanku*, which is also the Zulu name for the species."

10.3.3   Announcers of dawn and dusk

The cast of avian characters in this section includes larks (or pipits), night-hawks, goat-suckers, bush-pheasants, cuckoos, grass-birds and warblers.[58] There is an English expression though – 'up with the lark' – which is perfectly relevant to a discussion of birds in Zulu bird lore that mark the start of the day. Mynott reminds us that when Juliet and Romeo hear a bird call after a night-long tryst, Romeo says, "It was the lark, the herald of the morn . . .", showing that this notion of the lark as a 'day-breaker' goes back at least to Shakespearian times (2009: 37).[59]

Up with the lark!

Colenso is the earliest Zulu lexicographer to use the name **umngcelu**, and as is often the case, he is unsure of exactly what bird it is. His full entry for this word reads:

> **umNgcelu**: bird frequenting new grass; another bird (a sort of wagtail), which is one of the earliest to chatter in the morning = *umvemve*; generally any very early bird.
>
>   Ex. *ngiya'kuvuka imingcelu ingakakali*, I shall be up before the *imingcelu* begin to chirp = very early (1884: 363).

Bryant also refers to the name **umngcelu** used in the Zulu saying *Ngiyakuvuka imingcelu ingakakali*, which he also translates in the same way as Colenso (1905: 420). Bryant does not identify which bird this is either,[60] saying only

---

58. Not all of these names are in common use as English vernacular names today. The birder of the twenty-first century would likely be hard-pressed to identify a 'bush-pheasant'.

59. The similar role of the domestic cockerel goes back even earlier. In line 350 of his *Parlement of Foules* ('Parliament of Fowls', i.e. birds) Geoffrey Chaucer describes the cockerel as "The kok, that orloges is of thorpes lyte" (i.e. that is the clock (horloges) marking the [first] light (lyte) in a village (thorpe) (Robinson 1957: 314).

60. Woodward and Woodward have no Zulu names for the very few larks and pipits in their 1899 *Natal Birds*, so it is not surprising that Bryant is unable to identify this bird.

"certain veldt bird, frequenting new grass". Doke and Vilakazi identify the bird as "the road lark, *Anthus richardi*" (1958: 551). The 2013–2017 Zulu bird name workshops assigned the name to the Long-billed Pipit (*Anthus similis*). This bird, previously known as Nicholson's Pipit, is also referred to by the longer form of its Zulu name, namely **umngcelekeshu**.

At least one Zulu day-breaker bird has also become a colonial nickname: Carl Faye gives **uMngcelu** as the Zulu name of colonial magistrate G. Walker Wilson, and says it means 'the early riser' (1923: 25).

Earlier, in discussing birds that marked the seasons, an explanation was given of the name **uphezukomkhono** for the Red-chested Cuckoo, which among the Xhosa functions as a 'wake-up call'. This role of the bird is also echoed in traditional Xhosa beliefs about the Red-capped Lark, with an informant telling Godfrey: "It is the bird by which we are assured of the dawn. We hear it call and understand it is time to get up" (1941: 72). The Red-capped Lark (*Calandrella cinerea*) is widely distributed in KZN, where its Zulu name is **umntoli**. None of the sources that have contributed to this chapter have anything to say about the **umntoli** in Zulu bird lore, so either this belief is restricted to the Eastern Cape, or it is yet to be recorded among Zulu speakers.

Samuelson adds the nightjar into the mix of the dawn chorus. Introducing **uzavolo** as a name for a "small night-hawk, a goat-sucker",[61] he goes on to say:

N.B. – This is a small night-hawk living on beetles and such-like; it is generally seen just at dusk and a little after, and just before dawn, in the morning; it is very fond of frequenting roads; at daybreak it keeps out of sight; it utters a sweet gurgling, pathetic call, off and on, which natives have put into the words, viz:– *zavolo, zavolo, sengel' abantabakho*, which means *zavolo, zavolo*, milk for your children. One may be sure the dawn is breaking when one first hears the call of this bird (1923: 526).

In *Long, Long Ago*, Samuelson has much to say about various species of birds. Here is another contribution to the birds that 'rise with the lark':

---

61. The name is used generically for all five of the species of nightjar that occur in KZN.

Just after dawn one would be greeted, and often startled by "Kwehle-kwehle-kwehlehle: uttered by the speckled bush-pheasant,[62] suddenly starting up and moving from place to place, either alerted by an enemy or because it felt it should join in with the others in greeting the morn (1929: 419).

A few lines later he tracks more bird calls as the morning gets underway, and offers another role for the Red-chested Cuckoo that has already appeared in this chapter as a herald of spring:

Then as the sun gets into the sky, a pair of the birds called **Phezukwomkhono** entered into the vocal competition and uttered a call from which the name is derived, with a sweet, sad, ringing sound, something like the sound of the metal triangle rung in some distant land.

His interpretation of the call of this cuckoo is unusual, to say the least.[63]

Afrikaans folk names for birds can also refer to their role as heralds of the dawn. In his article on such names for birds in the Riversdale area in the Western Cape, John Muir (1940: 4) gives *dagbreker* (lit. 'day-breaker') as a name for the Familiar Chat (*Emarginata familiaris*). Unlike larks that rise higher and higher in the sky as they sing,[64] Muir's *dagbreker* sits on the roof and sings, whence its other folk name in Riversdale – *andriesdak* (lit. 'Andries's roof'). There is, however, another interpretation of *andriesdak*: Muir says of this name for the same bird that "the last syllable is an interesting substitute for '*dag*'". If this is the case, then the name belongs to the same group of names in Afrikaans as *jandiederik*, *klaasskaapwagter* and *groenpiet*, where personal names are used in the same way as in English names like Jackdaw and Magpie. In this interpretation, the name *andriesdak* for the Familiar Chat would be something along the lines of an English name such as Andy-the-dawn or perhaps simply Jackdawn (instead of Jackdaw).

---

62. From the imitation '*kwehle-kwehle-kwehle*' it is clear that Samuelson's "speckled bush-pheasant" is the bird **isikhwehle**, identified by Bryant as the "Natal Bush Partridge (*Francolinus Natalensis*)" (1905: 339), but now officially known as the Natal Spurfowl (*Pternistis natalensis*) (Chittenden, Davies and Weiersbye 2016: 48). In between it was known as the Natal Francolin.

63. I can hear the **uphezukomkhono** calling outside my window as I write this in November 2017, and there is no evidence of a metal triangle, distant land or not.

64. As seen in English poetry about larks in phrases like "mount up on high" and "sings hymns at Heaven's gate".

Returning again to the Congo, we find Hendrickx recording the name *kabikakutshe* for Heuglin's Robin (*Cossypha heuglini*)[65]. The name is derived from the verb *kubika* 'to cry', 'to call' and the noun *butshe* 'dawn' (1944: 197).

The coming of dusk

Samuelson (1929: 419) brings us towards the end of day in noting that "when the sun was leaving long shadows behind", the **intendele** (Red-winged Partridge) can be heard "with its cheerful 'kekelu-kekelu-kekelu'". And further south, in the Eastern Cape, Godfrey tells us that the Cape Grass-bird (Xhosa *udwetya* and Xulu *uvuze*) and the 'Drakensberg Wailing Warbler' finish off the day. Godfrey says that these birds call towards sunset so that even if it is overcast the boys know when to bring the cattle home (1941: 91, 96).

Having enlisted Samuelson to help start the day with his rendition of the call of the Red-chested Cuckoo as a "sweet, sad, ringing voice, something like the sound of the metal triangle rung in some distant land", it seems only fair to allow him to bring the day to a close as well, and this he does with the same lofty prose, using the Guineafowl (**impangele**) as his Muse:

> The **Impangele**, guinea-fowl . . . with its call "kerri-kerri-kerri" took turns in giving melody in the surroundings, as the bright day was dying in the West, and the glamour of night was fast approaching; so soon as it got dusk they hung up their harps, flew up into trees and slept in the care of their Creator (1929: 419).

And on that high note, we leave the announcers of dawn and dusk, and move on to the feathers of Chapter 11.

---

65. Now the White-browed Robin-Chat (but still *Cossypha heuglini*).

11 Feathers, food and fancies

In my 2015 book on Zulu plant names, in addition to a lengthy chapter on the medicinal use of plants, I devoted an entire chapter to the practical uses of plants in Zulu culture (Koopman 2015: 221–38). These range from firewood to plants used in construction and the making of utensils and implements, from providing food and drink to providing fibres, dyes and cosmetics. There is remarkably little overlap between the practical use of plants and the practical use of birds in traditional Zulu society. Plants contain a variety of chemical substances that make them not only useful in medicines, but also as tanning agents, insecticides, soaps, perfumes, and so on. Plants contain the sort of fibres useful for plaiting, weaving and twisting into yarns and twines; birds do not. On the other hand, birds have features that plants do not possess: they fly (most of them), they sing and make other noises, they nest and produce young, they lay eggs and they have feathers. It is, as one might expect, features like this that figure largely in the practical uses of birds in a traditional culture. Indeed, about the only overlap between plants and birds, when it comes to their practical use, is as a food resource, and even here, as is noted below, birds figure in the most minor way on the traditional Zulu menu, whereas plants constitute by far the bulk of what is eaten in the traditional Zulu diet.

In the following section, the major emphasis is on the use of bird plumage, partially as the result of the rich material contained in *The James Stuart Archive* (*JSA*), relating to the use of feathers as military insignia in the Zulu regiments of bygone years.

The cover illustration of Volume V
of *The James Stuart Archive*.

11.1   USE OF FEATHERS FOR DECORATION AND ORNAMENT

In the six volumes of *JSA*, mention is made of the use of feathers of ten different birds, some clearly identifiable species, such as the Ostrich and the Blue Crane; others are more generic references, for example to eagle and dove. In this section, we will discuss each of these ten birds in turn, indicating who (kings, for example, or chiefs, or particular regiments) characteristically wore each bird's feathers, adding details on how they were worn. At the end of this sub-section, the sources of the feathers will be discussed.

11.1.1   The Blue Crane (*Grus paradisea*), Z. indwa (also frequently indwe)

A single Blue Crane feather is indelibly associated with the Zulu King Shaka kaSenzangakhona, probably from J. Saunders King's drawing of the king, found in numerous publications (see Plates 57 and 58). The proliferation of this picture has led to a widely held belief that only Shaka, or only Zulu kings, could wear this bird's feather, and only a single feather should be worn at a time. However, sources in the literature make it clear that this was not so.

Godfrey says of the Blue Crane that "it was in the olden days distinctively a warrior's bird, whose feathers adorned the heads of the fighting men during drill or war" (1941: 44). Bryant says:

This feather was presented to a full-grown man by the king and was a preliminary sign that the recipient was about to be called to the honour of wearing the headring (1905: 123).

R.C.A. Samuelson emphasised that the feathers were a specific gift of the king:

This bird was one of the Zulu royal birds whose plumes were used by those to whom they [i.e. the feathers] had been specially gifted by the king (1923: 89).

Henry Francis Fynn recalls Shaka wearing a Blue Crane feather when the two first met in 1824, and recorded the following in his diary:

Round his forehead he wore a [headband] of otter skin with a feather of a crane erect in front, fully two feet long, and a wreath of scarlet feathers, formerly worn, only, by men of high rank . . .

Round the ring on the head, were a dozen tastefully arranged bunches of the loury feathers, neatly tied to thorns stuck in the hair (1969: 74).

JSA makes no reference of Shaka wearing Blue Crane feathers. In fact the single reference to Shaka and feather ornamentation in the six volumes simply confirms that "Tshaka . . . wore a bunch of loury feathers" (Vol. III: 72).[1] However, Cetshwayo kaMpande is mentioned: "Cetshwayo . . . had on his head a band of otter skin, with tassels of blue monkey skin, and a crane feather" (Vol. II: 223) and there is an interesting story about Dingiswayo kaJobe, king of the Mthethwa clan:

[Dingiswayo] ordered his war finery to be brought out. His crane feather was brought . . . While he was busy, a number of locusts suddenly settled on the feather which he had put on. The *izinduna* cried, 'Hau, Nkosi, what are those things on the feather?' Others exclaimed, 'They are locusts' . . . The feather came loose and fell on the ground . . . (Vol. II: 186).

---

1. The use of 'loury' or turaco, feathers is discussed below.

On the other hand, Delegorgue is not referring to a king but to any "fully dressed warrior" when he says:

> At the back of the head stands a Numidian Crane's feather,[2] its slender tip waving in the breeze and trembling as if from impatience, while behind it, attached to the back of the Cafre crown [i.e. the headring] flutters a bunch of feathers of many colours ([1847] 1997: 115).

According to various sources recorded in *JSA*, the following regiments wore crane feathers as part of their military identification: the amaMboza and Ndhlondhlo (Vol. II: 223);[3] the Tulwana and Dhlokwe (Vol. III: 318); the Ndabakawombe, Bulawayo, Siklebe, Dukuza, Ngwegwe and Mbelebele (Vol. IV: 119); and the iziMpohlo (Vol. VI: 291). Another source echoes what Bryant is quoted above as saying: "Crane feathers were distributed to a regiment that had put on the headring" (Vol. IV: 311). Another source gives an account of how the 'great men of Mpande', immediately prior to an encounter with the Boers, exchanged clothing with commoners (Vol. VI: 131):

> [At the time of the Boers coming to Maqongqo]:
> A man who was there at Maqongqo, one of Mpande's men – his name was Konjwayo of the Embo people – said that the next day the great men of Mpande's side took their headbands (*imiqele*) and their *amabeqe* of monkey-skin, together with their crane feathers, and made their ordinary men (*abafokazana*) put them on (*qilisa*) . . . For their part they took the garb of the ordinary men and put it on, and took the shields of the ordinary men.

The 'nodding blue plume' has become a stock cliché in novels about Zulu warriors in the nineteenth century. Here, for example, is George MacDonald

---

2. I am quoting here from Fleur Webb's translation of Delegorgue's 1847 French original so I do not know what term Delegorgue used for the Blue Crane. Collins English Dictionary (Hanks 1986: 1057) states that the term 'Numidian Crane' is a synonym for the Demoiselle Crane, a bird of the northern hemisphere that is very similar to the Blue Crane in appearance, suggesting that in Delegorgue's 1847 original he used the term *demoiselle* in referring to this bird.

3. Stuart's original spelling has been retained here and in the following names.

Fraser allowing his fictional character Flashman to describe an individual Zulu soldier in the aftermath of the 1879 Isandhlwana debacle:

> I turned and there . . . was a Zulu warrior. I could still tell you today every detail of him . . . the calf skin girdle, the white cow-tail garters, the ringed head with its nodding blue plume, even the little horn snuff-box swinging from his neck (2000: 282).

In one unusual case recorded in *JSA*, a single crane feather featured in the ceremonial dress of a woman, not a warrior who was about to receive the honour of the headring. After the assassination of Shaka in 1828, Mnkabayi, the daughter of Jama and aunt of Shaka, and a powerful political figure in the Zulu kingdom, helped to decide on who the next king of the Zulus would be. A woman, yes, but she "dressed like a man". The description of her outfit on this occasion is given in full, to give an idea of how a politically powerful man might be dressed at the time:

> [Mnkabayi] was summoned either to Dukuza or Bulawayo. She dressed as a man and came into the semi-circle . . . had on a skin skirt (*isidwaba*), not covered with black powder (*umsizi*) like others [worn by women], but left ruddy and simply covered with scent [??]. Over this she wore *umqubula*[4] of genet and blue monkey. She also had *imiklezo*,[5] i.e. *amatshoba*. When dressed, her identity could not be detected. She had a band of otter skin on her head, she had also *amabeqe* of monkey skin. She had also *imnyakanya*[6] of the widow-bird, also [a] long crane feather . . . (Vol. VI: 97).

It is interesting to note that the notion of the Blue Crane as distinctively a warrior's bird still has life in South Africa today. The box below gives information on an award made by the current ruling party, the African National Congress (ANC), for 'warriors' in the struggle against apartheid. Awards

---

4. *Umqubula*: "Dancing apparel, presented by the Zulu king to favourite warriors, consisting of three girdles of blue-monkey [*insimango*] tails" (Doke and Vilakazi 1958: 714).
5. *Imiklezo*: ox-tails worn suspended from the neck; *amashoba*: "bushy tails suspended from the upper arms and from the calves" (Doke and Vilakazi 1958: 743).
6. *Iminyakanya*: "Bunch of plumes of the long-tailed black finch [Long-tailed Widowbird]" (Doke and Vilakazi 1958: 617).

are still being made in the twenty-first century. Although the ANC website translates '*isitwalandwe*' as 'the one who wears the plumes of the rare bird', a more correct translation is '[the one] who carries (*thwala*) the blue crane (*indwe*) [feather]'.

Isitwalandwe/Seaparankwe Award

Isitwalandwe/Seaparankwe is the highest honour awarded by the people of South Africa, through the African National Congress, to those who have made an outstanding contribution and sacrifice to the liberation struggle.

Isitwalandwe, literally translated, means "the one who wears the plumes of the rare bird" and was traditionally bestowed only on the bravest warriors of the people, on those who distinguished themselves in the eyes of all the people for exceptional qualities of leadership and heroism.

Chief Albert Luthuli, Dr Yusuf Dadoo and Father Trevor Huddleston were the first to be proclaimed *Isitwalandwe/Seaparankwe*. This was at the Congress of the People in 1955. Since then many outstanding leaders have been honoured.

11.1.2   The Common Ostrich (*Struthio camelus*), Z. intshe

While ostrich feathers did not appear to feature as part of royal insignia, they were certainly commonly used as regimental insignia. Delegorgue noticed this on his travels and says, "I saw regiments which had adapted ostrich feathers as their insignia . . ." ([1847] 1997: 116). The six volumes of *James Stuart Archive* make seven references to ostrich feathers, and they were part of the dress of the Dhlokwe (Vol. II: 318), the izinGulube (Vol. III: 293), the Mxapo (Vol. III: 319), the imDhlenevu and iziNyosi (Vol. VI: 291) and the Ngobamakhosi (Vol. VI: 375) regiments.

This detail from a picture by G.G.F. Angas (Isaacs 1970) of a 'visiting Zulu youth' shows what may be an *iqholo*: a large ball of bunched up feathers worn on top of the head. This one would appear to have a single ox-tail protruding from it to go with the two tails under it.

Ostrich feathers could be worn in a number of ways: the simplest seems to have been the way in which the Dhloko regiment wore them – just "on the head . . . a circle of black and white ostrich feathers" (Vol. II: 242). They could be worn as *amaqholo* – a large bunch of feathers worn in a basket on top of the head, as the izinGulube (Vol. III: 293) and the imDhlenevu and iziNyosi (Vol. VI: 291) did. Or they could be worn in a very complex manner, as was the case with the Ngobamakhosi regiment:

> These had bunches (*amaqolo*) of grey and white ostrich feathers on the head, with one white ostrich feather stuck upright; this feather was known as *umbongo* and <was> snow-white (*ku mhlope kute ngqu!*). A man well off might stick in 2, 3 or more (Vol. VI: 375).

There is no indication that ostrich feathers were personally presented by the king, as was the case with crane feathers. They may, however, have been purchased with the king's funds, as indicated in:

> Crane feathers were distributed to a regiment that had put on the head-ring. They were bought with royal cattle from the amakhanda.[7] A bundle or two were given, but not sufficient. Ostrich feathers would

---

7. *Amakhanda*: military barracks.

be given out the same way – bought with the king's own cattle. Two small ostrich feathers were enough. No one could buy on his own account (Vol. IV: 311).[8]

The question arises as to where the ostrich feathers came from.

Source of ostrich feathers

It is generally accepted that ostriches have never occurred naturally in what is now KwaZulu-Natal (KZN). Hockey, Dean and Ryan say:

> Former distribution in s Africa not clear because of translocation of birds for domestication, but probably throughout semi-arid Namibia, Karoo and w S Africa, Botswana and sw Zimbabwe (2005: 60).

A source closer to the time when Zulu regiments were still wearing ostrich feathers, namely Layard, says:

> We would . . . remind our readers that the range of the wild Ostrich in South Africa is still the subject of much interest, and that the question of the number of species in the Northern regions of South Africa still remains unsettled . . .
> . . . The Ostrich is still found in most of the Karoo country, within the borders of the colony (1867: 791).

Hockey, Dean and Ryan's mention of an original population in south-western Zimbabwe is echoed in Layard's statement that "Mr T.E. Buckley states that it is still common in the Matabele country, but much hunted for the sake of its feathers" (1867: 792). Taking the previous comments on the earlier distribution of ostriches into account and the fact that they were hunted in 'Matabeleland', i.e. southwestern Zimbabwe as it is now, we can look at a further statement reported by Stuart:

---

8. However, the extract on page 375 of Vol. VI (shown above) states of ostrich feathers that "a man well off might stick in 2, 3 or more" suggesting that such feathers might indeed be purchased by individuals.

It was the Boers who used to trade in these feathers. Cetshwayo *shwaqa'd*,[9] i.e. purchased them in a wholesale manner (Vol. VI: 375).

The inference here is that the ostriches were being hunted in 'Matabeleland', either by the Boers themselves or by the amaNdebele (Matabele), and finding their way to the Zulu kingdom where the Boers used to exchange them for cattle. Given that at least six regiments were wearing ostrich feathers and a regiment may have had up to 500 men, and that many warriors were wearing at least three ostrich feathers each, this works out at 9 000 ostrich feathers, which presumably had to be renewed on a regular basis. And it was not only the warriors who were wearing ostrich feathers: an interesting anecdote comes from *JSA* (Vol. IV: 311) concerning a type of courting ritual, which makes it clear that even young men unattached to a regiment might be wearing ostrich feathers as well:

> During the *unomzimana* game, young men would come in a body and stand in a line where the girls were and a stick would be stuck in the ground. When a girl fancies a man she will grab the stick above where he is holding it, and thus signify that she has chosen him, or she may take whatever he may chance to be wearing, say a widow-bird feather, or ostrich feather, or a strip of [animal] skin.

Clearly the cattle-for-ostrich-feathers trade was extensive.

To end this section on the wearing of ostrich feathers, here is an interesting quote from Cocker, showing that ostrich feathers were worn in other parts of Africa, and not just as part of military regalia:

> For the Marakwet people of Western Kenya there is also a strong connection between ostrich and cattle, a traditional currency for measuring success. To hear the male bird making his strange booming note at dawn is a sign of good luck, predicting the ownership of a large herd. For the Nandi, near-neighbours to the Marakwet, the association was enshrined in an old riddle: What is the thing which, though so weak it is blown by the wind, is able to herd oxen? The answer was an

---

9. The verb *shwaqa* means 'collect together' in this context.

ostrich feather: in the tall vegetation his plumed headdress was often the only visible part of a man guarding his cattle (2013: 16).

11.1.3   Long-tailed Widowbird (*Euplectes progne*), Z. isakabuli

The famous late Victorian novelist Henry Rider Haggard's 1885 novel *King Solomon's Mines* traces the adventures of a party seeking these mines in the country 'Kukuanaland'. On first reaching the country of the Kukuanas and being met by regiments of warriors with "flashing spears and waving plumes", the narrator says:

> They wore upon their heads heavy black plumes of Sakaboola feathers, like those which adorned our guides ([1885] 1962: 82).

The Long-tailed Widowbird was universally known as the 'sakaboola' in Colonial Natal English, so much so that Clancey, in his 1964 *The Birds of Natal and Zululand* – a book that contains no Zulu bird names at all – gives 'Sakabula' as the English vernacular name for *Euplectes progne*.

The spelling of the bird's name was fluid, to say the least. Haggard wrote in his diary:

> [We came] . . . to a kraal where presently the [Zulu] dancers, male and female, arrived in all their finery . . . [difficult to persuade the dancers to let go of their umbrellas] . . . Umbrellas do not go well with shiny naked bodies, beads, saccaboola plumes and mottled oxhide shields (Haggard 2000: 177).

Delegorgue puts the plumes of the widow bird into their full ornamental and decorative context:

> The Amazoulus have their heads shaved . . . [the headring] is to hold the ceremonial plumes . . . the quills . . . the snuff boxes made from the cocoon of the bombyx . . . the plumes of war made from the tufts of touraco and widow bird tails . . . ([1847] 1997: 114).

There are six references to the use of the feathers of this bird in the six volumes of *JSA*. As we saw in the quote above, the long tail feathers of the male widowbird in breeding plumage were often worn by courting youths as part of their finery. They were also part of the regimental insignia of the Ngobamakhosi

(Vol. III: 318), Khandempemvu (Vol. IV: 83–4), and uDhlambedhlu regiments (Vol. VI: 291). Mnkabayi, as we saw above, added these feathers to her single crane feather when dressed as a man to impress the elders (Vol. VI: 97), and the Tulwana regiment likewise wore both widowbird and Blue Crane feathers.

> Tulwana: had Blue Crane feathers and feathers of the widow bird bound by a strip of otter skin – about 15 inches long; had headrings; had also brass armbands on right forearm, very many of them . . . (Vol. III: 318).

For Colenso, this bird was the quintessential provider of feathers for the Zulu military. In his dictionary, he glosses the word *isakabuli* as "name of a bird whose feathers make the plumes of Zulu soldiers" (1884: 500).

These birds are still common residents in KZN and would have been in the days of Mnkabayi and the Tulwana regiment, so there would have been no need to establish trade links and use cattle from the Zulu royal herd to purchase the feathers. As far back as 1867, Layard was noting of this bird[10] that "the [Xhosa] children stretch bird-limed lines across the fields of millet and sorghum, and snare great numbers of the males by their tails becoming entangled in the lines" (1867: 191). As it is only the males in their breeding plumage that would be caught this way and it is only the males in their breeding plumage whose feathers were and are wanted for ornamentation, this would seem to be an ideal method of trapping them. And a great saving in cattle, too.

The females, being much more drab creatures (and not sporting the long tail feathers), are referred to by the generic name *izintaka* (finches).[11] The word *intaka* is used in a Zulu saying that combines bird, plumage and traditional custom: *intaka ibekelwe amazolo kaSibanibani* (lit. 'the little finch is put out in the dew at So-and-so's place'). Bryant explains that this expression means that the wedding dance is tomorrow, and the expression is derived from the custom of putting feather head-dresses (often made of male finch feathers) out in the dew to remove any creases (1905: 607). The expression may be used of any imminent event requiring ceremonial finery, not just a wedding.

---

10. Which he called the Orange-shouldered Bunting or Kaffrarian Grosbeak, offering the following synonyms of the scientific name: *Chera Progne*, *Vidua Phoenicopterus* and *Emberiza longicauda*.

11. The word *intaka* simply means 'bird' in Xhosa.

11.1.4 Purple-crested Turaco[12] (*Tauraco porphyreolophus*), Z. igwalagwala and mousebirds[13] (indlazi and umtshivovo)

There is conflicting material about the feathers of this bird. One of Stuart's sources is reported as saying that among the things tabooed by the Zulu kings were leopard skins and "igwalagwala – the Lory [*sic*] . . .", suggesting that these were reserved for royal wear. And both Shaka (Vol. III: 72) and Dingane are reported to have worn **igwalagwala** feathers, with the latter having "a bunch of red loury feathers stuck in his headring" (Vol. I: 37). A particular episode involving Shaka and Sirayo and Mgabi, both contestants for the chieftaincy of the Nyuswa clan, describes the different status of **igwalagwala** feathers and those of the **indlazi** (mousebird). Shaka has apparently asked both Sirayo and Mgabi to put on their finery for an *umkhosi* (an important ceremony). The testimony was given to Stuart by Socwatsha kaPapu on 4 April 1916:

> Now all this took place in the presence of the King. Sirayo was wearing feathers of the loury. Mgabi was wearing feathers of the mousebird. Tshaka said, 'Has there ever been a chief who wore mousebird feathers? The chief is the one wearing feathers of the loury'. For the king had told them to put on the finery they wore for the *umkosi*, and they had done so (Vol. VI: 105).

The point here, namely that a chief wears 'loury' feathers, while a commoner wears mousebird feathers is clear in evidence that Socwatsha kaPapu actually gave to Stuart twenty years previously, on 10 January 1897:

> Don't you see by the very ornaments and dress that Sirayo has on that he is indeed a chief? Has there ever been a chief who dressed in mousebird feathers, who wore mousebird feathers on the head? (Vol. VI: 3)

There is no indication elsewhere in *JSA* that mousebird feathers were worn by other people, but Bryant seems to imply that this was quite common when he says "**Indhlazi** [**Indlazi**]: Mouse-bird . . . whose long tail-feathers are used as

---

12. Previously known as the Purple-crested 'Lourie', with a number of spelling variations, as seen in the extracts from *JSA*.
13. Presumably Stuart's reference to mousebirds includes both species found in KZN: the Speckled Mousebird (*Colius striatus*) and the Red-faced Mousebird (*Urocolius indicus*).

a head-ornament" (1905: 100). It is clear, however, that no one who wanted to be called a chief would be seen dead wearing them.

The turaco feathers were not necessarily always restricted to royalty and chiefs. *JSA* records that both the amaChunu people and the abaThembu people of that time wore "a bunch of loury feathers on the head" (Vol. I: 315), the former with long loin covers, and the latter with short loin covers, so that they could still be told apart from a distance.

Local hunting seems to have accounted for the supply of turaco feathers: "Louries are got in all parts – one got them at Pasiwe and at Qudeni. They frequent forests and especially the high trees therein" (Vol. I: 322).

Turaco feathers are still seen regularly today as head ornamentation: the Swazi King Mswati II always wears them in his hair when he wears traditional dress, as does the Zulu King Zwelithini kaBhekuzulu. Veteran politician Mangosuthu Buthelezi, a prince of the Zulu royal family on his mother's side, also does so.

As we saw in Chapter 8 (page 198), the bright red colour of the wing feathers, normally only seen fully when the bird is in flight, has given rise to the Zulu idiomatic expression *ukumthwesa umuntu igwalagwala* meaning to give a person a blow on the head so as to draw blood (lit. 'to make a person carry a gwalagwala') (Bryant 1905: 212).

Samuelson has an interesting, if debatable, idea about the etymology of the word **igwalagwala**, as well as yet another spelling of 'lourie':

> **Igwalagwala**, the Laurie, so named because it is a very shy bird, whose feathers used to be worn by kings only, and later by nobles; Igwala—coward (1929: 406).

11.1.5 Paradise Flycatcher (*Terpsiphone viridis*), Z. u(lu)ve or inzwece

The Paradise Flycatcher (Zulu **uluve** and **inzwece**) has long tail feathers as well, and these are equally prized as ornamental feathers (Bryant 1905: 677). The bird's Zulu name **uve** (or **uluve**) suggests that feathers of this bird were probably worn by one of Zulu King Cetshwayo kaMpande's regiments with the name Uluve, but from the two brief mentions of this bird in *JSA* (Vol. III: 328 and IV: 320) it seems that this was the one species of bird whose feathers were truly reserved for the king. One statement says that the Uve regiment did not wear the feathers of its namesake (Vol. III: 328); the other says more specifically:

Wearing feathers of the paradise flycatcher is prohibited; its feathers are put on by the king alone. No one was made a presentation of these feathers, or given them (Vol. IV: 320).

The tail feathers of this bird have given rise to the saying *uve ludl' isisilo salo* (lit. 'the flycatcher eats off its own tail', as the bird is said to do when closely pressed by boys hunting it). The expression may be used of a person whose bad conduct reacts harmfully on himself, as a father ill-treating his own children (Bryant 1905: 677).

Although there is no record of the use of the feathers of the Lilac-breasted Roller (*Coracias caudatus*, Z. **ifefe**) being used by Zulu royalty, these beautiful feathers were only allowed to be worn by Mzilikazi kaMashobane, the king of the amaNdebele. McLachlan and Liversidge state:

The blue wings are particularly beautiful in flight, hence the popular name 'Blue Jay'. Mosilikatze, king of the Matabele, reserved the feathers of this bird for his exclusive use, and it is often known as Mosilikatze's Roller (1978: 296).[14]

11.1.6   The Kori Bustard (*Ardeotis kori*), Z. umngqithi

There is a single reference in *JSA* to the Kori Bustard:

The iziMpohlo [regiment] wore *iziqova*, rolled into a ball about 8 or 9 inches in diameter; they covered the head . . . They were given *amagubela* of crane feathers, worked along their spines, made soft so that the feathers hang over, and *imnqiti* (15 inches high bird, with grey feathers) (Vol. VI: 291).

Current distribution maps of the Kori Bustard (see Chittenden, Davies and Weiersbye 2016) show that the bird occurs nowhere near KZN. This may not have been true at the time of the iziMpohlo regiment – a regiment of Shaka – or the bustard feathers may have been traded as some of the other feathers were.

---

14. Austin Roberts enters this bird as "Lilac-breasted Roller, Mosilikatzi's Bird" (1951: 169).

11.1.7 The dove (ihobhe, ijuba), the eagle (ukhozi) and the vulture (inqe)

There are single references to each of these, and in each case it is a generic reference.

Lunguza kaMpukane of the abaThembu is recorded as saying:

> My father wore his black sheepskin loin cover, with the front part made of genet skin. He might stick in his [head]ring a little bunch of feathers, possibly of dove feathers. He was otherwise naked (Vol. I: 34).

The abaThembu, as we saw above, also often wore a bunch of turaco feathers on their heads, so Mpukane (Lunguza's father) might have been a commoner. He could also have been a chief, although Lunguza does not say so. But the outfit seems a little simple for a chief. It may be worth mentioning here that in his discussion of N. Sotho bird names, Louwrens mentions that young men of that language group spend much time hunting doves, as their feathers are greatly favoured by young girls for decorative purposes (2004: 103). No similar reference has been found for the Zulu people.

The single reference to eagle feathers concerns Cetshwayo kaMpande at the Ndondakusuka battle with his half-brother Mbuyazi (Vol. II: 243):[15] "C[etshwayo] had on a bunch of eagle feathers. He had not put on the headring at that time." The Blue Crane, as we noted earlier, was only awarded to warriors who had put on the headring or were about to put on the headring. Perhaps this applied to royal princes as well (Cetshwayo's father Mpande was still alive at the time) and Cetshwayo chose a 'regal' bird that did not have to be gifted by the king. The 'eagle' feathers might indeed have been from a Bateleur, which as we have seen is a bird of great symbolic power.

Three regiments are recorded as having worn vulture feathers as part of their insignia: the Ihlaba, the Imkulutshane and the iziNyosi – all wore *izidlodlo*[16] of vulture feathers, although the iziNyosi regiment was later directed to leave off wearing the *isidlodlo* of vulture feathers and only wear ostrich feathers (Vol. VI: 291).

---

15. Two sons of the ageing King Mpande kaSenzangakhona, Cetshwayo and his uSuthu faction, and Mbuyazi, with his iziGqoza faction, fought in 1852 on the Ndondakusuka Ridge above the uThukela River.
16. *Isidlodlo* (also *isidlukula*): a bunch of feathers worn on the top or the back of the head (Doke and Vilakazi 1958: 159).

11.1.8   General plumage (unspecified birds)

Two extracts from *JSA* are worth looking at, even they do not mention a specific type of bird.

The first is about the regimental finery of the uNdabenkulu regiment:

> The uNdabenkulu [regiment] wore *izimbenge* on the head like the peak of a cap, with *umcono* (about 9–10 inches high) and stuck into the *imbenge*, and on top would be *isidhlodhlo* (feathers of various birds) about as large as one's hand. This finery (*vunula*) was called *ubuyungu* because it was said to have been obtained from *abelungu* though no one seemed to know what and where the *abelungu* were (Vol. VI: 290).

An *imbenge* is a small open basket of the type used as a lid for a beer-pot. The word *umcono* is untraceable, so it is not clear what the members of this regiment were sticking into the baskets on their heads, but it is interesting that when the bunches of feathers of the different birds was added the final result was called an *ubuyungu*.[17] Perhaps the upturned basket looked like the hats of white people, as reported by those who had come into contact with them.

The second extract concerns the famous madman that Shaka promised his mentor Dingiswayo he would deal with. The story has been told over and over again in popular accounts of Shaka's youth. The interest here lies in new details of the plumage cladding the madman, and the simile used for the clashing of their shields, which links directly to the Bateleur as an omen of war, as discussed in the previous chapter:

> [The madman] had on his finery, in the form of bird plumes, sewn to long threads, hanging down over his whole body; he flourished his shield as he went . . .
>
> . . . Tshaka went forward and confronted the madman. Their shields clashed against each other like the clashing of bateleur eagles (Vol. III: 197).

---

17. This is the '*thefuya*' pronunciation of the Lala people, first mentioned in Chapter 8, where the 'l' sound is pronounced as a 'y', turning the abstract noun *ubu-lungu* ('European-ness' < *abe-lungu* 'white people', 'Europeans') into *ubu-yungu*.

11.1.9   A lexicon of feather terms

It is generally accepted that if a particular entity (for example, cattle, fish, boats, snow, certain crops) are culturally important in a particular society, then there will be a highly specialised lexicon of words devoted specifically to that entity.

Certainly this is the case for terms dealing with feathers and plumage used for ornamental and decorative purposes in Zulu dress, and the following are some of the words found in Bryant (1905), Samuelson (1923) and Doke and Vilakazi (1958):

isAla: Bunch of crow or other feathers worn on the back of the head by young men and boys when out courting, though originally only at royal festivities.

imBhangayiya: An ostrich plume given as a gift.

umBhongo: Ostrich plume.

isiDlodlo: Bunch of feathers worn on the top or back of the head.

isiDlukula: As above.

inDwa: The Blue Crane, *Tetrapteryx paradisea*. Zulu royal bird whose plumes could only be worn by permission of the king.

umGomba: Long tail plume of a bird, often worn as a head-dress.

iGubela: Plume, long decorative feather.

ubuLumbuza: Bunch of feathers worn dangling from the back of the neck.

uNkonkowane: Plume worn on the forehead.

iNtshe: 1. Ostrich; 2. Ostrich-feather plume.

umNyakanya: Bunch of plumes of the long-tailed black finch, worn by young warriors at the feast of the first fruits.

iQholo: Large bunch of feathers set in a basket frame and worn on the head by young Zulu warriors at the feast of first fruits. *Ukuthwala iqholo*: to carry plumes = to be conceited.

isiQhova: 1. Crest of a bird; 2. Ornamental crest of feathers worn by young men.

isiSaka: Plume of black finch feathers (cf. *idlokolo*).

uSoyaka: The Father of flowing plumes: a praise-name for a warrior (cf. *umyaka*).

ubuThekwane: Bunch of feathers worn behind the head (resembling the hammer-kop's crest).

u(lu)Ve: 1. Species of bird, Paradise fly-catcher, *Terpsiphone perspiculata*. *Uve ludlé isisila salo*: The fly-catcher has devoured its own tail = hoist with his own petard; 2. Long tail-feathers of the Fly-catcher, used for ornaments; 3. *u(lu)Ve*: one of Cetshwayo's regiments.

isiYaya: Necklet of tufted feathers.

11.2 USE OF BIRDS FOR FOOD

In the section on taboos in Chapter 10 (page 289), I quoted Gcumisa and Ntombela (1993) as listing several birds, including the Blue Crane, the vulture, the Bateleur and the Southern Ground Hornbill, whose flesh is not eaten. The unstated implication here is that the flesh of other birds is eaten. A number of other food taboos linked to birds were also mentioned, such as Godfrey saying that boys do not eat the flesh of owls. As pointed out earlier, this is ambiguous: it could mean that the flesh of owls is indeed eaten, only not by boys, or it could mean that boys, who are the only [main?] eaters of birds, do not eat owls. Similar implications underlie Bryant's statements that girls do not eat the Grey-backed Bush Warbler [now the Grey-backed Camaroptera] (but every other member of society eats it) and that only boys and old women eat Burchell's Coucal (but all members of society eat other birds).

Some birds clearly are eaten: Wilson says that the Chewa name *nkomatsabola* for the Black-bellied Bustard means 'good to eat' (2011: 48). According to him, the name is a compound of *ku-koma* ('be beautiful', i.e. 'tasty') and *tsabola* ('taste like pepper').

However, despite this tasty bustard in Malawi, the narrative in the literature is almost entirely a negative one. For example, almost every statement made about birds as food in traditional Zulu society has to do with what people do not eat: there is little if anything about what birds people do eat. Indeed, the following statement by Nancy Jacobs for sub-Saharan Africa generally seems to hold true for traditional Zulu society:

> Across the continent [of Africa] in general . . . wild birds have not been a leading food source. In some places, people do not eat birds or are not supposed to eat them, including all Maasai and Nuer twins and adults in the Sudan. The people who ate the most birds were those who hunted them the most – small boys – who spent their days in pasture and forest with time to kill and animals to hunt. None other than Nelson Mandela recalled bird hunting among his boyhood activities. Boys hunted for something to do as much as to get food, and in the process they learned skills (2016: 52).

Ichikawa (1998), as we have noted earlier, writes about the Mbuti hunter-gatherers of the Ituri Forest in the Congo, and we might expect such people to regularly eat wild birds. But in fact Ichikawa states that "[w]hile the Mbuti consider almost all of the birds in the area as their food, [the] actual proportion of birds in their diet is not high". He further states that while traps are set for small birds like greenbuls and other bulbuls, these birds are mainly eaten by young children (1998: 109).

As Jacobs says ("or are not supposed to eat them") and as pointed out above, much of the bird-food paradigm has to do with taboos.

The point she makes about young boys being the major consumers of birds seems to hold true for traditional Nguni society, if the limited references in the literature are anything to go by. One such reference is an early statement by Godfrey:[18]

> The tinky is a favourite with everybody, even with the little herd-boys, who are lost in wonder at his power of mounting in the sky until he disappears from sight; yet this respect for his powers does not prevent [its] figuring as the commonest item on these same herd-boys menu card (1919: 202).

Biyela speaks from her own personal experiences growing up in rural KZN as well as conducting research in the rural areas when she says:

> In the past, boys used to spend much of the day in the veld where they were taught advanced life skills such as herding, hunting and protecting livestock . . . According to the informants, bird hunting was the most popular activity that boys enjoyed in the wild (2009: 38).

As mentioned in Chapter 9, she remembers how her brothers and her cousins used to bring their caught birds home to show their grandfather:

> Each of the boys who came with a bird would receive the head of his bird and my grandfather would take the rest and divide it amongst other small children who were not yet capable of hunting (2009: 39).

---

18. Most of the quotes from Godfrey in this book are from his 1941 book *Bird-Lore of the Eastern Cape Province*. This one is from an earlier (1919) article in a journal.

Krige (1950: 78) makes a very basic statement about herdboys trapping and killing birds, with no detail at all apart from a reference to *isife* and *ingcokovane* traps. The latter word cannot be found in either Bryant (1905) or in Doke and Vilakazi (1958), but Bryant (1905: 140) explains *isife* as a "bird trap, as commonly built of a stone resting on a stick". The bait used in these traps would have been *incombo* or *inganga*, both explained by Bryant as small white termites used by boys as bait in bird traps (1905: 79, 170). We saw a reference to the use of this bait in the last line of the praises of the *inqomfi* lark in Chapter 9 above (page 240), and we also noted in the same chapter the number of Zulu proverbs based on this and other methods of hunting birds (pages 249–50).

Trapping by 'stone-and-stick' seems to be universal, and Cocker notes that even in the twenty-first century this is still legal in France:

> One method is the *tendelle*, a stone trap, which involves a bird coming to a fruit bait and then dislodging the support twigs that hold up a large flat stone.
>
> Environmentalists today regard this as 'Stone Age poaching'; French hunters say it is the "maintenance of an ancient tradition which dates back to Ancient Greece" (2013: 468).

Slightly more detail about Zulu trapping traditions is given by Kidd, writing in the condescending and patronising tone often found among writers of his time, in connection with small boys making spring traps to catch small birds:

> [A]s soon as a bird is caught they run up and kill it and eat it, either raw or slightly cooked over the fire. Of course, they eat the feathers, entrails and all; these things are too good to be wasted, and improve the flavour (1904: 311).

Kidd has other details about how small birds are hunted by herdboys. He remarks, on the way they use the 'kerrie' to knock over birds: "I have seen them knock flying birds over at forty or fifty yards . . ." (1904: 313), and on their use of bow and arrow he has this to say:

> When boys wish to shoot small birds with arrows, they put a wooden knob on the point of their arrow, and so the bird is not broken to pieces, but simply stunned (1904: 314).

Bryant adds a further method of catching birds, in his entry for the verb *kisila*:

> **kisila**: hiss as a snake. N.B. Children are accustomed to go among the reeds of a river in the evening time when the birds are in their nests, hissing softly as they go in order to rouse out the birds, which, however, do not fly far – being no doubt unaccustomed to night-travelling – and are easily caught by other children waiting in readiness (1905: 306).

As we saw in the section on feathers above, Layard refers to Xhosa children stretching bird-limed lines across fields to catch on the long, dragging tails of the Long-tailed Widowbirds (1867: 191). Once they had removed the feathers for trade or sale, they no doubt ate the birds.

To round off this section on birds as food, here are some food taboos from the Shona people of Zimbabwe, as given by Renata Coetzee (1982: 113):

DO NOT eat the flesh or the eggs of a nightjar; you might become habitually sleepy during the day.

DO NOT eat the flesh of a kingfisher; you might have bad luck.

DO NOT eat the flesh of an owl; it consorts with a witch and you may invoke the wrath of the witch.

DO NOT eat the flesh of a swallow as it does not eat grain.

DO NOT eat the flesh of doves; you may develop epilepsy.

DO NOT eat birds' eggs; you may have fits of giddiness.

11.3   MISCELLANEOUS BELIEFS AND USES

In this final section we look at (a) beliefs that relate to the perceived relationship between humans, cattle and birds; (b) beliefs about where birds go for the winter; (c) two unrelated belief 'curiosities' that defy categorisation elsewhere; and then (d) some practical uses of birds unrelated to the uses as feathers or food described above.

11.3.1   Humans, cattle and birds

Given the extremely important role played by cattle in Nguni societies, both economically and culturally, it is perhaps not surprising that Nguni bird lore links cattle and birds in a number of ways. Chapter 5 on the meanings of bird names already at looked how words like *ilunga* and *inkwazi* can refer to both birds (the Southern Fiscal and the African Fish Eagle respectively) as well as to cattle with colour patterns resembling those of the birds. In this section, we look primarily at cattle herding: the role of certain birds as cattle herders themselves, and as aids to the finding of lost cattle.

Birds as cattle herders

In the case of the Fork-tailed Drongo (**intengu** in both Zulu and Xhosa) and the Cape Wagtail (*umcelu* in Xhosa, and **umvemve** in both Xhosa and Zulu), the bird acts as a cattle-herder, leaving the boys who ought to be in charge of the cattle to go off and sleep, hunt or play games. Godfrey provides much of the detail of this (1941: 101ff.). His notes on the role of both the drongo and the wagtail prompted American historian Nancy Jacobs to write an article on 'herding birds', with some interesting observations on interspecific communication, including communication between birds and mammals, and between birds and humans. After some discussion, she concludes:

> The birds are known, across a wide area, to interact with cattle. A single ecological explanation is possible. These birds eat the insects flushed up by stock. Perhaps they associate with stock because of the feeding advantages (Jacobs 2015: n.p.).

The **umvemve**, a name given generically to all wagtails in KwaZulu-Natal, plays a number of roles in traditional Zulu culture.
© Adrian Koopman

Godfrey's schoolboys, whose essays provided him with so much information about bird lore, indicate that the wagtail does not only look after cattle:

> The wagtail may be trusted to look after sheep, for it blows a whistle (just like a herd[boy]) . . . When boys are herding, it is especially fond of being among the sheep. It likes to be among horses, cattle and sheep, whistling all the time (1941: 102).

In Chapter 3 above, in an investigation of the underlying meanings of bird names from a variety of different African languages, those that referred to birds as 'herders' were discussed. Among the names detailed were:

- *klaasperdewagter* (Nicholas the horse-watcher) for the African Stonechat, and *skaapwagter* (sheep-watcher) for the Capped Wheatear (Afrikaans);
- *kadima mbuzi* (goat-herd) for a species of wagtail (Tanzanian dialect);
- *kitàndry* (the watchman) and *langòroaomby* (ox-heron) for a species of cattle egret, *vórontàniómby* (attendant on cattle) for the cuckooshrike, and *fitìlibèngy* (goat-watchman) and *manàranòsy* (goat-ibis) for a species of ibis (Malagasy);
- *molisalipela* (herder of dassies) for the Drakensberg Rockjumper (S. Sotho).

And let us not forget the Zulu names **ilindankomo** (what waits on cattle) and **ilanda** (the one that follows) for egret species. In the 2013–2017 Zulu bird name workshops, the name **umalusinkomo** (what herds cattle) was coined for the Eastern Nicator, the Zulu-speaking bird guides all concurring that this bird is a great follower of cattle.

Birds as aids to finding lost cattle

There are many references in the literature to the **isiphungumangathi**, a name both in Zulu and Xhosa that refers not only to the Long-crested Eagle, but also to the chrysalis of a certain insect. Bryant says:

> **isi-Pungumangati**: . . . African Crested Eagle (*Lophoaetus occipitalis*); chrysalis (of any 'moving' kind) – both of these are applied to by herd-boys to know whereabouts the cattle are, or by a child when its mother is late coming home in the evening, the motion of the crest or the extremity giving the direction.

Ex. *we! 'sipungumangati, umame ngapi?* I say, sipungumangati, whereabouts is my mother? (1905: 517)

To enlarge on what Bryant has said, and summarising from different sources, the herdboy or herdsman who has lost cattle approaches the bird (or the chrysalis) and asks "*Siphungumangathi, Siphungumangathi, zikuphi izinkomo zikababa?*" (Siphungumangathi, Siphungumangathi, where are the cattle of my father?), upon which the eagle inclines its head and points with its crest to the correct direction. If the question is asked of the chrysalis, it is picked up and placed on the palm of the hand of the enquirer whereupon it wriggles around until its 'sharp' end is pointing in a certain direction – the direction in which the lost cattle will be found.

Samuelson provides further details in his two publications (1923 and 1929). In the first, his dictionary, it is a simple explanation of the name and the way the bird is used:

isiPhungumangati: the South African crested eagle.
N.B. – The natives apply to this bird for directions as to the whereabouts of strayed cattle, and it is said to point out directions by its crest (1923: 381).

In the second, he adds his own (unsuccessful) experiences with this bird:

isiPhungumangathi, the Crested Hawk Eagle . . . "*Siphungumangathi, Siphungumangathi zingaphi izinkomo na?*" . . . When in Zululand as a boy, when herding cattle, whenever the herd-boys had made themselves scarce, I used to appeal to these hawks when the cattle had strayed, and would see the hawk drooping his crest as if pointing, but I do not remember one instance where I found the cattle by the hawk's help (1929: 408).

Samuelson does not record whether or not he had ever applied successfully to the **isiphungumangathi** to know when his mother might be coming home.

It is interesting that all sources that have mentioned the role of the **isiphungumangathi** (and only two out of the several that are available have been mentioned here) need the enquirer after lost cattle to address the eagle directly, repeating its name. The opposite is the case in the Ituri Forest among the Mbuti people, where this bird is also present: Ichikawa says that its name (*sonbuoko*) must never be mentioned out aloud by anyone (1998: 110).

Other bird-cattle links

The Bokmakierie is a good-luck omen where cattle are concerned, with the belief among the Xhosa that the calling of the bird at a cattle kraal indicates that the cattle will increase (Godfrey 1941: 111). And Godfrey, quoting schoolboy essays, says much the same about wagtails: "When we see this bird at the cattle kraal, we say the cattle will increase; it is the 'bird of good fortune' [*yintaka yamathamsanqa*]" and "The wagtail is not killed, for it is called 'the bird of the cattle' [*yintaka yeenkomo*]" (1941: 101), Bryant does not mention any similar beliefs about the wagtail in Zulu culture (1905), but it is of interest that in addition to being a generic term for wagtails, the word **umvemve** also means a young feeble calf of only a few days old – yet another kind of link between bird and cattle (Bryant 1905: 678).

The reference to nightjars as 'goat-suckers' is too well known to need further elaboration here. But perhaps worth recording is Samuelson's variant on the nightjar/goatsucker theme, when he says:

> **umKhonya**: a large locust, beautifully coloured with a bladder-shaped stomach, which is also transparent, found during summer in certain parts of Natal, Zululand and East Griqualand; it travels about at night and utters a deep-toned call, something like this, *voyo voyo*, with acute accent on the last *o*.[19]
> N.B. The natives detest it and kill it when they can, as they say it sucks the milking cows and causes them to cease giving milk (1923: 228).

11.3.2   Where birds winter

Birds that indicate the onset of a particular season (like spring) are invariably migratory birds that disappear for the South African winter, and then reappear when winter is over. Godfrey lists two traditional beliefs among the Xhosa about where these birds go and what they do during winter:

Quails are not migratory birds, but keep hidden during winter and are not easily seen. The reason for this, according to traditional Xhosa belief, is that the quail turns into a frog during the winter (Godfrey 1941: 40).

---

19. Bryant says that this locust's call resembles "the sound of a London tram-conductor's whistle" (1905: 317).

When the Yellow-billed Kite disappears during the winter months, it is believed, says Godfrey, that it retires to some safe place in the rocks "whither it has conveyed a large number of chickens to serve as winter provisions" (1941: 29). How does it keep this chicken meat fresh for all these months? Very simple, says one of Godfrey's informants: "*Wenza imiqwayitho kanti ukuze aphila ngayo ebusika*" (it makes biltong [out of the chicken flesh] so that it can live on it during winter). We see here a human activity (that of making biltong) assigned to a bird species.

Samuelson, in his notes on the lore of the Yellow-billed Kite, records that he discovered for himself that it is true what the Zulus have always said, namely that the kite 'hibernates' in a cave with its mate during the winter, living off food that they have stored up during the year, and having shed their wing feathers (1929: 409). It is interesting that he finds this belief to be 'true', but was not able personally to confirm the cattle-finding powers of the *isiphungumangathi*.

11.3.3   Curious miscellanea

There are two curious bird-related beliefs recorded in the literature that cannot be categorised elsewhere. The first appears in Samuelson (1929) and then again in Godfrey (1941).

The kite as a 'tooth fairy'

Samuelson, in connection with the Yellow-billed Kite, says:

> The Zulus have a superstition about this hawk: they hold that if a child has shed a tooth and throws it away between his legs backwards and calls to the said hawk, while doing so, "Nhloile, Nhloile, thatha izinyo lami ungiphe elitsha," meaning "Nhloile, Nhloile, take my tooth and give me a new one," it will soon get a new tooth.

He goes on to say, and it is difficult to say whether this is tongue-in-cheek or serious moralising:

> It is a pity that such procedure does not apply to older people too, for then the world would not have so many dentists and old people would have a sporting chance (1929: 418).

Godfrey (1941: 30–1) records exactly the same belief relating to the Yellow-billed Kite, saying that it is one of three notes relating to the kite given to him by teacher John Sotashe (i.e. Godfrey has not derived this story from Samuelson's *Long, Long Ago*). The following is Godfrey's translation from the original Xhosa of Sotashe's note:

> When a child is losing its milk-teeth, it is told to throw its old tooth to the kite and beg for a new one, so that the new teeth may come out. The child says 'Kite! Kite! Take that old tooth of yours ['mine', perhaps?][20], and bring me my new one!' The child at the same time throws it [the tooth] away between its legs, without looking at where it is going to fall.

The similarity between the details in the two extracts given by Samuelson and Godfrey are striking, and, given that these were recorded separately from Zulu-speaking and Xhosa-speaking communities, it must be assumed that this is a very old Nguni belief. It is interesting to note that this belief is still held among Xhosa speakers and Zulu speakers today, according to the participants of the June 2017 Zulu bird name workshop.[21]

The other belief under the heading 'Curious miscellanea' comes from Samuelson alone (1929). It has not been found elsewhere in the literature.

The partridge in the thumb

Samuelson, in a loosely assorted collection of odd notes about birds (many of which have already been quoted in this chapter), gives the following:

> **Inswempe**, 'the white-winged partridge' which the Amatonga tribe ventriloquists have learned to imitate and use when they hypnotised the Zulus to believe that they hear it in their thumbs, when they hold out their thumbs on the order of these ventriloquists (1929: 405).

---

20. Although the original Xhosa does read '*thabatha izinyo lakho elidala*' (take your old tooth).
21. See section 13.6.2. 'The (virtually) unchanged survival of a meme' in the last chapter.

Inswempe is the Zulu name for the Coqui Francolin (*Peliperdix coqui*); 'Amatonga' in this context refers to the people of the erstwhile 'Tongaland' – the far north-eastern corner of KZN where it borders Mozambique; and 'ventriloquists' usually refers to a particular sub-category of *izangoma* (diviners) who are said to be able to call up 'whistling spirits' (*abalozi*) to their aid when divining. Samuelson's curious note on the **inswempe** is the only mention of the imitation of bird calls in *abalozi*-related divination that we have found and the only mention of hypnotism. The notion that the imitated sounds of this francolin are heard in the thumbs of the hypnotised Zulus is unique. Unlike the case with the role of the Yellow-billed Kite as a 'tooth fairy', the participants of the June 2017 Zulu bird name workshop, when asked about this, said they had never heard of this before.

11.3.4 Practical uses with some modern implications

The Lanner Falcon, says Godfrey is used in 'natural falconry' when it attends boys who are on a quail hunt (1941: 28). The boys welcome the assistance of the falcon, but it is not immune from being killed by the boys' sticks if it gets too close.

In the case of the Long-crested Eagle finding lost cattle and the drongo and the wagtail helping to herd cattle and other livestock while the boys relax, these birds are helping people in a practical way. In a different vein is the use of a swallow's wing, fed with milk to dogs to help make them as swift in the hunt as the swallow is on the wing (Godfrey 1941: 74). Ichikawa records a similar belief among the Mbuti when he says swallows and swifts are not eaten because:

> [T]hey are used as ritual medicines in hunting; in particular their feathers are fastened to a hunter so that he may move in the forest as fast as swallows (1998: 106).

Linked to the Xhosa belief of the swallow's wing fed to dogs to make them swift is the belief that the head of a coucal is preserved and given to pups for the purpose of making them expert hunters (Godfrey 1941: 58). The thought-patterns are perhaps not so easy to follow here, one can easily understand the cognitive link between the swiftness of a swallow's flight and the anticipated swiftness of a dog in a hunt, but surely there are birds other than a coucal that suggest expert hunting? Falcons and hawks come easily to mind. Perhaps relevant here is the S. Sotho name for the Lanner Falcon – *phakoe-ea-balisa* (falcon of the herders), with Meyer explaining that hunters drive out birds with sticks and the falcons then catch them.[22]

Godfrey only gives one reference to the practical (and economic) use of the Black-shouldered Kite (now officially the Black-winged Kite), but it is worth recording this if only for the nicknames that arise out of this rodent eradication:

> The Black-shouldered kite is well-furnished with names. Its economic value is recognised in the names *umdlampuku* (mouse-eater) and *unoxwil'impuku* (mouse-catcher) . . . elsewhere known as *umlungwana* (little Whiteman) (1941: 31).

The nicknames *umdlampuku* and *unoxwil'impuku* are readily explained, but Godfrey does not explain *umlungwana*. However, the Black-winged Kite has a considerable amount of white on its front, and this is probably the reason.

There are certainly many birds that fill this same role of keeping the population of rodents and other pests down. McLachlan and Liversidge say for example of the Mountain Buzzard that its diet, which includes insects, mice and rats, makes it "on balance a most valuable species to man" (1957: 81). And in April 2016, Jim Taylor of the Wildlife and Environment Society of South Africa told me of a project to encourage 'owl boxes' in urban townships to encourage owls to nest in urban areas and so deal with the burgeoning rat population in some South African townships and their concomitant health issues. It would be interesting to see what Zulu nicknames develop for owls in such contemporary urban settings.

11.4 CONCLUSION: ATTITUDES TOWARDS TRADITIONAL BELIEFS

The material covered in this chapter was accessed from publications ranging from Layard's 1867 *Birds of South Africa* to material published in 2017, a span of 150 years. During this time, the way in which writers (almost invariably white males from a Euro-Western background) wrote about Zulu (and other African) culture(s) has changed completely. Besides the terminology issues discussed in Chapter 10, section 10.1.1 (for instance, *Kafir* or *kaffir* for black

---

22. Johan Meyer, personal communication, January 2017.

South Africans, especially the Xhosa), the tone and the attitude of the writers has changed.

As pointed out in Chapter 10, many of the earlier writers referred to traditional beliefs about birds as 'superstitions'. Perhaps this was not so when those writers were writing, but today the term carries with it connotations and 'baggage', such as 'superstitious nonsense', 'uneducated' and generally a feeling of superiority on the part of the writer (who invariably is not a member of the society he or she is writing about).

This sense of being superior is seen in the quote on page 331 given by Dudley Kidd, writing in 1904 about young Zulu boys trappings birds for the purpose of eating them. I described it earlier as being in a "condescending and patronising tone". Let us look at it again before we analyse it a little further:

> [A]s soon as a bird is caught they run up and kill it and eat it, either raw or slightly cooked over the fire. Of course, they eat the feathers, entrails and all; these things are too good to be wasted and improve the flavour (Kidd 1904: 311).

On the surface, Kidd is giving details about accessing birds as a food source, which is why the quote was incorporated into the section on birds as food. But there is a subtext: effectively, what Kidd is doing here is contrasting the way in which these Zulu boys prepare and eat the birds they have trapped with the Western ways of doing things. He does not need to spell out the Western ways because he knows that his readers do not eat birds raw or even "slightly cooked". He knows that that in Western cuisine, birds are plucked and the entrails removed before cooking. By providing the details of the boys eating birds raw, with their feathers and entrails intact, Kidd is painting them as 'savages', as 'barbarians' who have not learnt the 'proper' way of doing this. The phrase 'of course', inserted into this description, can easily be translated as 'What can one expect?', and goes hand in hand with the sarcasm of "these things are too good to be wasted" and "improve the flavour".

Another way of expressing "what can one expect?" is through the rhetorical question "Can you believe it?", reflecting astonishment and surprise at the way in which these ignorant, uneducated 'savages' are behaving. This can be expressed very subtly, such as in the addition of an exclamation mark. Even someone as respected an ethnographer and lexicographer as Father Bryant is guilty of this on occasion, as seen in the quote about the use of a Southern Ground Hornbill as a medical cure, given in Chapter 10, on page 293:

A person suffering from unusual prominence of the eyeball – which is said to be due to an *umtakati* – may have the defect removed by the application of a little *inTsingizi* eyeball, whereupon the offending organ will return to its normal size! (Bryant 1905: 376)

Over the last few decades, the tone of writing about such beliefs has changed radically. In my opinion, one publication that marks a change in the thinking of those writing about Zulu cultural beliefs is Axel-Ivar Berglund's 1976 *Zulu Thought-patterns and Symbolism*. As I have pointed out in an earlier publication:

For Berglund (1976), the manipulation of material for the purposes of love charms, protection against evil, warding off lightning, and so on, are all acts that fall into Zulu thought patterns that are both systematic and intelligible (Koopman 2013: 88).

Many, if not most, writers before Berglund's time described Zulu cultural practices from <u>outside</u> the culture they were describing, seeing it through Western eyes, and explicitly or implicitly comparing it to Western cultural patterns. Berglund based his writing on years of consultation with *izangoma*, *izinyanga* and *izinyanga zezulu*, discussing with them the interpretation of cultural practices. The result is a finely nuanced view, where cultural beliefs are no longer the "superstitious nonsense" of uneducated people, but systematic, intelligible, thought-out beliefs.

It may be too simplistic to regard Berglund as the single projector of new ways of perceiving and describing Zulu traditional beliefs, but certainly over the last 30 or so years the idea of traditional beliefs, not just in Zulu society, but in other African societies, as being those of ignorant or uneducated people has changed completely. From 'ignorance', traditional beliefs of societies all over the world have for the last several decades become 'knowledge'. In a parallel development, this 'indigenous knowledge'[23] has been aligned with environmental studies, producing terms like 'traditional ecological knowledge' often abbreviated to 'eco-knowledge'. So beliefs that the Hamerkop is an

---

23. A phrase that has become decidedly familiar in South Africa, particularly among politicians and academics, and usually found in extended form as "Indigenous Knowledge Systems" or "IKS".

omen of evil, or that the **inqomfi** lark is lucky, or that the Bateleur should be drowned in a river if one wants rain, are no longer the superstitions of the uneducated. Today the study of these is 'ecological anthropology' or the like, and if specifically to do with birds, it is 'ethno-ornithology'. Such traditional beliefs are no longer irrational, but regarded as scientific. Consider, for example, Ichikawa's phrasing in the introduction to his article on the 'folk beliefs' of the Mbuti hunter-gatherers (my underlining, Ichikawa's italics):

> One of the major fields of our interest is _ethnoscience_, that is, a study of traditional <u>knowledge</u> on the plants and animals in the forest (1998: 105).

In a very recent article on the importance of 'indigenous knowledge' in issues related to both language loss and biodiversity, Wilder et al. begin with the opening paragraph:

> With the accelerating losses of biodiversity, habitats, and native languages, indigenous knowledge – including the study of traditional ecological knowledge of species and landscapes maintained by native languages – has become ever more significant (2016: 499).

After talking about language and biodiversity loss they go on to say:

> Fortunately, there are growing efforts to incorporate indigenous cultures into projects that restore habitats of declining species and resuscitate lost practices and knowledge.

And on the same page:

> These efforts are complemented by drawing attention to the wisdom embedded in _traditional ecological knowledge_ (TEK) (2016: 499).

They continue: "Accommodating their knowledge should be of interest to those scholars working to bridge the environmental sciences with social sciences, arts and religion." (2016: 500). A statement two pages later concerns an indigenous culture from Mexico, but could hold equally true of traditional Zulu society:

As could be said of many place-based societies, the way Yucatec
Maya people [of the Yucatan Peninsula, Southern Mexico] relate with
their surrounding landscapes pivots on a system of values, beliefs and
symbolic representations of the natural world.

. . .

The social and ecological resilience of the Maya is intertwined with the
sacred, in which the interweaving of the spiritual and practice realms
is encoded in a healing ritual with the land that sustains a biologically
diverse landscape (2016: 502).

The following, from the introduction to Deikumah et al.'s recent paper on bird
naming systems among the Akan people of Ghana, is also relevant:

Native and traditional wisdom has historically provided the basis of
much of what scientists have documented. Moreover, there is a large
body of knowledge about the environment contained in indigenous
culture. Indigenous knowledge is based on people's experience[s]
with the environment that are passed from generations to generations
usually by word of mouth and cultural ritual. The dynamic nature of
indigenous knowledge has assisted communities to survive in their
changing environment but its application in biodiversity conservation,
monitoring and management has not been fully explored (2015: 1).

The traditional bird lore explored in this chapter has been about how birds figure
in the worldview (*weltanschauung*) of the Zulu people and other communities in
sub-Saharan cultures. Some of these beliefs may be explained away rationally,
as Jacobs does for the closeness of drongos and wagtails to livestock; others
are obvious facts: the Yellow-billed Kite <u>does</u> come back to South Africa when
the *Scadoxus* plant rises from the ground, and the Red-chested Cuckoo and the
inkanku cuckoo <u>do</u> return when the spring rains have made the soil suitable
for ploughing. Other beliefs are aetiological: they seek to explain observed
facts like the disappearance of the kite during the winter months. Many beliefs
are what was earlier called 'sympathetic magic': if a swallow is swift then
applying a part of the swallow to dogs will make them swift; if a hornbill's
eye protrudes, it must be suitable for treating protruding eyes, and so on. But
all of these are what Berglund explains as 'thought-patterns' – perhaps not
immediately rational in a Western world, and certainly not to earlier writers,

but completely explainable and rational in the cognitive world in which these beliefs develop and hold sway.

Finally, we should note that traditional lore of the sort explored in this chapter is part of a cultural heritage, another notion that has taken root in South Africa over the past 25 years. Socio-political thinking over this period has come to realise that Western-origin education is all to the good for developing the economy of the country, but not at the expense of alienating people from their traditional beliefs. The notion of 'heritage' has now been allied to traditional beliefs, and this is often expressed in phrases like "heritage beliefs". Heritage beliefs are seen to be part of the cultural identity of a nation or a society such as Zulu society.

These notions are expressed in practical ways as well as being part of political rhetoric. When an organisation like the Women's Leadership and Training Programme asks for material on traditional bird lore for the Youth Bird Clubs they are setting up in KZN they say that they expect this to make birding activities far more relevant to Zulu youth.[24] Such 'traditional' material, it is suggested, will help bridge the gap between the environment in which these children have been raised, and the Western notion of 'birding' as an activity.

Developments such as these, and the way in which traditional Zulu bird lore can be integrated into birding programmes of all sorts, including educational programmes, is explored in more detail in the final chapter of this book.

---

24. See section 13.4.2 'Bird clubs and the Women's Leadership and Training Programme (WLTP)'.

12 New names and new identities

12.1 INTRODUCTION

The African Oystercatcher (*Haematopus moquini*) was the next to flash on the screen (see Plate 16). There it was, clearly, uncompromisingly – black. I recall that there was no discussion about what the salient feature of the bird was. The blackness was undoubtedly its most obvious characteristic. But how to convey this in a name? For some reason or another, no one wanted to use the Zulu adjective *–mnyama* (black). Someone suggested using the lesser-used alternative *–mpisholo* (black, dark-coloured). Others wanted metaphorical references to mourning and mourning clothes. Eventually three names became the main contenders: *umfelokazi* (widow), *unompisholo* (using the species formative *–no–* with *–mpisholo*), and *unozila* (*–no–* with the verb *zila* 'grieve', 'mourn'). It came down to a vote in the end, and **unozila** was declared the new Zulu 'book' name for the previously unnamed African Oystercatcher. All participants in the exercise agreed, however, that the names *umfelokazi* and *unompisholo* were too good to waste, and should be kept in mind when the next really black bird came up for naming.[1]

The discussion above took place during the first of what would be several workshops on revising and extending the Zulu names of birds: addressing the 'problem' identified by Gordon Maclean in 1984 (as discussed extensively in Chapter 2). Professor Noleen Turner had brought together a number of Zulu-speaking bird experts to discuss Zulu bird names, and I had been asked

---

1. At a later workshop, **umfelokazi** was chosen (quite suitably!) as a generic name for the widowbirds.

along to help facilitate discussion, to take notes, and to give a brief outline on the linguistic aspects of name formation. We met in September 2013 in a conference room at Phinda Private Game Reserve, kindly loaned to us by the Phinda management. The first day was spent 'finding our feet'. We started off with a dozen or so Zulu-speaking participants, whose knowledge of birds we had been assured was exceptional.

In that first workshop approximately one-third of the bird species on our list of KwaZulu-Natal (KZN) birds had been discussed, and we realised that follow-up workshops would have to follow. These could only be held once a year as a result of funding constraints. Accordingly, in 2014 four of the participants who had proved their mettle in 2013 were invited again, together with five new members, to a workshop held in St Lucia near the iSimangaliso Wetlands Park. A third workshop, held in October 2015, enabled us to complete the list of bird species, and so a draft revision and extension of the list of Zulu names for KZN species could be drawn up.

12.2   BACKGROUND TO THE WORKSHOPS

I had been working independently on Zulu bird names in the early 1980s. I had been doing fieldwork during this period on Zulu names of all sorts, particularly place names. The Zulu place name research was a direct consequence of requests from the then Department of Forestry, which had produced drafts of a series of what they called 'recreational maps' of the whole of the Drakensberg. I had seen these drafts, and reported to Bill Bainbridge, who was in charge of producing these maps, that the Zulu names were inconsistently and frequently incorrectly spelt. Under the auspices of the Department of Forestry, and with the help of the then Natal Parks Board, I was able to spend many weekends in the different Drakensberg areas and speak to the older Zulu-speaking members of staff about the place names in their area.

Later, but still in the early 1980s, I used the contacts I had established to interview other older Zulu-speaking members of the Natal Parks Board, in all the KZN areas under their control, about Zulu bird names. When University of Natal (Pietermaritzburg) colleague, Professor Gordon Maclean, an ornithologist in the Zoology Department, was asked to do the fifth edition of *Roberts' Birds of Southern Africa*, I was able to use my dictionary trawling and field research to show Maclean that the Zulu names in previous editions, like the Zulu place names on earlier maps, were inconsistent and incorrect for many species. Working with me on the Zulu names inspired Maclean to seek assistance from bird-loving experts in the other Bantu languages of South Africa.[2] This exercise had greatly improved the situation as regards the Zulu names in previous editions of Roberts, but they were still problematic, as noted in Chapter 2 of this book, in talking about the 'Maclean problem'.

In 1989 I was asked to be a consulting editor for the translation into Zulu of the 1980 publication *A Beginner's Guide to Our Birds*, written by Jo Oliver and published by the Natal Branch of the Wildlife and Environment Society of South Africa (WESSA). The 1989 translation, by Mduduzi T-G Mchunu and B.G. Nkwanyana, was published as *Izinyoni Ezingamashumi Amane Nantathu ZakwelaKwa Zulu* (Forty-three Birds from KwaZulu) by uPhiko LweNgcebo Yemvelo KwaZulu, the then governmental wildlife and conservation authority of KwaZulu (see Plates 62 and 63).

My earlier fieldwork research was not published (apart from my contributions to the 1984 *Roberts' Birds of Southern Africa*), but I did publish an article on how bird calls are frequently captured in English, Afrikaans and Zulu bird names (Koopman 1990) and I included a chapter on Zulu bird names in my 2002 book *Zulu Names*. By the time of the first workshop in 2013, my knowledge of bird species and their Zulu names had grown somewhat rusty, but as I was at that stage working on *Zulu Plant Names*,[3] I was at least familiar with the naming of biological entities in Zulu, and the problem of reflecting either a genus or a species in the naming of plants. I had also become familiar with the various creative linguistic strategies employed by the Zulu language in forming names for plants.

It was because of all the above background that Turner invited me to attend the first Zulu bird name workshop in 2013.

The 1989 Zulu translation of Jo Oliver's 1980 *A Beginner's Guide to Our Birds* seems to have gone unnoticed by the birding association BirdLife South Africa, as in 2003 they approached Turner, then a senior lecturer in the Department of Zulu at the former University of Durban-Westville, to translate exactly the same book into Zulu. She was equally unaware of the existence of the earlier translation, and together with Doris Kumalo, she translated Jo Oliver's book all over again. It was published by WESSA and BirdLife South Africa in 2003 as *Ibhuku Lokucathulisa Abasaqala Ulwazi Ngezinyoni Zethu*

---

2. See Maclean (1984: xliii) for the names of the various linguists involved.

3. Koopman (2015).

(loosely: 'A book of taking first steps about the knowledge of Our Birds' – clearly a version of Jo Oliver's original title, *A Beginner's Guide to Our Birds*) (see Plate 64).

Turner's association with both bird names and BirdLife South Africa, through her translation of this beginner's bird guide, led BirdLife South Africa to approach her in 2011 about the possibility of a project that could lead to a situation of an individual Zulu name for each discrete species of bird occurring in KZN. It was not long before she came to the conclusion that a project of this nature required considerably more input from mother-tongue Zulu speakers who were bird experts. As a keen birder herself, who had been on numerous birding outings over the years, and as someone who had many times visited the game reserves of KZN – both those of the provincial government as well as the private reserves – Turner knew several Zulu-speaking bird guides and a number of the Zulu-speaking game guards who accompanied game-spotting vehicles at the different reserves. Thus she was able to move her enquiries from the more confined environment of her office at the University of Durban-Westville to the more open spaces, such as eShowe and the Dlinza Forest, where bird guide Sakhamuzi Mhlongo was based, or the estuarine environment of the iSimangaliso Wetlands Park, where Themba Mthembu ran a birding safari company, or to the bushveld of Phinda Game Reserve, where Benson Ngubane had long been showing visitors the birdlife of this area. These three experts were, of course, present at the first workshop in 2013, and were also able to suggest other Zulu speakers with greater or lesser knowledge of birds and their Zulu names.

12.3   PREPARATIONS FOR THE FIRST WORKSHOP

12.3.1   Logistics of the workshop

In choosing the venue, it was important to take into account geographical and financial constraints: the venue had to be reasonably central so that participants from various places in Zululand had to travel approximately the same distance, and it had to be at a place on a recognised taxi route. The venue had to be able to provide a conference room big enough for up to twelve persons, with accommodation at the same venue or within close walking distance. Turner managed all the financial arrangements to facilitate this process.

12.3.2   Collating of existing material

Turner and I shared the duty of providing documentary material for all participants.

I provided existing lists of Zulu bird names from Bryant's and Doke and Vilakazi's dictionaries; together with separate lists of names from these two dictionaries such as "**imBucu**: species of bird", "**isiCibilili**: species of brown bird with red beak", and "**isiBulalambiza** (pot killer): species of small bird which draws people from its nest through feigning inability to fly."[4] The idea behind this list was to be able to assign one of these unassigned names to a bird without a name if the unassigned description seemed to fit the bird.

Turner had received from BirdLife South Africa a list of all birds occurring in KZN and had converted these into worksheets: entering the existing recorded name (where these were known), and leaving space for participants to enter the names that were decided on at each workshop.

Turner had also brought a laptop with the Roberts VII Multimedia program installed, so all the existing names in the 2005 edition of *Roberts Birds* could be accessed, including names in other Bantu languages. The program included several photographs of each species of bird and, most importantly, the vocalisations of each bird.

12.4   MODUS OPERANDI OF THE FIRST WORKSHOP

The format established at that first workshop in September 2013 was to be followed in subsequent years. The only difference is that at the first workshop Turner and Koopman each explained different aspects of the aims and goals of the exercise, and because at least four of the participants of this initial workshop attended the next workshop in 2014 and again in 2015, there was little need to explain again. At the first workshop Turner explained the background and the eventual goal of 'one bird one name'. I went over Zulu derivational grammar, reminding the participants about the name formatives *–so–*, *–no–* and *–ma–*, about reduplication of stems, about compounding of different lexical elements, and about literary devices such as metaphor and onomatopoeia. Turner then explained the use of the worksheet issued to each participant, the list of unassigned names from Doke and Vilakazi's dictionary, and the way in which the *Roberts* multimedia program was to be used during the workshops. She

---

4. Sadly, we were never able to identify this bird, so this wonderful name remains unused.

then divided the twelve participants present on the first day into three groups of four each and gave each group six unnamed birds and asked them to think up names. Each group then presented their 'findings' and then discussed these with the other groups.

The group as a whole then settled down to the first birds on the list: albatrosses, pelicans, gannets, cormorants, and then on to the herons. For each group of birds, Turner called up each species found in KZN, reviewed the existing names, enumerated the problems (the 'Maclean problem') and displayed the characteristics of each bird on a screen: their appearance (plumage, colour, markings, distinctive stripes, spots, crests, etc.), their call, and their movements (flight, gait, diving, and so on). Details about habitat, distribution and diet were also available when needed.

During this first session, Turner and I played three roles: operator of the multimedia program (Turner), facilitator and scribe. All the participants were asked to write on their worksheet what name decisions had been reached for each species but when six or seven people were talking at the same time, ideas were flying around, and birds kept calling from the laptop speaker, it was not always easy to remember to do this. Consequently, when participants were asked to return their completed worksheets, there were often large gaps next to the English names of the birds.

On the second day of the workshop, as the aims, goals and methodology started to clarify, the pace quickened, and by the end of the workshop we were often finding names of groups of six or seven birds every hour. By the end of the first workshop in 2013, names had been revised or created for about a third of the birds found in KZN, and Turner and I estimated that we would be able to complete the entire list if we held similar workshops in 2014 and 2015.

This proved correct. A three-day workshop in St Lucia in November 2014, and another three-day workshop near Mkhuze in November 2015 enabled us to complete the first draft of a complete list of birds found in KZN, each one with a confirmed, a revised, or a newly coined name.

After that it was a case of checking for inconsistences, gaps, duplication and other queries. This is discussed below in section 12.7.

12.5   A CASE STUDY OF SHORE BIRDS AND WADERS

In order to give some sort of idea of how a 'cluster' or loose group of birds was tackled in a typical workshop session, I give here a case study of the names of a group of fifteen waders and shore birds: sandpipers, stints, turnstone, curlew and similar birds. I begin here by looking at what Zulu names (or even what African names) were recorded by Maclean in his 1984 edition of *Roberts' Birds*:

Turnstone (*Arenaria interpres*): No African names

Terek Sandpiper (*Xenus cinereus*): No African names

Common Sandpiper (*Tringa hypoleucos*): No Zulu name; Xhosa *uthuthula* and a S. Sotho and Tsonga name

Wood Sandpiper (*Tringa glareola*): No Zulu name; Xhosa *uthuthula*; S. Sotho and two Tsonga names

Marsh Sandpiper (*Tringa stagnatilis*): No African names

Greenshank (*Tringa nebularia*): No Zulu name; Xhosa *uphendu*; S. Sotho and two Tsonga names (S. Sotho and Tsonga names the same as for Wood Sandpiper)

Knot (*Calidris canutus*): No African names

Curlew Sandpiper (*Calidris ferruginea*): No African names

Little Stint (*Calidris minuta*): No African names

Sanderling (*Calidris alba*): No African names

Ruff (♂), Reeve (♀) (*Philomachus pugnax*): No Zulu name, one S. Sotho name (same as for Wood Sandpiper and Greenshank)

Ethiopian Snipe (*Gallinago nigripennis*): Zulu name: **unununde** (Maclean 1984); [**ubuklekle** (D&V)]; Xhosa *umnquduluthi*; two S. Sotho names, one Kwangali name

Bartailed Godwit (*Limosa lapponica*): No African names

Whimbrel (*Numenius phaeopus*): No Zulu name; Xhosa *ingoyi-ngoyi*

Avocet (*Recurvirostra avosetta*): No African names

The African language name situation for these fifteen birds, as shown here, is dire: nine of them have no African name at all (at least, as recorded in Maclean); five have a Xhosa name (two sharing the name *uthuthula*); five have one or more shared S. Sotho names; three have a shared Tsonga name; one has a Kwangali name; and only one of the fifteen has a Zulu name (two names, with only one reflected in Maclean, the other from Doke and Vilakazi, 1958: 432).

At the end of the 2013–2017 workshop, all these birds had distinct species-specific Zulu names:

Ruddy Turnstone (*Arenaria interpres*): The consensus was that we choose a name reflecting the bird's habit of turning over stones to seek food, as does the English name. The first choice was **umajikamatshe** (<*jika* 'turn' and *amatshe*

'stones', but at a subsequent workshop it was realised that *jika* is intransitive, and is used for 'turn' when, for example, a car turns the corner. The verb *phendula* (turn something over) was substituted and the name **umaphendulamatshe** (that which characteristically turns over stones) used instead.

Terek Sandpiper (*Xenus cinereus*): The Zulu name **unopheshwana** was coined, meaning 'the little one with the upturned beak'. The name comes from the same root as the name *usipheshula* for the Pied Avocet (see below).

Common Sandpiper (*Actitis hypoleucos*): The name **ucijomhlophe** was coined, derived from *cija* 'be pointed' (or from *umcijo* 'diamond in pack of playing card') + *omhlophe* 'white', a reference to the narrow white pointed marking in front of the bird's wing.

Wood Sandpiper (*Tringa glareola*): The name **unogqabhakazi** was coined, derived from –*no*– + *gqaba* 'make incisions' + -*kazi* 'great(er)', a reference to the markings on the face and body. At a subsequent workshop the same name was given to the Ruff. Much later, in 2018, this duplication was resolved by changing the name of the Wood Sandpiper to **umakhwifikhwifi**, from a Zulu adjective meaning 'speckled'.

Marsh Sandpiper (*Tringa stagnatilis*): The workshop participants chose the swinging gait of this bird as the base of the name **unothwayiza**. It is derived from –*no*– + the verb *thwayiza* 'walk with swinging gait'.

Common Greenshank (*Tringa nebularia*): The name **unompempe** was chosen for this bird. Usually meaning a referee in a football or rugby match, the word is derived from –*no*– + the noun *impempe* (whistle) and refers to the bird's loud ringing *tew-tew-tew* on take-off.

Red Knot (*Calidris canutus*): The name **unovimba** is one of the more intriguing names coined in the workshops. The participants were greatly interested in the story of King Canute and how he tried to push back the waves (as reflected in the scientific epithet *canutus*). They felt that a Zulu name should have the same underlying meaning. The name **unovimba** is derived from –*no*– + the verb *vimba* 'block off', and of course, like the Latin *canutus*, is a direct reference to the bird feeding on the shore at the very edge of the waves.

Curlew Sandpiper (*Calidris ferruginea*): The coined name **ungozwana** is a diminutive form of the noun *ingozo* (the tiny elephant shrew with a long snout).

Little Stint (*Calidris minuta*): This bird is perceived as a smaller version of the Marsh Sandpiper and it has the same gait. The diminutive suffix *–ana* is added to **unothwayiza** to form **unothwayizana**.

Sanderling (*Calidris alba*): As with the Marsh Sandpiper, the workshop participants felt that the behaviour of this bird – its haphazard dashing about on the wave edges – was its defining feature. The name decided upon – **umaphithizela** – is derived from *–ma–* 'characteristically' and the verb *phithizela* 'move about in a haphazard, confused and disorderly manner'. Maclean (1984: 251) describes its movements as follows:

> Forages on wave-washed beach, running after receding wave, ploughing bill through sand; as next wave breaks, flock turns and runs upshore; movements quick and lively.

Ruff (♂), **Reeve** (♀) (*Philomachus pugnax*): as noted above under Wood Sandpiper, the name **unogqabhakazi** was erroneously assigned to both these birds. It was retained for this species when the name for the Wood Sandpiper was changed.

African Snipe (*Gallinago nigripennis*): Zulu name: **unununde** (Maclean 1984); **ubuklekle** (D&V, 1958): The well-known name **unununde** was confirmed for this bird. Doke and Vilakazi's clearly onomatopoeic name **ubuklekle** will be retained as well.

Bar-tailed Godwit (*Limosa lapponica*): The coined name **unodaka** is derived from *–no–* and *udaka* 'mud', and is another name that refers to a bird's foraging habits, this time using the place where the bird forages, in shallow water or on exposed mud.

Whimbrel (*Numenius phaeopus*): The coined name **unokhifi** is derived from the ideophone *khifi* 'of drizzling rain'. The fine spots and speckles of the plumage suggest a light spattering of rain.

Pied Avocet (*Recurvirostra avosetta*): At an earlier workshop, the name **umhloshana** (the little white one) was coined, but at a later workshop it was agreed that the long upturned beak was a far more salient feature of the bird, and the name was changed to **usipheshula**, derived from the noun *isipheshula* which refers to any turned up or reverted thing.

*   *   *

This group of fifteen birds was tackled in one long sitting of the relevant workshop. Only one species of this loose group of marshbirds and shorebirds had a Zulu name, the African (formerly Ethiopian) Snipe, with the long-established name **unununde**, together with an onomatopoeic name (**ubuklekle**), which is recorded in the pages of Doke and Vilakazi's dictionary, but was not known to the workshop participants. With fourteen species of bird with no Zulu name, and indeed in most case with no African name at all, this was a challenge to the creative thinking of the participants. This was not a case like the largely unnamed members of the duck family, where the generic word **idada** could be adapted in a number of ways to distinguish between the various species of duck. The Zulu language has no generic name that equates to English 'wader' or 'shorebird'.[5]
What was required was innovative thinking based entirely on the bird itself: its appearance, its call, its behaviour, or its diet and habitat.

As pointed out earlier, the workshop participants were all bird experts, many of them trained bird guides. Nonetheless, in the process of deciding which of the salient features of each bird was the most desirable for naming purposes, each bird was called up on the screen. The bird was looked at from each angle, videos were played of its flight or gait. Its diet and habitat were noted, and recordings were played of its call. For each bird, there was discussion, not only on which of the features of the bird was the most suitable for naming, but which linguistic strategies could be used for naming it: onomatopoeia, metaphor, adaptation of a suitable verb or noun, or a compounding of these.

It should be noted that salient features per se do not predict which naming strategy is likely to be used, apart from the link between a definitive call and the strategy of onomatopoeia. These two notions (salience and linguistic strategy) must therefore be looked at separately in any summary.

In the case of the 15 marsh and shore birds above, the following salient features were identified:

---

5. Much later, in 2018, a generic name equivalent to English 'wader' was coined. The name, **usoxhaphozi**, is derived from –so– and *ixhaphozi* 'marsh'. The intention was to have a word suitable for school biology textbooks.

Appearance: The upturned beaks of the Terek Sandpiper and the Pied Avocet were captured in the names **unopheshwana** and **usipheshula** with the noun root /*pheshu*/ the basis of each. The white mark on the neck of the Common Sandpiper is reflected in **ucijomhlophe**. The appearance of facial incisions gave the name **unogqabhakazi** to both the Wood Sandpiper and the Ruff/Reeve. The speckled appearance of the Whimbrel is what gave rise to the name **unokhifi**. And it is the small size of the Curlew Sandpiper that is reflected in **ingozwana**.

Behaviour: I include here movement as well as foraging behaviour as they are linked. Foraging behaviour is also of course linked to the environment in which the bird forages, thus some of the names here include a reference to local habitat, whether overt or covert. The name **umaphendulamatshe** (what turns over stones) may seem to be a direct translation of the English name (Ruddy) Turnstone, but it was reached after quite some discussion about which features of the bird should be the peg on which to hang a name. The Red Knot's Zulu name **unovimba** places it at the very edge of the wave line on the seashore. The name for the Sanderling – **umaphithizela** – refers to its foraging behaviour, but is also a covert reference to the same wave edges where this bird seeks food. The apparently haphazard running back and forth is a direct result of the bird seeking food as the water of the waves retreat. Similarly, the Bar-tailed Godwit also receives the name **unodaka** because of the way it forages for food in mudflats.

It is only how the bird walks that gives rise to the name **unothwayiza** for the Marsh Sandpiper, and there may be a similar reference to this swinging wide-striding gait in the name **unothwayizana**, or it may simply be that the Little Stint is seen as a kind of 'lesser Marsh Sandpiper'.

Voice: only two names refer to the call of the bird concerned. One is the apparently forgotten name **ubuklekle** for the African Snipe, the other is the name **unompempe** ('referee', i.e. the one with the whistle) for the Common Greenshank.

When it comes to linguistic strategies, we can note a wide range of grammatical complexities:

The name-forming prefixes –*no*– and –*ma*– are evident in **umaphendulamatshe**, **unopheshwana**, **unogqabhakazi**, **unothwayiza**, **unovimba**, **unothwayizana**, **umaphithizela**, **unodaka**, **unokhifi** and **unompempe**, showing how frequently the workshop participants made use of these word-forming devices. The verbs *gqaba*, *vimba*, *thwayiza* and *phithizela* are used directly to form names, while the verb *phendula* is compounded with the noun *amatshe*. The nouns *ingozo*, *udaka*, *isipheshula*, *ikhifi* and *impempe* are used directly to form names, and the noun *ucijo* is compounded with the adjective *omhlophe*. Two names are metaphorical and, curiously enough, both are borrowed metaphors: the name **unovimba** for the Red Knot borrows the metaphor of King Canute from the Latin specific epithet *canutus*, while the name **unompempe** for the whistling Common Greenshank borrows the metaphor of a referee with its shrill whistle from the world of team sports. During the five years of workshops, the workshop participants frequently used onomatopoeia for the naming of birds with distinctive calls, but they did not do so for this group of waders and shore birds. The only onomatopoeic name was **ubuklekle** for the African Snipe, taken from a dictionary and unfamiliar to those present.

This 'case study' of fifteen waders and shore birds gives an idea of the creative thinking that was produced in the session. Similar case studies could, of course, be produced for any group of birds largely lacking individual names. On the other hand, where a generic name was known, as with **ukhozi** for the eagles, or **idada** for the ducks, workshop thinking tended to be more prosaic and concentrate on the linguistic process of extending, rather as English has done when a generic name such as 'eagle' or 'duck' was required to be differentiated. The creative muses were most obviously present where no generic name existed, as with the terns and many other avian clusters.

At the end of this chapter, there is an appendix of the kingfishers' names. This is also a case study, but additionally it situates the naming of the kingfishers during the relevant workshop within the context of Zulu kingfisher names from 1940 to today (i.e. over the last 80 or so years).

12.6   A SUMMARY OF THE LINGUISTIC STRATEGIES AND PROCESSES USED

In this section the following 'strategies' are discussed: 'confirmation', 'selection', 'relegation', (previously 'archiving'), 'redirection', 'extension', 'adaptation' and 'coining'. These are not necessarily always mutually exclusive processes: adapting and extending existing bird names produces new lexemes (new words in a bird name list), so in a way they are coinages as well.

12.6.1   Confirmation

In the 2013–2017 workshops, the most common process, and the one that took the least amount of time and thought, was the process of confirmation. An excellent example here is the ostrich (even if this is said not to occur naturally in KZN). Every dictionary records the Zulu name **intshe** for ostrich. Research has shown cognates in a number of southern African Bantu languages so the word is old. The bird has no other known names, and all workshop participants had no hesitation in confirming the name **intshe** for the ostrich.

Confirmation of a name was easy if three factors coincided: the name was well-attested in the literature, the name was well known by the participants, and the name was the only name that particular bird was known by.

Examples of other names that were confirmed with no hesitation are **ilongwe** (Egyptian Goose), **ukhozilwentshebe** (Bearded Vulture), **inkwazi** (African Fish Eagle), **indwe** (Blue Crane), **impangele** (Helmeted Guineafowl), **isikhwenene** (Cape Parrot) and **uphezukomkhono** (Red-chested Cuckoo). There were dozens more.

12.6.2   Selection and relegation

The process of selection involved a choice. Two different types of selection were carried out, one type involving a group of closely related birds (a 'cluster' or 'folk genus'), the other involving a particular species.

Let us take the first one: <u>selection among a group</u>. The eagles discussed in Chapter 2 make a good starting point. In Maclean (1984) many eagles have the name **ukhozi**, many have no name at all, and a few have individual names like **ingqungqulu** (Bateleur), **isiphungumangathi** (Long-crested Eagle) and **inkwazi** (African Fish Eagle). For the last three it was a case of <u>confirming</u> the name. At the first workshop in 2013, the name **ukhozi** was <u>selected</u> as the species-specific name for the Martial Eagle, as this bird was regarded as the 'king of the eagles'. In the last workshop in 2017, when it was realised that generic names for folk genera were also important, the name **ukhozi** was <u>selected</u> as the generic name. This meant, of course, that the Martial Eagle required a discrete name of its own. The workshop participants saw no reason to change their thinking about this species being the 'king of the eagles' and renamed it **inkosiyezinkozi** (king of the eagles). Similar changes in thinking saw an earlier <u>selection</u> of the word **idada** as a species-specific name for the African Black Duck to then selecting the name as a generic name for all ducks.

<u>Selection as applied to individual birds</u>: here we refer to the choice made when a bird had more than one name, or a single name had been recorded in different forms. The Secretarybird has been equally recorded as **intinginono** and **intungunono**. Clearly this is the same word, but which should be the first choice in a system of 'one bird one name'? In other words, which name should be selected as the Zulu vernacular book name? On balance, more workshop participants were familiar with **intinginono**, and so this form was selected.

The selection becomes more difficult when there are several names for one bird: the Black-headed Oriole is commonly recorded as **umbhicongo** and **umqoqongo**, less commonly as **umgongolozi**. The name **umqoqongo** appears as **umgoqongo** and **umqaqongo**. To make things more complicated, most participants said that the more commonly used name in KZN today was the onomatopoeic **usibó**. After considerable discussion on the names of this bird, the recorded name **umqoqongo** and the well-known (but unrecorded in the literature) name **usibó** were both selected as 'book' names for the Black-headed Oriole. The names **umbhicongo** and **umgongolozi** were <u>relegated</u> to a lower status: not to be forgotten, but given the status of 'alternative names with historical reference'.

In a way, these become 'yesterday's names' – the fate that awaits 'lourie' (in all its spellings) now that the Knysna and Purple-crested Louries are officially 'turacos'. Who remembers that these birds were once officially known as 'plantain-eaters'?

Relegation came up when I was working with Maclean in the early 1980s on the Zulu names for the fifth edition of *Roberts' Birds*. Scouring the existing dictionaries, and doing field research that turned up a number of regionally different names for one species, often produced a situation where a given bird had three or four names. Maclean worked on the 'one bird one name' principle as much as he could, given that this was the prevailing principle for the English and Afrikaans names – not to mention the scientific names. However, he would accept two Zulu names for a species if these were equally well known and neither was obviously used more than the other. He drew the line at three Zulu names for one bird, however, and the ghost of Maclean seems to have been hovering in the background when the workshop participants made their own selections. For example, Doke and Vilakazi (1958) have three names for the Yellow-billed Kite: **isikhokhwane**, **ukholo** and **unhloyile**. Maclean and I selected **isikhokhwane** and **uhloyile** as the *Roberts* book names for this bird, and the 2013 workshop whittled this down to just **unhloyile**.

There was one notable exception to this: the names of the Black-collared Barbet. This bird has so many different names, and these are all so well known that no consensus was reached on which to select as the main 'book' name. The bird currently carries, therefore, the 'official' book names **usibagwebe**, **isiqonqotho**, **isandondwane** and **ingongoni**. It is the only bird in the final list that has four names.

Other examples of selection and the corresponding relegation are:

- The name **isikhombazane sehlathi** selected as the 'book' name for the Tambourine Dove, while the well-attested **isibhelu** is relegated to second level status.
- The name **ijubalaphansi** selected as the 'book' name of the Lemon Dove (whose earlier English name 'Cinnamon Dove' has likewise been relegated to 'an earlier name once used as the book name'), while the well-attested name **isaqgukwe** goes into the historical records.
- The Grass Owl loses its original unique name of **umshwelele**, and takes on the extended generic **isikhova sotshani** (owl of the grass). The reasoning behind this highly disputed choice was that '**isikhova sotshani**' would be 'easier to remember'.
- Maclean accepted all three names **ihlabankomo**, **ihlolamvula** and **ijankomo** for the Black Swift. The workshop process selected **ihlolamvula** as the sole name for this bird; **ihlabankomo** fell by the wayside, and **ijankomo** was 'redirected' (see below).
- The White-fronted Bee-eater has long had the name **inkotha**. This was confirmed by the workshop participants, but they selected the apparently well-known oral name **uswenka** to be a 'book' name as well. The name appears to be from the use of the English adjective 'swanky' as applied by Zulu speakers to brightly coloured clothing.
- The Greater Honeyguide has been recorded with the names **unomtsheketshe**, **inhlavebizelayo** and **ingede**. An earlier selection relegated the name **unomtsheketshe** to the lower level of 'earlier recorded for this bird but now not used', while **ingede** was selected at a later workshop as the generic term for honeyguides. By default, as it were, **inhlavebizelayo** becomes the 'book' name for this species.

A particularly interesting case of selection of names was that of the Rufous-naped Lark. This bird is well known in the literature as **ungqangendlela** (what goes straight along the path). Indeed, this name is developed into an extended image in the historical praises of this bird. The bird has also been extensively recorded in different forms of the root /*ngqwashi*/: **unongqwashi**, **umangqwashi**, **usangqwashi** and more. Yet, despite this plethora of existing names, the workshop participants selected the names **untilontilo** and **uqaqashe** for this bird. There was considerable, and lengthy, discussion about this, but eventually the consensus opinion was that these were the two names by which the bird was currently known among the Zulu-speaking people of KZN.

Selection, and its strategic corollary relegation, often seem quite arbitrary – a flip of the coin, as it were – but in cases like the Rufous-naped Lark the issue has been rigorously debated, and as with so many well-known and well-loved bird names used by English-speaking birders, the old is replaced by the new.

Turner and I have struggled to come up with appropriate terminology to describe names that were previously recorded and are almost certain to still be in oral use, but which have not been selected by the workshop processes to be the standardised Zulu 'book names'. We originally used the term 'archived' but discarded this with its negative connotations of being consigned to a dark repository where the names could only be accessed by specialists. As seen above, we are currently using the concept of 'relegation to a lower status', a concept that is not only vague, but has its own negative connotations. An idea comes here from Skead. Although he is talking of scientific nomenclature rather than of vernacular naming, there are some interesting points that could well be valid for Zulu names:

> Indications of the changes in the names of the various species are to be found in what is known as the Synonymy, a term used to denote prior names applied to the same bird, but which are now *discarded*. A future worker may well resuscitate one of these names as a result of further research, in which case the name used in this book will then be relegated to the synonymy [my italics] (Skead 1960: 6).

There are a number of interesting points about this brief paragraph. Firstly, the term 'synonomy', within the specific framework of biological nomenclature, means 'alternative, usually older, scientific names used for a particular species'. Skead has noted that although these names may be discarded (a term with stronger negative overtones than 'archived', although meaning much the same in this context), they may also later be resuscitated. Skead may well have had in mind the English vernacular name Emeraldspotted Wood Dove, discarded (or archived) by Maclean in 1984 in favour of 'Greenspotted Dove', only to be resuscitated later as 'Emerald-spotted Wood Dove'. Secondly, the same discarding and later resuscitation has indeed happened several times with scientific nomenclature. It does not take much imagination to replace "the name . . . will then be relegated to the synonymy" with "the name will then be placed in the Dustbin of History".

A third point is the way in which Skead uses the term 'Synonymy', almost as if this abstract linguistic and nomenclatural notion is a real place.

A final point of interest is that although Skead talks of placing unused names in the 'synonymy', his book is one of the few that does actually record past and/or alternative names. Modern books on birds, such as the 2016 *Roberts Bird Guide* tend to give the current name for a species, whether the vernacular name or the scientific name, and no other names. Skead, on the other hand, despite remarks about 'discarding' older names, actually retains them in his publication. An example can be given of how he treats the nomenclature of the Bully Seedeater (Skead 1960: 70):

THE BULLY SEEDEATER: **SERINUS SULPHURATUS** LINNAEUS
Loxia sulphurata, Linnaeus. Syst. Nat. 12th Ed. i, p. 305, 1766. *Cape of Good Hope.* No holotype extant.

Synonomy and References. *Crithagra sulphurata*, Swains. Zool. Journ., iii, p. 348, 1828; Layard Bds. S.A., p. 218, 1867; Sharpe ed. Layard, Bds. S.A., p. 486, 1884; Roberts Ann. Tvl. Mus., X, p. 186, 1924; Roberts, Bds. S.A., p. 367, 1940 and p. 462, 1957. *Serinus sulphuratus*, Sharpe, Cat. B.M., xii, p. 352, 1887; Butler, Foreign Finches, p. 29, pl. 1894; Shelley, B. Afr., i, p. 21, 1896; Stark & Sclater, Bds. S.A., i, p. 169, 1900; Sclater, Ann. S.A. Mus., iii, p. 313, 1903; Gunning & Haagner, Chk. List. Bds. S.A., p. 54, 1910; Sclater, Syst. Av. Aeth., pt. 2, p. 816, 1930; Gill, Fst. Guide S.A. Bds., p. 19, 1936; Clancey, Prelim. Lst. Bds. Natal & Zululand, 1953; Smithers *et al.* Chk. Lst. Bds. S. Rhod., p. 159, 1957. *Serinus sulphurata* Vincent, Chk. Lst. Bds. S.A., p. 114, 1952.

Local Names. ENGLISH: Brimstone Canary, Bully Seedeater; AFRIKAANS: Dikbeksysie, Geeldikbeksysie, Dubbele Geelsysie, Blokbek; AFRICAN: Indweza eluhlaza (Xhosa), nzwiyi (Rhodesian Shona).
The name *sulphuratus* is from Latin = Yellow.

We can see here his chosen 'book names' for this bird are the English vernacular name Bully Seedeater and the scientific name *Serinus sulphuratus.* The English vernacular name has been selected from the options 'Bully Seedeater' and 'Brimstone Canary', both of which are listed below the main entry. The scientific name '*Serinus sulphuratus*' has been selected from the options *Loxia sulphurata* (a Linnaean name from 1766), *Crithagra sulphurata* (supported by a host of named taxonomists, with dates and references), *Serinus sulphuratus* (supported by an even greater number of named taxonomists), and the variant *Serinus sulphurata.*

Although there is no main Afrikaans entry, Skead offers four possibilities.[6] He also offers the only 'African' names he has been able to lay his hands on: a Xhosa name and a Shona name.

In a separate book that Turner, Porter and I are compiling, which lists all the final Zulu bird names decided on in the 2103–2017 workshops, we do not give the full lists of scientific synonyms as Skead and other writers (e.g. Clancey 1964) do, but we do give recent English vernacular and scientific names if these have changed in the last twenty or so years, and we do attempt to give all the older Zulu names that we have been able to find. See, for example, our proposed entry for **iwabayi** (the White-necked Raven *Corvus albicollis*):

---

340 Corvus albicollis   **iwabayi**   White-necked Raven
[ihub[h]ulu, iwabayi, ihlungulu]

All three names recorded in ROB5† are well known and well attested, but *iwabayi* has the widest use in KZN. D&V 339 have: **ihlungulu**: "White-necked raven, *Corvultur albicollis*" and give the Ur-Bantu form *likuŋguvû*, indicating that this name is very old, and has cognates in other Bantu languages. Nyanja, for example has *khungubwi* for a large black and white crow, and Shona has *chikhunguo* for a black crow. For **ihubhulu** [note correct spelling], D&V346 give "White-necked raven, *Corvultur albicollis*", and for **iwabayi**, D&V847 have "White-necked raven, *Corvultur albicollis*."

---

† ROB5 stands for '*Roberts* Fifth Edition' in this particular book.

Here we see that although we have selected **iwabayi** as our 'book name' (i.e. the main entry), and have relegated **ihubhulu** and **ihlungulu** to 'non-book name' status, they have still been retained in the entry. Whether or not we could say that we have 'placed them in the Synonymy' is debatable.

---

6. The latest authority (Chittenden, Davies and Weiersbye 2016) has selected *dikbekkanarie* as the 'official' Afrikaans book name (not one of Skead's options). They have also selected 'Brimstone Canary' as the English vernacular name, and *Crithagra sulphurata* as the scientific name.

Note also the reference to a possible much older Ur-Bantu name, and probable cognates in Nyanja and Shona.

12.6.3 Redirection

The White Stork is another bird with a number of previously recorded names. Doke and Vilakazi give, for example, the names **unogolantethe**, **unowanga** and **unoyenge** for this bird. In the workshop session that dealt with storks, the name **unogolantethe** was selected for the White Stork, **unoyenge** was relegated to lower level status, and **unowanga** was <u>redirected</u> to the Black Stork, which at the time had no previously recorded or known name. Redirection is in effect a kind of onomastic 'wealth sharing': such and such a species has three or four names while two or three other species in the same folk genus have no names at all, so why not redistribute or redirect the unneeded extra names?

Above I mentioned the three names recorded for the Black Swift: **ihlabankomo**, **ihlolamvula** and **ijankomo**, and pointed out that while **ihlabankomo** had been relegated to lower status, and **ihlolamvula** selected as the book name for this bird, **ijankomo** was redirected. It was redirected to the Horus Swift, which at that time had no known individual name.

In similar fashion, the workshop participants felt that the Green Wood Hoopoe (formerly the Red-billed Woodhoopoe) did not really need both the names **inhlekabafazi** and **unukani**, while the closely related Common Scimitarbill (formerly the Scimitar-billed Woodhoopoe) had none, so the name **unukani** was redirected to the Common Scimitarbill (later **unosungulo** was coined for this bird).

In the case of the Cape Batis, with the two names **umnqube** and **udokotela**, the name **umnqube** was redirected to become the generic name for the three species of batis found in KZN. The name **udokotela** (< English doctor),

The Cape Batis and the Chinspot Batis both share the name **umnqube**, but in many parts of KwaZulu-Natal these birds are called **udokotela** because the chest patch resembles the skin bag of medicines worn by a traditional Zulu doctor.
© Adrian Koopman

incidentally, refers to the markings around the neck, which appear similar to the medicinal charms often worn around the neck by traditional healers (*izinyanga*). Such charms are known as *incweba*, and it is a regional form of this word – **incwaba** – that has become the name of the Chinspot Batis.

A similar situation is found among the helmetshrikes. The best-known of these birds is the White-crested Helmetshrike, with the two recorded names **iphemvu** and **uthimbakazane**. The name **uthimbakazane** has been confirmed for this bird, but the name **iphemvu**, corrected to **impevu**, has been redirected to become a generic name. The White-crested Helmetshrike has also had the name **abayeni** (husbands) confirmed as a well-known and widely used name, which has not previously been recorded in the literature.

12.6.4 Assignment

Turner and I have used the term 'assignment' for a particular process: during the workshops it often happened that no decision could be reached on what features of a particular bird should be used as the base of a name, or, having decided on a salient feature, we could not come to a decision on how to frame a name for that feature. On such occasions, we then checked the list of dictionary names without designated species, names such as **isambatha** (species of yellowish-brown bird resembling a plover), **isantinti** (species of small waterhen), **imbekle** (species of small bird with green head, and red tail and beak) and **ubholoba** (species of small bird found in the grass). If the bird for which we could not decide a name was in fact a yellowish-brown bird resembling a plover, then the name **isambatha** was <u>assigned</u> to it, and if our unnamed bird was a type of small bird often found in grass, then the name **ubholoba** might be <u>assigned</u> to it.

In this way, the Purple Heron (*Ardea purpurea*) was assigned the name **unoxhongo**, glossed in Doke and Vilakazi as "species of heron" (1958: 588), and the Black Crake, with no previously recorded Zulu name, was assigned the name **isiqhanazana**, an abbreviated version of the name *isiqhananazana* glossed by Doke and Vilakazi as "species of water-bird" (1958: 687).

The Three-banded Plover (*Charadrius tricollaris*) had no previously recorded Zulu name, and as no new name came readily to mind, the name **igwigwi** was assigned to it. Doke and Vilakazi give the second meaning of this word as "species of water bird" (1958: 287). They also give the name *igwigwigwi*, glossing it as "species of bird, destructive in the sorghum fields (so called because of its cry)" (1958: 287). Maclean gives the habitat as

". . . shorelines of lakes, dams, pans, rivers, sewage ponds. Marshes . . ." (1984: 224), so it seems reasonable to see this as a 'waterbird'. As a bird that eats "insects, crustaceans, molluscs, worms", it seems unlikely to be a bird that is destructive in sorghum fields, and its cry as described by Maclean does not suggest '*gwigwigwi*'. The name **igwigwi** may well be suitable, but its close resemblance to the name **igwigwigwi**, with its unconnected connotations, is somewhat disturbing.

In the relevant workshop, the name **isishishi** was given to the Southern Black Tit, with no previously recorded name. Turner's notes taken at the workshop suggest that this was an onomatopoeia-based coinage, but the name **isishishishi** does occur in Doke and Vilakazi with the meaning "species of small forest bird" (1958: 741), so this looks more like assignment than coinage. Curiously, Doke and Vilakazi also give the name **isicukujeje** specifically for the Southern Black Tit, but this name was not known by the workshop participants, and I had somehow neglected to include it when assisting with the Zulu names for Maclean's 1984 edition of *Roberts' Birds*.

12.6.5   Coinage

Strictly speaking, the term 'coinage' refers to the creating of new words, a process that can include both adapting and extending existing words as well as creating entirely new words that have no resemblance to any existing words. In the case of bird names, adaptation and extension usually involve the reworking of an existing bird name, often a generic name. Creating new bird names is always the manipulation of an existing word in the Zulu lexicon, which is not a bird name. The exception to this is an onomatopoeic name, which usually has no resemblance or connection to any existing Zulu word.

Let us take the following Zulu bird names as examples:

- The name **usantiyane** for the Green-winged Pytilia (formerly Melba Finch) means 'something like the **intiyane**' and it was an <u>adaptation</u> of the name of the Common Waxbill.
- The name **intiyaneluhlaza** (green *intiyane*) for the Green Twinspot is an <u>extension</u> of the same word (**intiyane**).
- The name **insukakude** is a completely new coinage for the Arctic Tern, but it is derived from the existing words *suka* (come from) and *kude* (far away).
- The onomatopoeia-derived name **inzwinzwi** for the White-backed Duck is imitative of the vocalisation of the bird, and thus not derived from any other existing word in Zulu.

Adaptation

Adaptation was one of the linguistic strategies used when the workshop tackled the Zulu names of thrushes. Maclean had given **umunswi** as the name of the Olive Thrush and the Spotted Ground Thrush. The Orange Ground Thrush and the Groundscraper Thrush had no recorded Zulu names, but were seen as part of the same 'thrush cluster'. The workshop participants decided to assign the name **insansa** to the Groundscraper Thrush, taking the name from Doke and Vilakazi, where it is glossed as "species of small bird, speckled black and white" (1958: 508). They wanted to keep the Olive Thrush, Spotted Ground Thrush and Orange Ground Thrush together, so kept the name **umunswi** for the Spotted Ground Thrush.[7] They then <u>adapted</u> it by adding an extra syllable to make **umunswili** for the Olive Thrush, and reduplicating the noun stem –*nswi* (and changing the noun class) to make **inswinswi** for the Orange Ground Thrush.

Adaptation was also a strategy adopted for two tern names. The process began with the Lesser Crested Tern, which had no previously recorded Zulu name. The workshop participants decided on a unique coinage here, the onomatopoeic name **ukliyo**. This was then adapted with the prefix –*no*– and the suffix –*ane* to produce the name **unonklilwane**, given to the Little Tern. A different onomatopoeic element /*bhakla*/ was then prefixed to –*kliyo* to produce the name **ubhaklakliyo** for the Caspian Tern. The repetition of the rasping sound /kl/ in this name captures perfectly the call, described by Maclean (1984: 283) as "Raucous *kraka-kraaa* and *kraaak*; rapid *kak-kak-kak-kak* . . .".

An interesting type of adaptation occurred in the session when names for various species of swift were discussed. Maclean had given for the Black Swift the names **ihlabankomo** (what stabs the bovine), **ihlolamvula** (what predicts the rain) and **ijankomo** (a shortened form of **ijiyankomo**, loosely 'what follows the cattle'). The participants selected the name **ihlolamvula** here, as traditional Zulu beliefs linked this bird to the coming of rain. The Little Swift and the Alpine Swift lacked names, so it was decided to adapt the notion of the coming of rain and create the names **inhlolazulu** (what predicts the weather) for the Alpine Swift and **imvuliyeza** (the rain is coming) for the Little Swift.

---

7. Later, in 2018, the name **umunswi** was adopted as a generic for thrushes, and the name was extended to **umunswi wehlathi** (forest *umunswi*) for the Spotted Ground Thrush.

Martins are also perceived in Zulu traditional ethno-biological classification to belong to the same folk-genus as swifts, and Maclean has **inhlolamvula**[8] as the name of the Rock Martin (later extended in the workshops as **inhlolamvula yamadawala** 'of the rocks'). In the same way, the House Martin received the name **inhlolamvula yasekhaya** (of the home). For the Sand Martin, however, the workshop participants adapted the name **inhlolamvula**, changing only one consonant to form **inhlolamfula** (literally 'what predicts the river'), a reference to this bird building its nests in river banks and, of course, more than just a nudging reference to the scientific name *Riparia riparia* (where 'riparia' means 'associated with rivers'). The name **inhlolamanzi** ('what predicts the water', i.e. rain) chosen for the Brown-throated Martin was also an adaptation of **inhlolamvula**.

Extension

Extension was one of the most common ways of forming names for birds that had no previously recorded Zulu names. This is the method used by the committees that decided on species-specific English names for South African birds in the first half of the twentieth century. The process is simple: one takes a generic name such as 'duck', finds a group of a dozen or so ducks, and then qualifies the word 'duck' with whatever feature seems most salient. It was this process of extension that created names like White-faced Whistling Duck, Fulvous Whistling Duck,[9] White-backed Duck, Yellow-billed Duck and African Black Duck. The workshop participants used the same linguistic strategy of extension in creating **idadelimlomophuzi** (duck which is yellow-billed), **idadelincane** (little duck = Hottentot Teal) and **idadelimnyama** (duck which is black = African Black Duck).

Other ducks, as a matter of interest, received new names that were a combination of various strategies. The previously unnamed White-backed Duck was perceived to have a distinctive voice, reflected in the onomatopoeic name **inzwinzwi**, as mentioned above. The White-faced Whistling Duck

---

8. Note that the name for the swifts is **ihlolamvula** while that for the martins is **inhlolamvula**. The first is in noun class 5, where there is no nasal in the noun class prefix; the second is in noun class nine, where there is always a nasal in the prefix. The difference is insignificant and has no bearing on the meaning. Bird names are often found in different noun classes, depending on regional forms, and often on the individual who uses the name.

9. For those not familiar with the word 'fulvous', Collins Dictionary (Hanks 1986: 613) defines it as "of a dull brownish-yellow colour; tawny".

was perceived to have a similar voice, so **inzwinzwi** was adapted to create **inzwinzwinzwi** for this bird. The name **inzwinzwi** was then treated as a kind of generic and extended with *–ebomvu* (which is red) to make the name **inzwinzwebomvu** for the Fulvous Whistling Duck. The Cape Shoveler simply received the name **unofosholo** (*–no–* + *ifosholo* 'shovel'), and the Knob-billed Duck was named **unosimila** (*–no–* + *isimila* 'horny growth').

Extension has also produced the names **ifefeluhlaza** ('green roller' = European Roller), **ifefemidwa** ('striped roller' = Purple Roller), **ifefebomvu** ('red roller' = Broad-billed Roller), all three extended from the name **ifefe**, a long-established name for the Lilac-breasted Roller (which has now become **ifefelihle** 'beautiful roller').

And as we saw in Chapter 2, where the generic and specific Zulu names of eagle were discussed, it is this strategy that produced **ukhozolumabala** (spotted eagle) for the Lesser Spotted Eagle, **ukhozolunsundu** (brown eagle) for Wahlberg's Eagle, **ukholumidwayidwa** (streaked eagle) for Ayres's Hawk-Eagle and many more. All are extensions of the generic **ukhozi** (eagle) just as 'Lesser Spotted Eagle', 'Wahlberg's Eagle' and the others are extensions of the generic 'eagle'.

'True' coinages

The 'true' coinages, i.e. those that are not extensions or adaptations of previously existing bird names, are perhaps the most interesting of the new Zulu names for birds. The onomatopoeic ones simply try to recreate in combinations of Zulu phonemes the sounds that the birds themselves make. Many existing names are onomatopoeic in nature: **ingududu** for the Southern Ground Hornbill with its *du-du du-du-du* call, **uphezukomkhono** for the bird that has the equally onomatopoeic name *Piet-my-vrou* to reflect the same call, and **unohhemu** for the Crowned Crane that calls *ma-hem, ma-hem, mahemu-hemu,* and is known in Afrikaans as the *mahem*. Equally onomatopoeic names were coined in the different workshop sessions, including **umcwicwicwi** for the Green Malkoha (formerly the Green Coucal), **isipopopo** for the Yellow-rumped Tinkerbird, **iklosi** for the Grey Penduline Tit, **usibó** for the Black-headed Oriole and **isicivó** for the Black-backed Puffback (formerly the Puffback Shrike).

Other coinages were adaptations of words that exist in the Zulu lexicon, but not as bird names. Many of these also ended up as names referring to song, often using metaphor and simile as in **isangulube** ([sounds] like a pig) for the grunting alarm call of the Gorgeous Bushshrike, a bird that has also long had the onomatopoeic name **ingongoni**. The newly coined name **isipoki** (spook), like the Afrikaans name *spookvoël*, refers to the mournful hooting whistle of the Grey-headed Bushshrike.

A separate book could be written on the creative thinking that lay behind many of the coinages created out of existing words in Zulu. The following are fifteen names that I found to be particularly creative:

- The name **isicibamanzi** is derived from the verb *ciba* (move like an arrow) and the noun *amanzi* (water) and is a perfect description of the way the Cape Gannet plunges into the water.
- The workshop participants felt that the most salient feature of the Southern Pochard was the way its call sounded like the alarm call of a vervet monkey, and so coined the name **isankawu** (like a vervet monkey).
- It is surprising that the highly distinctive African Harrier-Hawk (or Gymnogene, as many still continue to call it) had no previously recorded Zulu name. The workshop participants were unanimous that the distinctive feature of this bird was the way it can bend its foot to get into crevices where prey birds have laid their nests. The result was the name **ijikanyawo** (what bends the foot).
- Because of the way the Brown Skua robs other birds of their food, it was felt that this bird deserved the name **impisiyolwande** (hyena of the ocean).
- The Rose-ringed Parakeet is a recent invader species, so it is not surprising that it had no previously recorded Zulu name. The decorative ring around the neck was held to be a distinguishing characteristic, and for this the participants settled on the Zulu word *ucu*, a bead necklace given by a girl to a young man she fancies. The word is prefixed with –*no*– to form the name **unocu** (see Plate 45).
- More beadwork appeared in the name **umagumejana** for the beautifully coloured Pink-throated Twinspot. One of the participants recalled a girl of his acquaintance from the Gumede clan, famous for her highly-decorative bead work. The name consists of –*ma*– (characteristically) with a diminutive form of the clan name Gumede (see Plate 43).
- Based on a place name is the name **unongoyana** for the Green Barbet (Woodwards' Barbet), endemic to the Ngoye Forest and a few other highly restricted locations. The name is derived from –*no*– and the diminutive form of Ngoye, and so means 'the little one of the Ngoye Forest'.

- The Southern Yellow-billed Hornbill has long had the name **umkholwane**, which it has shared with the Southern Red-billed and the Crowned Hornbill. This name was selected as a generic name, and extended to give **umkholomphunga** (yellow [billed] hornbill), but the participants were insistent that the most commonly used name for this bird all over KZN was **uzazu**. This is, of course, not so much a coinage as a loanword, borrowed from the Disney film *The Lion King*, as discussed in Chapter 4 (see Plate 36).
- The salient feature of the African Broadbill, decided the workshop participants, was its screw-like display flight, and they used the word *isikulufu* (from English *screw*) to form the name **umasikulufu** (that which characteristically [flies like a] screw).
- The Red-capped Robin Chat (still called by many the Natal Robin) has a beautiful call, reflected in the coined name **unonkositini**, derived from –*no*– and the word *inkositini*, derived from the English word 'concertina'.
- The word *umathithibala* is used for a well-known species of medicinal aloe. It is 'borrowed' intact from the world of botany to serve as the name **umathithibala** for the Red-backed Shrike, the female of which has markings underneath which are exactly like those markings on the leaves of the aloe (see Plate 22).
- Discussion on the name of the invader Common (Indian) Myna centred on the way it seemed to be taking over the entire country, and this was reflected in the coined name **usothathizwe** (from –*so*– 'master of' + *thatha* (take [over]) + *izwe* (country).
- The Grey Waxbill is an overall grey bird, but has bright crimson feathers around the rump. These bright spots of crimson inspired the workshop participants to coin the name **ivuzigazi** (what leaks blood).
- At first the European Honey Buzzard received the name **umanyosini** (characteristically found among bees) but when Porter pointed out that this bird actually ate wasps, not bees, it was agreed to change just one letter to form **umanyovini** (characteristically found among wasps).
- The African Finfoot has the Afrikaans name *watertrapper* (watertreader), but it was not this name that inspired the coining of the Zulu name **igwedlamanzi**, but rather its habits of paddling (*gwedla*) the water (*amanzi*).

These fifteen names, out of the more than 100 names coined during several workshop sessions spread over three years, give an idea of the creative thinking exhibited by the Zulu-speaking birders who contributed to each workshop. It can be seen that all sorts of distinctive features gave rise to names: plumage, diet, behaviour and song, and at the same time a wide variety of linguistic strategies were adopted. Chapter 6 of this book explained the grammar of Zulu bird names in great detail, but one point should be made here. The Zulu language is rich in derivational possibilities, far more so than English, and the Zulu birders at the workshops showed an easy familiarity with the grammatical potentials. This kind of skill: creating new words and modes of expression through derivational grammar, as well as through literary devices such as metaphor and onomatopoeia, is usually something ascribed to the professional creators of traditional oral praise poetry – the *izimbongi* (bards) who create and perform praise poems (*izibongo*) in honour of kings and chiefs. But the composing of oral poems in honour of this or that person, or indeed of this favourite dog or ox, is still common among the ordinary (non-professional) Zulu citizen, male or female, young or old, and so it is no surprise to find the Zulu-speaking birders at the 2013–2017 workshops employing the same skills in wordplay. The process of following these discussions, and occasionally contributing to them, was at times exhilarating, and it was decidedly a privilege to be present when so many creative ideas and thoughts were whirling around the room.

12.7   INTERROGATION OF THE RESULTS OF THE 2013–2015 WORKSHOPS

After the three workshops of 2013–2015, I produced an 'Annotated Proceedings' in February 2016, listing all the names chosen for the KZN birds, the old, the revised and the newly coined, as well as identifying a few problem names. A one-day 'mini-workshop' with only Sakhamuzi Mhlongo, Themba Mthembu, Turner and myself was held at Turner's house in Durban to clarify and correct these problems.

In order to check whether or not the names passed scientific muster, Koopman and Turner passed the 'Annotated Proceedings' on to Porter, as mentioned already, a highly knowledgeable birder. In sessions with him in the second half of 2016 and the first half of 2017, Porter identified a considerable number of problems. For example:

- birds recorded that in fact did not occur anywhere near KZN and the opposite: birds not recorded in the workshops that did in fact occur in KZN;
- names that suggested characteristics that the birds did not in fact have;

- names that suggested that certain birds were related in some way when they were not;
- frequent identification of distinguishing characteristics of certain birds that were markedly more salient than the ones chosen in the workshops as the basis of names.

As I look through my copy of the 'Annotated Proceedings', some of Porter's suggestions for querying and amendments stand out. The following are examples:

The 'Annotated Proceedings' show the names for Francolins and Spurfowl as sharing various forms of **inswempe**, **ithendele** and **isikhwehle**. For example, the Coqui Francolin is named **inswempe**, the Crested Francolin **isikhwehlesiqhova** (crested *sikhwehle*), Shelley's Francolin **ithendele**, the Grey-winged Francolin **ithendelomlotha** (grey *thendele*), and so on. Porter's suggestion was that the names **inswempe**, **ithendele** and **isikhwehle** should all be confirmed as generic names for this cluster of birds and where one of these names had been used as a species-specific name, such as **inswempe** for the Coqui Francolin, this be extended to make a unique name for this francolin. Discussion on this point made it clear that we needed to abandon the earlier practice of using a generic name for just one species of the genus.

When it came to the names of the korhaans and bustards, Porter stressed that all these should be uniquely named. We looked at the confusion arising out of naming the Red-crested Korhaan '**ufuba**' (perhaps in error for Doke and Vilakazi's *ufumba*) and how this could be confused with the word **ifuba** used for the pelican. In the 2017 follow-up workshops, the Red-crested Korhaan was renamed **umngqithi** and the name **ufumba** transferred to the Black-bellied Bustard.

I indicated under the heading 'Assignment' that I had misgivings about assigning the name **igwigwi** to the Three-banded Plover. The name came up again when the whole issue of plovers and lapwings was discussed with Porter. Effectively, it was a case of determining which of these birds had a call that sounded like *gwi gwi* and which had a call that sound like *ti ti ti hoye*. Porter suggested, inter alia, that the name **ititihoye** be removed from the Senegal Lapwing (formerly the Lesser Black-winged Plover) and that the name of the Grey Plover be changed from **ititihoyelimlotha** (grey 'titihoye') to **igwigwelimlotha** (grey 'gwigwi').

We had originally coined the name **umhloshana** (the little white one) for the Pied Avocet (and later discovered that we had 're-coined' the same name for the Acacia Pied Barbet). On the advice of Porter, the avocet was given the name **usipheshula,** which referred to its distinctive long up-curved beak, and the barbet was renamed **unomunga** (from *–no–* and *umunga* 'acacia tree').

An earlier workshop had coined the name **unocilongo** (from *–no–* and *ichilongo* 'trumpet') in reference to the call of the White-browed Robin-Chat (previously Heuglin's Robin). Porter disagreed: the bird's call sounded far more like a flute than a trumpet. Participants at the July 2017 workshop at eShowe agreed, and the name was changed to **unomtshingo** (from *–no–* and *umtshingo* 'reed-flute').

Several more pages of examples could be given of Porter's expert advice on the Zulu names of birds and how his suggestions were almost invariably heeded at subsequent workshops.

It was interesting working with someone who knew the birds as well as the Zulu-speaking workshop participants who had revised the old and coined the new Zulu bird names, but who approached the topic of bird naming from a different perspective to that of the scientist. Occasionally science and knowledge of Zulu meaning coincided to produce new revelations. I recall working with the names of the thick-knees (dikkops) with Porter, when we came to the name **umbangaqhwa**, a long-established name for the Spotted Thick-knee (earlier Spotted Dikkop) and I explained to Porter that the underlying meaning was 'what causes frost' and that I had never been able to understand why. Porter knew the bird and had seen the Zulu name but did not know its underlying meaning, but once he knew it, the scientist instantly saw the connection: the Spotted Thick-knee does not run away from danger but crouches low, hoping to be taken for a stone or rock. With the white flecks on its back, it looks just like a stone covered in frost. Hence the name (see Plate 28).

12.8   THE SECOND ROUND OF WORKSHOPS

Once interrogation of the 'Annotated Proceedings' was completed, it was clear that it was time for another workshop, with as many as the original participants as possible to be included, in order to answer the questions raised about certain names from the first series of workshops. Accordingly, Turner organised another three-day workshop in June 2017, again to be held near Mkhuze where the 2015 workshop had been held. This was to be a slightly different workshop as traditional beliefs about birds (omens, portents, taboos, etc.) were to be discussed as well. For the first time, Porter was invited to be a participant.

The first day was scheduled for discussion about the name problems, with the second and third day scheduled for discussions about traditional bird beliefs:

- On the first day we were able to complete less than a third of the list of problems to be addressed. Certain problems required far more discussion than the original revision or coining of a name had needed. After the first day, discussion of problematic names was halted, with the decision that this would continue at a second workshop to be held as soon as possible within that year.

- On the second day the workshop was divided into three groups of four people each and each group was given a list of 'belief categories'. One group discussed birds of evil omen, of bad luck, witchcraft and death; the second discussed birds related to predictions of rain and other weather, including 'thunder birds' and 'lightning birds'; and the third group discussed harbingers: birds that marked the seasons or the times of day. Discussion was lively and it was clear that many of the traditional beliefs recorded in the literature were still valid among the Zulu participants (and the Zulu population) seventeen years into the twenty-first century. After lunch each group appointed a spokesperson to report back in a plenary session. It was agreed that the discussion had been very productive, but had only scratched the surface of current Zulu bird beliefs. A follow-up workshop on this topic would be held in 2018.

- On the third day, discussion centred mainly on the use of birds in the education of Zulu-speaking children. Different speakers spoke about giving talks on birds to schoolchildren, either at the schools themselves or when the children went on 'bird safaris', on the setting up of bird clubs for Zulu-speaking youth, and on the role of bird quizzes, bird booklets and bird apps. Further discussion centred on education for the bird guides themselves, including the merits of various training courses on bird guiding, the organisations that ran these courses, the problems involved and the qualifications that could be gained. This conversation would be followed up via email and online discussion groups.

Given that the name problems had not been completed by the end of the first day of the June workshop, Turner scheduled a second workshop, this time a two-day workshop at the end of July 2017, to be held in eShowe. This time two days were adequate to complete the work, and all the questions raised by Porter during his interrogation of the 'Annotated Proceedings' were answered.

It was at this July workshop that the question of generic names arose. Turner, Porter and I had earlier decided that Zulu needed generic names for birds as well as species-specific names and these needed to be confirmed.

As I look through my notes from the eShowe workshop of 31 July to 1 August 2017, a number of points jump out at me.

The name of the African Darter (*Anhinga rufa*): I had identified problems with the existing name **ivuzi** as far back as the first workshop in 2013. It was given as a Xhosa name in Maclean's 1984 *Roberts' Birds*, but could be confused with **umvuzi** for the Fan-tailed Warbler (formerly the Broad-tailed Warbler) and **uvuze** for the Natal Grassbird (now known as the Cape Grassbird).

A name based on *inyoka* (snake) and *inyoni* (bird) had been suggested at the June 2017 workshop but there had been no enthusiasm for this idea, no support for any other ideas and lengthy discussion preceded the decision to continue at a later workshop. When this difficult question of a new name for the African Darter came up again at the eShowe workshop, it took less than two minutes for complete acceptance of the name **inyoninyoka** (bird-snake). Such are the different dynamics among the same actors at different times and in different places.

The eShowe workshop was also able to solve with little difficulty the issue of a name for the Cape Teal – a seemingly unsolvable problem only two months earlier. The name **unosikhutha** (from *–no–* and *isikhutha* 'mould or mildew, as on bread') was quickly found suitable as reflecting the pale mottled plumage of this duck.

The difficulties about the names of the smaller raptors, though, took quite a bit longer. One of the problems is that we had earlier used the name **uheshe** as a generic for a loose group of hawks, goshawks and falcons, and used the same name as a species-specific name for the Lanner Falcon. We had coined the diminutive form **uheshana** (little *heshe*) for the Pygmy Falcon but simultaneously used the name **uheshane** as a generic for sparrowhawks. The end result was total confusion. Porter helped to distinguish between hawks, goshawks, harriers, and falcons and it was a big step forward when a decision was taken to use **uheshane** as a generic for sparrowhawks, **uklebe** (an old spare name for the Lizard Buzzard) as a generic name for falcons, and **umamhlangeni** as a generic for harriers. After this decision, it was easy to add extensions so as to distinguish between, for example, **uklebemawa** (cliff falcon) for the Lanner Falcon and **uklebe osankonjane** (falcon which is like a swallow) for the Eurasian Hobby.

Problems that Porter had raised earlier about the names of cisticolas were also dealt with decisively at the eShowe workshop, using both **uncede** and **intinga** as equal generic names. This enabled us to distinguish between, for example, the Desert Cisticola with the name **uncede womhlane** (*ncede* of the desert) and the Cloud Cisticola, with the name **intingamafu** (*intinga* [of the] clouds). The eShowe workshop decided to retain the generic name **uncede** as a species-specific name for the Neddicky, as it is this Zulu word, via Xhosa, Dutch and Afrikaans, which is the basis of the word *neddicky*.[10]

The major issue on the second morning of this workshop, was the discussion on the kingfishers, and particularly on what to choose as a generic name. For a full hour, the participants could not reach agreement. As there was no distinguishing characteristic that they could see that covered all nine species of kingfisher that occur in KZN, it was impossible to coin a name. And as no single kingfisher seemed to represent the whole group, it was not possible to agree on one of the species names as being suitable for the generic name. Eventually it came down to a vote, and one of the four names long recorded to the Brown-hooded Kingfisher – **indwazela** – was assigned generic status. This name has connotations of sitting motionless while staring intently forward, and a majority of the participants felt that it was suitable for most of the kingfishers. That meant that the name **indwazela**, previously taken away from the name-rich Brown-hooded Kingfisher and assigned to the previously unnamed Mangrove Kingfisher, could no longer be used for the latter. So what to call it? Again discussion went on and on, mainly between a camp that favoured a word based on *inkalankala* (crab) as this was the main food source for this bird, and what could loosely be called the 'anything but crab' camp. Eventually three names were put to a vote, and the 'crab camp' won: the name **unonkalankala** is now the 'official' book name for the Mangrove Kingfisher.

What actually makes a bird name 'official'? Who decides, or rather, who has the power or authority to decide? These are questions I try to answer in this last section of the chapter.

---

10. See Chapter 9, footnote 16, on the origin of the word Neddicky.

12.9   FINAL LIST OF ZULU BIRD NAMES: THE PROCESS OF ACCEPTANCE

Desmond Cole, as someone who grew up on a remote farm in Botswana, was delighted to find that the epithet in the scientific name of the Groundscraper Thrush – *litsitsirupa* – was the same Tswana name he had known since young. Not only did he know the Tswana names of the birds he grew up with:

> In some cases, too, I used English names of my own devising. I suspected that this bird was some kind of thrush, for it resembled a bird thus named and illustrated in a tattered old book on the fauna and flora of England. However, I named it the "Salute Bird", from its habit of flicking a wing after running a short distance and stopping, smartly to attention, on a stone or dry cow pat. So it was Salute Bird, or *letsetseropa* as, years later, I learned to spell it, this being the official rendering in Tswana (Cole 1984: 11).

Anyone is entitled to give a name to a species of bird as Cole did. When I moved from Durban in 1978 to live in a wooded outlying suburb of Pietermaritzburg, I met up with a number of bird species I had never known in Durban. Purchasing my first ever copy of *Roberts Birds*, I was able to identify many of them by sight. Others remained elusive, represented only by an unknown call coming from the bush. One of these was the Orange-breasted Bushshrike, which, for lack of a better name, I named the *Sweety-Sweety-Woo*. I used this name for quite a few years before someone identified this bird for me by its proper name.

Cole and I can make up these individual names for different species of birds, but neither of us have any thoughts other than that these are probably short lasting, highly individual, and, of course, have not the slightest air of officialdom about them. When, however, during the earlier years of the twentieth century, the List Committee of the South African Ornithology Union met regularly to revise existing English bird names, and in so doing discard many of the old names and replace them with new ones, they had every intention that the new names should become general in use, last for a long time and, above all, become official.

The question is: when one is revising, changing, and making up new bird names, what are the processes that should be followed if one wishes the names to be used 'officially'? By 'officially', I mean in all published bird guides, in scientific reports, in academic articles, dissertations and theses, in governmental literature, including texts approved for use in education, and in all such texts and contexts where the reader would expect that the bird names given are generally and formally accepted.

In South Africa, the Pan South African Language Board (PANSALB) is an official, government-linked body that has 'official' control over the eleven official languages of South Africa. Under PANSALB are the various Language Boards of these eleven languages. The Zulu Language Board,[11] for example, makes decisions about spelling revisions and terminology development among other things. The business of the Zulu Language Board was the development of terminology in particular disciplines, (for example macro-economics, mathematics or dentistry). If the latter, the Board would spend two or three days, just like the participants of the various Zulu bird name workshops, in deciding what the appropriate Zulu words or phrases should be for dental terminology.

Such terminology as developed by this Board, (or now by other structures within PANSALB?), has weight and authority, and is often published in 'terminology handbooks' published by the government printing works in Pretoria.

Neither PANSALB nor any of its subsidiary structures dealing with Zulu (whether still alive or long since dead) have been involved in any of the 2013–2017 Zulu bird name workshops, so the associated weight and authority is absent from the final list of birds. In seeking such weight and authority, Turner and I have had two choices: either to seek this authority from PANSALB itself, or not to do so and go some other route.

---

11. I suspect that the Zulu Language Board does not still exist. A UNESCO World Languages Report Survey Questionnaire completed in 2002 by a certain Buyisiwe Phyllis Mngadi of the National Language Service, Department of Arts, Culture, Science and Technology says (my emphasis):

> One of the major contributions of the Zulu Language Board (<u>now disbanded</u>) was the standardisation of Zulu orthography rules, now taken over by the Pan South African Language Board, on issues such as where to use capital letters in days of the week, e.g. UMgqibelo.

On completion of this manuscript I again attempted to find information about this board, but was unsuccessful. PANSALB, however, has its own website www.pansalb.org and is also linked to 'Brandsouthafrica'.

Going through PANSALB

Going via PANSALB means sending the list to PANSALB with a request for official approval. The following are possible responses:

- PANSALB refuses to grant authority and 'official' blessing to the revised list of Zulu names on the grounds that the research work and the workshops were not carried out through official PANSALB channels and procedure.
- PANSALB recognises the high academic standards of the research processes and is happy to give the revised list of bird names its blessing.
- PANSALB recognises the high academic standards of the research processes and takes over the project as its own, including adding the names of its own officials to the full report (or even replacing the names of Turner and Koopman), and insisting on publishing the report under its own aegis and imprint.

Not going though PANSALB

Turner and I understand that if recognition, leading to acceptance (with or without 'official' government approval), is to be obtained, then publication is necessary. Doing so through a publisher recognised and supported by BirdLife South Africa accrues its own kind of authority and moreover ensures dispersal to the widest group of interested parties: birders particularly, but also linguists and those interested in Zulu culture and heritage, whether they are Zulu speaking or not. Accordingly, we were delighted when Porter suggested sending the final list to the John Voelcker Bird Book Fund in Cape Town. This publisher not only publishes the various editions of *Roberts Birds*, but also other specialist books, such as Richard Dean's recent *Warriors, Dilettantes & Businessmen: Bird Collectors During the Mid-19th to Mid-20th Centuries in Southern Africa.*

In June 2018, the manuscript of the projected book was nearing completion. It contains a completed list of Zulu bird names, with references to all past Zulu names for birds in the KZN area, and a 30-page 'Introductory Essay', which, like this chapter, explains the 'mechanics' of the 2013–2017 workshops.

At the same time, Turner and I have published articles in academic journals on various aspects of our bird research, such as an article on the morphology (i.e. grammar) of Zulu bird names, and another on the linguistic strategies (see section 12.6) involved in revising and creating Zulu bird names.

As to whether these names will eventually gain acceptance as standard Zulu names for birds occurring in KZN, let me quote once again from Maclean's introduction to his 1984 fifth edition of *Roberts' Birds of Southern Africa*, where he speaks about the African language bird names that he took such pains to improve upon:

> In most cases I accepted the linguistic experts' opinions without question, but in others I was able to discuss certain points with them, or at least to query a few individual names that did not seem appropriate. Only time will tell with what degree of cooperative accuracy we have succeeded (1984: xxx).

In the Zulu bird name workshops Turner and I also accepted the Zulu-speaking bird experts' opinions, but were fortunate that we were able to discuss all of them where necessary. At the end we too queried a few individual names, and held two further workshops to discuss these queries with the experts.

Whether we were successful or not, only time will tell.

12.10 APPENDIX: THE KINGFISHER'S TALE

12.10.1   Nine birds need nine names

The names of the kingfishers present an interesting problem. There are ten species of kingfisher that occur in southern Africa, and all of them except the Grey-headed Kingfisher occur in KZN (see Plates 65–72). For the nine species that are found in KZN, nine distinct and different names have been recorded in the literature, suggesting the 'ideal' situation where each discrete species has its own discrete Zulu name. However, this is not the case as can easily be checked in Maclean, which at the time of the 2013–2017 workshops was still the 'official' representative of the bird names in African languages. This is how he sees the naming of kingfishers (1984: 373–81), as compared to the three previous editions of *Roberts Birds* edited by McLachlan and Liversidge (1957, 1970, 1978):

As can be seen from Table 12.1, taken from McLachlan and Liversidge's second edition of 1957, the Zulu names for kingfishers in the first four editions of *Roberts' Birds* were few and far between: one name each for the Pied Kingfisher, Giant Kingfisher, African Pygmy Kingfisher and the Brown-hooded Kingfisher. This was the case right up until, and including, the 1978 fourth edition of *Roberts*, 30 years after Doke and Vilakazi had published the first

Table 12.1   Zulu names for kingfishers in McLachlan and Liversidge (1957) and
Maclean (1984).

English and Latin names	McLachlan and Liversidge's Zulu names (1957)	Maclean's Zulu names (1984)
Pied Kingfisher (*Ceryle rudis*)	isiquba	isiquba, isixula
Giant Kingfisher (*Megaceryle maxima*)	isivuba	isivuba
Half-collared Kingfisher (*Alcedo semitorquata*)	No Zulu name	isixula
Malachite Kingfisher (*Corythornis cristata*)	No Zulu name	inhlunuyamanzi, uzangozolo
African Pygmy Kingfisher (*Ispidina picta*)	isipigileni (Natal Kingfisher)	isiphikeleli, inhlunuyamanzi, uzangozolo
Woodland Kingfisher (*Halcyon senegalensis*)	No Zulu name (Angola Kingfisher)	unongozolo
Mangrove Kingfisher (*Halcyon senegaloides*)	No Zulu name	unongozolo
Brown-hooded Kingfisher (*Halcyon albiventris*)	unongozolo	indwazela, unongozolo, unongobotsha
Striped Kingfisher (*Halcyon chelicuti*)	No Zulu name	No Zulu name

edition of their *Zulu-English Dictionary*. When Maclean published the fifth
edition in 1984, the following names were available in Doke and Vilakazi and
were the basis of his addition to the Zulu kingfisher names: The names below
are as they are given in the 1958 revised edition of the dictionary:[12]

It was from these names that Maclean and I selected names to go into the
fifth edition of *Roberts*, where although nine names for kingfishers were now
entered into the pages of the famous bird guide, there were still 'problems'.

---

12. 'D&V 542' refers to page 542 in the Doke and Vilakazi *Dictionary*.

D&V 542: *indwazel*a: Brown-hooded Kingfisher, *Halcyon albiventris*

D&V 339: *inhlunu-yamanzi*: Natal Kingfisher, *Ispidina natalensis*; also Malachite Kingfisher, *Corythornis cyanostigma*

D&V 661: *isiphikeleli*: Natal Kingfisher, *Ispidina natalensis*

D&V 714: *isiquba*: Pied Kingfisher, *Ceryle rudis*. [vl *isiqula*]

D&V 715: *isiqula* (Not in Maclean): Pied Kingfisher

D&V 839: *isivuba*: Great African Kingfisher, *Ceryle maxima*. [cf. *unongozolo*]
       [See also D&V 839: *ivuba*: 1. Pelican; 2. Button quail.]

D&V 869: *isixula*: 1. Pied Kingfisher, *Ceryle rudis*; 2. Blue Kingfisher, *Alcedo semitorquata* [= Half-collared Kingfisher]

D&V 586: *unongozolo*: Brown-hooded Kingfisher, *Halcyon albiventris;* Angola Kingfisher, *H. cyanoleucus* [= Woodland Kingfisher, *H. senegalensis*]; Mangrove Kingfisher, *H. irroratus*.

D&V 887: *uzangozolo*: Natal Kingfisher [cf. *isiphikeleli*]

Table 12.1 shows that far from the situation of 'one bird one name', we have a situation where one kingfisher has no Zulu name, four have one Zulu name, two have two Zulu names and two have three Zulu names (nine birds altogether).

This situation is fairly typical of a folk taxonomy except there is no generic name for all kingfishers (as with 'kingfisher' in English). Rather we have generics with a smaller number of birds: **unongozolo**, for example, is a generic for the Woodland, Mangrove and Brown-hooded Kingfishers.

As can be seen in the following list, various renaming strategies were applied in the workshop that dealt with kingfishers:

Pied Kingfisher: The recorded name **isixula** was kept for the sole use of the Half-collared Kingfisher and the recorded name **isiquba** was relegated to historical status. Workshop participants were clear that this bird was well known as **ihlabahlabane**, a name consisting of the reduplicated verb *hlaba* (prod, stab) and the species-forming suffix *–ane*. The reduplication of the verb indicates a continual prodding or stabbing.

Giant Kingfisher: The name **isivuba** was confirmed.

Half-collared Kingfisher: The name **isixula**, given by Doke and Vilakazi for this species as well as for the Pied Kingfisher, was confirmed for the use of this species alone.

Malachite Kingfisher: The workshop participants regarded the name **inhlunuyamanzi**, which this bird shares with the African Pygmy Kingfisher, to be offensive, consisting as it does of the offensive term *inhlunu* (vulva) and the noun *amanzi*, suggesting an image of the bird plunging into water. It was accordingly relegated to historical status. The name **uzangozolo**, which Doke and Vilakazi give for the sole use of the similar African Pygmy Kingfisher, was re-assigned to the Malachite Kingfisher instead.

African Pygmy Kingfisher: Doke and Vilakazi have three names for this kingfisher (previously Natal Kingfisher): **isiphikeleli**, **inhlunuyamanzi** and **uzangozolo**. As we have seen immediately above, the supposedly offensive name **inhlunuyamanzi** has been sent to the archives, and the name **uzangozolo** given to the Malachite Kingfisher. The remaining name **isiphikeleli** was re-assigned to the Brown-hooded Kingfisher. The previously unrecorded, but apparently well-known onomatopoeic name **isikilothi** was confirmed for this bird instead.

Woodland Kingfisher (previously Angola Kingfisher): This species was one of three kingfishers that Doke and Vilakazi had included under the name **unongozolo**. As it was not unique to this bird, a new name was coined: **imbuyelelo**. This name is a noun derived from the verb *buyelela* (persistently return to) and was coined for this bird because after hawking its insect prey it always returns to the same branch.

Mangrove Kingfisher: The name **unongozolo** could have been used for this bird, as it was one of the three kingfishers included by Doke and Vilakazi under this name. For reasons that are not clear, the workshop participants re-assigned the name **indwazela** (used for the Brown-hooded Kingfisher) to this species.

Brown-hooded Kingfisher: This bird had the three names **indwazela**, **unongozolo**, and **unongobotsha** in Maclean. The name **indwazela** was given to the Mangrove Kingfisher, the name **unongozolo** was given to the previously unnamed Striped Kingfisher, and the name **unongobotsha**, which was completely unknown to the workshop participants, was relegated to historical status. This left what is certainly one of the best-known kingfishers without a name, so the name **isiphikeleli** was re-assigned from the African Pygmy Kingfisher.

Striped Kingfisher: This kingfisher species, the only one not to receive a name in Doke and Vilakazi's dictionary, was given the name **unongozolo**, previously used as a generic name for the Woodland, Mangrove and the Woodland, Mangrove and Brown-hooded kingfishers.

Decisions made at the 2013 workshop

The position at the end of the 2013 workshop, where the names of the kingfishers were discussed, is shown in the column on the right of the table below. The Maclean names are included for comparative purposes.

As can be seen from Table 12.2, the situation of Zulu names for kingfishers has become much more orderly. There are nine names for nine birds, as was the case before, but now each of the species found in KZN has its own discrete Zulu name. In the process, much moving about of names has taken place, the names *inhlunuyamanzi*, *isiquba* and *unongobotsha* are now in the same situation as the names Plantain-eater and Loerie/Lourie for the turaco cluster, or the name Gymnogene (now African Harrier-Hawk). Three new, previously unrecorded names, have taken their place: **ihlabahlabane**, **imbuyelelo** and **isikilothi**. The new arrangement of names for the kingfisher cluster has emulated the scientific, English and Afrikaans names for kingfishers in more ways than one: not only is this new arrangement a much tidier 'one bird one name' situation, the Zulu names have, in the process of revision, involved a throwing out of old, well-known names, and a replacement with new names. If these new names are in fact names that have simply never made it into print before (despite being well known to many Zulu speakers), then the Zulu-speaking birding public should be much happier than the English-speaking birders who voice their unhappiness at every major name-changing exercise.

Table 12.2   Zulu names for kingfishers in Maclean (1984) and as approved in the 2013 first Zulu bird name workshop.

English and Latin names	Maclean's Zulu names (1984)	2013 workshop-approved Zulu names
Pied Kingfisher (*Ceryle rudis*)	isiquba, isixula	ihlabahlabane
Giant Kingfisher (*Megaceryle maxima*)	Isivuba	isivuba
Half-collared Kingfisher (*Alcedo semitorquata*)	Isixula	isixula
Malachite Kingfisher (*Corythornis cristata*)	inhlunuyamanzi, uzangozolo	uzangozolo
African Pygmy Kingfisher (*Ispidina picta*)	isiphikeleli, inhlunuyamanzi, uzangozolo	isikilothi
Woodland Kingfisher (*Halcyon senegalensis*)	unongozolo	imbuyelelo
Mangrove Kingfisher (*Halcyon senegaloides*)	unongozolo	indwazela
Brown-hooded Kingfisher (*Halcyon albiventris*)	indwazela, unongozolo, unongobotsha	isiphikeleli
Striped Kingfisher (*Halcyon chelicuti*)	No Zulu name	unongozolo

At the end of the 2013 workshop, when this neater 'one bird one name' status was achieved for kingfishers, there remained one problem: English and Afrikaans had retained a generic name for the cluster (kingfisher and *visvanger* respectively) but there was no generic name for the kingfisher cluster in Zulu. This would have to wait for the 2017 workshop in eShowe.

At this workshop, most of the discussion centred on a generic name for kingfishers, and then, having chosen the name **indwazela** for this purpose, what then to call the Mangrove Kingfisher. Below are the final decisions, and the underlying meanings and reason for names chosen at the 2013 workshop.

Decisions made at the 2017 workshop

<u>Generic name</u>: Use **indwazela** as a generic for kingfishers. The word connotes sitting motionless while staring intently forward and is suitable for all kingfishers. The word was originally used only for the Brown-hooded Kingfisher, but was moved in the 2013 workshop to the Mangrove Kingfisher.

<u>Pied Kingfisher</u>: retain the name **ihlabahlabane**: it refers to the continually prodding of this bird while fishing.

<u>Giant Kingfisher</u>: retain the well-attested and old name **isivuba**: it refers to the fact that this bird resembles the pelican in general appearance.

<u>Half-collared Kingfisher</u>: retain the old and well-attested name **isixula**.

<u>Malachite Kingfisher</u>: retain both the old and well-attested names: **uzangozolo** and **inhlunuyamanzi**. The former name refers to the shape of the bird's beak, and is reflected in the saying "*lo zangozolo ngomlomo*" ('don't give me that pursed-up sullen look', said to a mutinous child); the latter name literally means 'vulva of the water' (using an obscene word) and refers to the plunging of the bird into water.

<u>African Pygmy Kingfisher</u>: retain the old and well-known name **isikilothi**. It is onomatopoeic.

<u>Woodland Kingfisher</u>: retain the coined name **imbuyelelo**. It refers to this bird continually coming back to exactly the same perch.

<u>Mangrove Kingfisher</u>: delete the previously assigned name *indwazela*; replace with **unonkalankala** (< *inkalankala* 'crab') to refer to this bird's specific diet of crabs.

<u>Brown-hooded Kingfisher</u>: retain the well-known name **isiphikeleli**, one of four names originally assigned to this bird.

<u>Striped Kingfisher</u>: add the diminutive/species suffix *–ane* to **unongozolo**: **unongozolwane**.

12.10.2   Nine names for nine birds

We can now look at three sets of Zulu names for the kingfisher cluster: (1) the few names given in the first four editions of *Roberts Birds*, from 1940 to 1978; (2) the names given by Maclean in the 1984 edition after I had helped insert names from Doke and Vilakazi's dictionary, names that were still 'in print' in 2005; and (3) the names after the 2017 eShowe workshop:

Table 12.3   Three different sets of Zulu kingfisher names.

English vernacular name	*Roberts Birds* (1940–1978)	*Roberts Birds* (1984–2005)	After the 2017 Workshop
Generic name (*kingfisher* and *visvanger*)			indwazela
Pied Kingfisher	isiquba	isiquba, isiqula	ihlabahlabane
Giant Kingfisher	isivuba	isivuba	isivuba
Half-collared Kingfisher		isixula	isixula
Malachite Kingfisher		inhlunuyamanzi, uzangozolo	inhlunuyamanzi, uzangozolo
African Pygmy Kingfisher	isipigileni	isiphikeleli, inhlunuyamanzi, uzangozolo	isikilothi
Woodland (Angola) Kingfisher		unongozolo	imbuyelelo
Mangrove Kingfisher		unongozolo	unonkalankala
Brown-hooded Kingfisher		indwazela, unongozolo, unongobotsha	usiphikeleli
Striped Kingfisher			unongozolwane

The 'Zulu kingfishers' are now ready to present themselves to a Zulu-speaking birding public.

13 Change is in the air

13.1 YESTERDAY, TODAY AND TOMORROW

The basic theme throughout this book has been the intersections between birds, names and people. People interact with birds in a number of ways: they see them, admire them and weave them into their literature and their traditional beliefs. They eat them and, in turn, the birds eat their grain. In order to distinguish between one type of bird and another, people name birds.

Chapters 1 to 11 of this book have looked at the human-bird-language connections from a number of different perspectives:

- Chapter 1 considered the question of whether the 'names' of birds are really 'proper names' in an onomastic sense.
- Chapter 2 explored the issue of generic and specific names, both from a scientific perspective as well as from an ethnobiological perspective.
- Chapters 3, 4 and 5 all examined the underlying meanings of bird names, all of which showed the different ways in which humans have perceived birds.
- Chapter 6 looked at the derivational aspects of Zulu grammar that allowed for a wide range of different possibilities for the adaptation, extension and creation of bird names.
- Chapter 7 discussed how bird calls are 'verbalised' into human language.
- Chapters 8 and 9 explored how birds are woven into different forms of Zulu traditional oral poetry, from the comparatively lengthy praise poems of kings and chiefs, to the more pithy expressions of proverbs and riddles.
- Chapters 10 and 11 looked at a wide variety of traditional beliefs relating to birds: birds as omens and portents, as harbingers of seasons, and a

variety of other beliefs that reflected the deep impact birds have had on human consciousness.

All of these chapters examined bird names, literary expression and beliefs typical of Zulu culture since it first started being recorded by colonial diarists, ethnographers and lexicographers. It was only in Chapter 12 that we looked at new developments in Zulu bird names, sparked off in part by a perceived need to situate Zulu bird names within the context of birding, a social activity that requires a unique name for each distinct species of bird, as became the case for English and Afrikaans during the twentieth century. This perceived need for such unique species-specific names led to the Zulu bird name workshops of 2013–2017, and the adaptation, extension and allocation of existing Zulu bird names as well as the coining of totally new names.

Besides the changes in Zulu bird nomenclature, other changes have taken place in the subtle interrelationships between people and birds in Zulu society. In the past, for example, barefooted herdboys spent much of their free time hunting and trapping birds, taking them home to eat, and listening to their elders talking about bird lore in the evenings around the fire in the evenings. Today these children are no longer herding cattle for they are at school, where some of them may learn about birds from an environmental programme that includes the use of the internet, laptops, mobile phones and associated technologies.

This final chapter now looks at these changes, firstly in the context of the past: interrogating the process in which Zulu bird names were recorded by colonial diarists, ethnographers and lexicographers who converted traditional oral Zulu knowledge into writing. I look at the problems involved in this process of conversion, and then go on to discuss some of the major figures who gathered bird names as part of other research into Zulu language, literature and culture. From this exploration of 'yesterday' I go on to look at some of the major Zulu figures who are involved today in Zulu bird names within the wider context of birding, avitourism, conservation and environmental education.

In 1923 R.C.A Samuelson recorded in his dictionary the traditional children's game *Bhula 'Ntsentse Bo!* where children competed to see who knew the most bird names. Today schoolchildren have a *Roberts'* bird quiz app on their mobile phones and undoubtedly have just as much fun as the children of three generations ago. This is but one of the changes that are explored in this chapter.

13.2   ZULU BIRD NAMES IN THEIR HISTORICAL CONTEXT

Here I look specifically at the recording of Zulu bird names in the second half of the nineteenth century and the first half of the twentieth century. The section begins by looking at the issues involved in transcribing oral knowledge as written knowledge and then goes on to examine the known and the (generally) unknown people who did the actual collecting of the knowledge.

13.2.1   Oral contributions and their transcription

The road between an orally supplied bird name and that same name in a later publication is one that is fraught with difficulties of two kinds: the first is the difference between oral knowledge and written knowledge; the second concerns the transition between two languages as well as between two or more different means/methods of recording the message.

Oral knowledge and written knowledge

Oral knowledge is held in common by an entire linguistic community; written knowledge is usually authored by one person. In any item of oral knowledge (a proverb, clan praises or *izinganekwane* [folktales], bird names) there are many variations depending on the person speaking out the 'knowledge item', the context of such speaking, and the place or region where such speaking out takes place. The praise singer who recites the praises of the clan chief will do so differently at each recitation. The grandmother who tells the story of the mongoose and the lion will not tell it in the same way as the grandmother in the next village, even if it is the same story. The person who gives you a particular name for a species of bird may give it in a different form the next day; a person giving you the name of the same bird in a neighbouring district may give a completely different name. Yet Western authors insist on the 'correct' name' and the 'correct' form of that name. Let us take the Zulu name(s) of the Rufous-naped Lark, for example. Depending on which author has collected the name from which source, the names of this bird are given as **ungqangendlela**, **uhuye, untilontilo, uqaqashe, iqabathule** and **ungqwashi** – six completely different names. In addition, the last of these has been recorded with four different forms of the same word: **ungqwashi, unongqwashi, umangqwashi** and **usangqwashi**. To the Western mind, it is unacceptable to have four different spellings: there has to be a 'correct' form. Yet in an orally based knowledge system, these names are all the same: they consist of the class 3a noun prefix *u–*, a name-forming prefix like *–no–* or *–ma–*[1] and the root *–ngqwashi*.

In a more extreme example of this, I noted in an earlier publication that the Zulu name for the Bushveld Arum (*Stylochiton natalensis*) is given in ten different forms in various publications.[2] We can select **umfana-kaNozihlanjana** (the little boy of the marshes), for example, to represent all ten variants, and then break this name down into its various elements, each of which (apart from */fana/* and */hlamb/*) can either be present or absent. There are sixteen permutations possible here, each of which would be considered correct by a Zulu speaker. And those sixteen variations are the <u>oral</u> forms. To put this name into writing, however, there are more decisions to be made: write it as one word, two words or hyphenated (as above)? Put the capital letter on the first letter, the first letter after the vowel or the first letter of the noun stem? (All are in fact acceptable.) Or don't use a capital letter at all? Adding in these permutations allows for a splendid 384 possible variations, all of which would be acceptable spellings. To a Zulu speaker all sixteen of the oral forms would be essentially the same 'name'. To the Western author, a choice must be made between 384 possibilities, and only one can be 'correct'.

Transition from one language to another; from one medium to another

In the recording of bird names since the time of Adulphe Delegorgue, the 'giver' of the name has invariably been a Zulu speaker, the person recording the name invariably not. People like Colenso, Bryant and Gerstner[3] were highly competent in Zulu, but were not mother-tongue speakers. Others who have recorded Zulu names of birds were undoubtedly less competent. Even a competent listener cannot always hear clearly, particularly if the speaker has some kind of speech impediment, or if there is background noise going on, such as hadedas calling.

Once the name has been recorded for the first time, a whole new set of problems arises. In their travel narrative, the Woodward brothers tell of the

---

1. Or in the case of **ungqwashi**, a zero manifestation of such a prefix.
2. Koopman (2015: 37–8).
3. See section 13.2.2.

problems when one is crammed in a tiny leaky tent during heavy rain and attempting to prepare specimens by the light of a single candle.[4] The same applies to writing down notes under cramped or rainy conditions, particularly if one if using a pencil or an old-fashioned nib/quill dipped in ink. Later, in the course of travelling, this notebook may get mud or spilled tea on it, and be folded up and put in a back pocket. Later again – sometimes considerably later – the recorder of the original information retrieves the stained and many-times folded notes and tries to make sense of them. Worse still, the notes may be given to someone else to be transcribed, often someone unfamiliar with both the handwriting and the Zulu language.

This is, of course, speculation, and it has not been possible to get hold of the fieldnotes or the notebooks of the Bryants and Godfreys of early bird nomenclature. Evidence of this problem did come to light, however, when in 2012 a Norwegian publisher published, in two volumes, the notebooks of Dr Henrik Greve Blessing.[5]

Blessing qualified as a doctor in Norway in 1893 and came to Zululand in 1901, spending four years learning about Zulu medicinal plants from local traditional healers, and writing down what he had learnt in his notebooks. What is unique about this publication is that while Volume I is a transcription with additional notes, chemical analyses of the medicinal plants and a quantity of high-quality photographs, Volume II is a facsimile of Blessing's original two notebooks in his own handwriting, allowing a glimpse into the kind of problems mentioned above. There is no space for a full analysis here, but I can give just two examples:[6]

- On page 16 of Notebook 1, Blessing has given the names *Umhlonhlo*, *Umhlonhlwana*, *Umsolasola* and *Umsololu* for a type of *Euphorbia*. The notebook reveals a peculiarity about Blessing's handwriting: he often writes his 'a's and his 'o's in the same way, so that the names *umhlonhlo* and *umhlonhlwana* could be transcribed by someone who knows no Zulu (or does not know Zulu medicinal plant names) as 'umhlanhla' and 'umhlanhlwana' or even as 'umhlanhlo', 'umhlonhla', 'umhlanhlwona', and so on. In the end, the non-Zulu-speaking editors

4. Woodward and Woodward (1897).
5. Paulsen et al. (2012).
6. See Koopman's book review of these two volumes (2014c).

have ended up with 'umhlonhlo' and 'umhlanhlwana', a decision that makes no sense given that *umhlonhlwana* is derived from *umhlonhlo* via the diminutive suffix *–ana*.

- On page 168 of the edited Volume I, the editors have recorded 'ISI-FULL' as one of two names for the Red Beech (*Protorhus longifolia*). Anyone with a rudimentary knowledge of Zulu would recognises this as a non-Zulu form. Checking the original handwriting in Volume II allows us to see that Blessing has written the correct Zulu name *isifuce* in a very ambiguous manner: a Zulu speaker would immediately see the word as '*isi-fuce*', while an English speaker would just as immediately see the word 'isi-full'.

While we do not have the Woodward brothers' original fieldnotes to hand, there are plenty of examples in their 1899 *Natal Birds* that at first glance are just uncorrected typos but when looked at together show a misreading of handwriting. Two examples are '**umch**wlane' for '**umehlwane**' (Cape White Eye) and 'hl**eba**bafazi' for '**ihlekabafazi**' (Green Wood Hoopoe). Many similar errors have also been copied faithfully into the relevant volumes of Stark and Sclater's 1900–1906 *Birds of South Africa*, and this is an excellent example of one of the difficulties involved in converting oral Zulu knowledge into published English knowledge: the perpetuation of errors once they enter into print.

Below we see the word **inhlekabafazi** (the correct name for the Green Wood Hoopoe) written in longhand by a person who consistently does not bring the pen back to the line on the down-stroke. After writing this name in a small notebook, the page is folded twice and put in a back pocket for several weeks, where it comes into contact with mud. Much later the person who wrote the name down opens the folded piece of paper and tries to make sense of what he has written. Is there any wonder it comes out as 'inhlebabafazi'?

13.2.2   Earlier contributors to the Zulu ornithonomasticon

Under this heading I only include those figures who have (or seem to have) collected Zulu bird names themselves from the Zulu-speaking inhabitants of the former colony of Natal and neighbouring Zululand. I am not specifically interested in publications such as the four volumes of Stark and Sclater between 1900 and 1906,[7] or the several revised editions of *Roberts Birds of South Africa* that have appeared over the years since Austin Roberts published his first version of this classic in 1940. These books do indeed contain a number of Zulu names for birds, but they were not collected in the field by the authors themselves.

Delegorgue

The French traveller, explorer and naturalist Adulphe Delegorgue (1814–50) can perhaps be considered the pioneer of collecting Zulu bird names and recording them in print. Between 1837 and 1844 he travelled through Natal and Zululand hunting and collecting specimens of natural history, and on his return to his native France he published two volumes about his travels.[8]

Details in his travel narratives show the young Delegorgue[9] to be an accomplished linguist who picked up a great deal of Dutch at the Cape on his arrival in South Africa, and a considerable amount of Zulu in his subsequent travels in the Zulu-speaking country. Indeed, his second volume contains a *Vocabulaire de la Langue Zoulouse* with over 800 lexical items, the earliest such published lexicon of Zulu.[10] There are only ten Zulu bird names in this vocabulary list, so although Delegorgue can be considered the first contributor to record Zulu bird names, he was certainly not a major contributor. He spelt the Zulu names as if they were French, so we have *omoucé* for **umunswi** (thrush) and the slightly less recognisable *ikoé* for **intshe** (ostrich). The words *izikova* for 'owl', *omkoloani* for 'hornbill' and *landa* for 'egret' are instantly recognisable as the modern forms **isikhova**, **umkholwane** and **ilanda**.

---

7. *The Birds of South Africa*, volumes I to IV.
8. Delegorgue (1847), published in English translation in 1990 and 1997.
9. He was only 23 when he started his Natal and Zululand travels and he died at sea at the untimely age of 36.
10. See Koopman and Davey (2000) for a detailed analysis of this *Vocabulaire*.

Delegorgue's *ikoalakoala* is also immediately recognisable as the modern **igwalagwala** (turaco) even though he mistakenly assigns this name to the pheasant or francolin.

Colenso

The Right Reverend John W. Colenso (1814–83), Bishop of Natal from 1853 to 1883, is well known for his dispute with the Church of England over an interpretation of the Pentateuch, and perhaps less well known for his contributions to the writing of Zulu. He published many ecclesiastical works in Zulu and a Zulu grammar as well, but it is his 1861 dictionary that we look at here.

There are 126 bird names in the dictionary, and a considerable number of entries resemble the following:

> **isAnzwili**: a small bird of the table-land, which makes a whistling sound
> **isiBelu**: bird with brownish wings and red breast, about the size of a dove
> **isiCelegu**: small bird with white spots

Two things become clear when we look carefully at this dictionary: (1) Colenso was well-versed in the Zulu language but he was no ornithologist, with little or no knowledge of the bird species in the wider area where Zulu was spoken;[11] (2) he collected the names from oral sources, very probably by asking a number of informants to give him the names of as many birds as possible, and then to describe them.

The second and third editions of this dictionary (1878 and 1884) contained the same number of bird names with the same lack of species identification, but the fourth edition, revised by Colenso's daughter Harriette Colenso and published in 1905, contained far more names and species identification. She had, by then, had access to the Woodward brothers' 1899 publication.

---

11. To be fair, Layard's *Birds of South Africa*, the first ornithological compilation of South African birds, only appeared in 1867. Woodward and Woodward, writing in the 1890s, continually refer to "Dr Layard" as their authority for identifying species; Colenso was unable to do so.

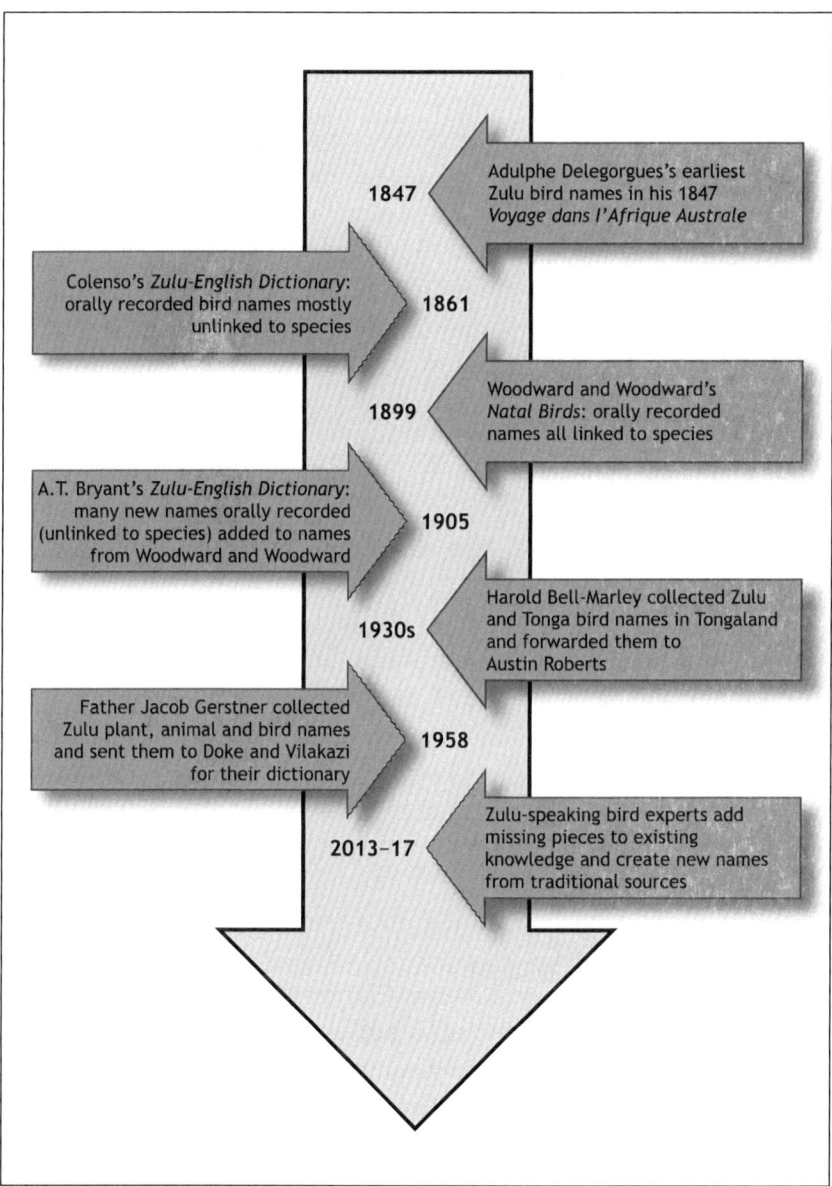

Figure 13.1   Early collectors of Zulu bird names.

Woodward and Woodward

The brothers Richard and John Woodward came out to the colony of Natal in the 1880s,[12] ostensibly as missionaries, but they also did a little farming on the side and spent a considerable amount of time travelling through Natal and Zululand collecting birds and butterflies and writing about birds. They published three articles on their travels in the British ornithological journal *Ibis*,[13] but it is their 1899 book *Natal Birds* that concerns us here.

This book contains 386 entries for specific species of KwaZulu-Natal (KZN) birds, although only 107 are assigned what Woodward and Woodward call the "native" name. This figure is less than the 126 names in Colenso's dictionary but at least each name is assigned to a particular species of bird.

In their travel narratives, Woodward and Woodward speak occasionally of their "native guides" or the porters that accompanied them on the march, but give no indication of accessing Zulu names, apart from the giving of the odd Zulu name for a species of bird. Nevertheless, we must assume that Woodward and Woodward recorded their Zulu names directly from oral Zulu sources, simply because there is nowhere else they could have got them. They came to Natal as missionaries and nineteenth-century missionaries were renowned for their ability to pick up a good command of the Zulu language very quickly. There is no reason to suppose that these brothers were any different, and with their abiding interest in birds, it seems highly likely they would have asked the Zulu names of birds from the Zulus with whom they came into contact.

Bryant

Father A.T. Bryant (1865–1953) was a British-born Catholic priest and missionary. He is well known for his 1929 historical work *Olden Times in Zululand and Natal* and his 1949 ethnographic work *The Zulu People*, but it is his much earlier 1905 *Zulu-English Dictionary* that concerns us here.

Bryant's dictionary contains 211 entries for birds, considerably more than the 126 names in Colenso and the 107 in Woodward and Woodward. Bryant can only have got these extra names from oral sources and, as he is known to have done considerable field research for historical and ethnographic publications, it is more than likely that he collected Zulu bird names 'on the side'. He certainly

---

12. Details of their lives, including dates of birth and death, are very sketchy.
13. Woodward and Woodward (1897, 1898, 1900).

collected the names of medicinal plants and information on their usage, as reflected in his lesser-known *Zulu Medicine and Medicine-Men*.[14]

His dictionary has been extremely useful in the writing of this book because Bryant frequently does not just identify a bird, but has much more to say about it. Consider the following entry:

> **uMabengwane**: Woodford's Owl (*Syrnium Woodfordi*) whose peculiar hoot is often heard in the woods at night, saying *Woza! woza! 'mabengwana!* (Come! come! 'Mabengwana – its mate, of course). Cp. **isiKova**.
>
> N.B. The fat of this bird, mixed with *iSokalakwazulu* (common washing-soda) is used as an *ihabiya* (1905: 371).

Bell-Marley

Austin Roberts published the first edition of the well-known series of *Roberts Birds of South Africa* in 1940. This immensely successful field guide had names for birds in a number of southern African Bantu languages, including, of course, Zulu. From the incomplete number of names, it was clear that Roberts had not consulted either Bryant's 1905 dictionary or Samuelson's 1923 *King Cetywayo Zulu Dictionary* (which also contains a number of Zulu bird names). The question then was – where did Roberts get his Zulu names from? An answer to this question is suggested by Richard Dean in his 2017 work on bird collectors in South Africa.[15]

According to the S2A3 Council, which, in 2002, launched a project to compile a biographical database of southern African science,[16] Harold Bell-Marley (1873–1945) was "an ardent naturalist, an enthusiastic collector, and a generous contributor of specimens to museums in southern Africa and overseas". He spent six weeks each year in the Ubombo area of northern Zululand, where both Thonga and Zulu were spoken. During this time he apparently lived with the local population,[17] where he would certainly have learnt the local names that he wrote on his specimen labels. Dean tells us:

---

14. Bryant ([1906] 1970).
15. Dean (2017).
16. See http://www.s2a3.org.za/bio/Biograph_final.php?serial=215 (accessed 26 April 2017).
17. See https://www.biodiversityexplorer.org/people/bell-Marley-hw.htm (accessed 30 April 2017).

One of Bell-Marley's minor interests was in indigenous bird names, and he wrote the isiZulu and Tonga names of birds on the data labels for his egg collection. This was almost certainly one of the sources for the African names of birds that were included by Roberts in the first edition of *Birds of South Africa* (2017: 91).

From almost continuous references in the writing of the Woodward brothers (1899), Stark and Sclater (1900–06) and Godfrey (1922, 1941), it is clear that in the latter half of the nineteenth century and the first half of the twentieth century, the colony of Natal was filled with amateur naturalists who sent notes to museum-based scientists on bird sightings and bird behaviour, often following this up with specimens they had collected of nests with eggs, or of the birds themselves. Many of these ardent amateur ornithologists would have had a good command of Zulu and some of them are certain to have compiled lists of Zulu bird names. Harold Bell-Marley just happens to be one whose name has been recorded in print as recording Zulu bird names.

Gerstner

Clement M. Doke and Benedict W. Vilakazi published the first edition of their *Zulu-English Dictionary* in 1948. They had the benefit of Woodward and Woodward (1899), Bryant (1905) and Samuelson (1923) to draw on, but also managed to collect considerably more Zulu names than their predecessors, entering a total of 388 bird names. The same question arises: where did they get their additional Zulu names from? There is a clue in the 'Acknowledgements' of the second edition:

> Special mention must be made of the help of the Rev. J. Gerstner, Ph.D., in placing at our disposal his published and unpublished collections of botanical terms, a large number of which have been incorporated in the dictionary . . . *In addition Gerstner supplied considerable lists of natural history terms* (Doke and Vilakazi 1958: xii; my italics).

Father Jacob Gerstner (1888–1948) was a Bavarian-born Catholic priest who came to South Africa in 1924 and was for many years (1928–42) the Superior of the Catholic mission farms in Zululand, giving him ample opportunity to indulge his love of botany and other aspects of natural history. It is extremely likely that the "considerable lists of natural history terms" mentioned by Doke and Vilakazi included lists of Zulu bird names.

The original Zulu-speaking sources: unnamed and named

All of the above-mentioned names – Colenso, the Woodward brothers, Harold Bell-Marley, and so on – are of course the white, mainly English-speaking colonialists who collected the bird names. The source of the information, however, was the whole body of Zulu speakers who had learnt their bird names and bird lore through oral transmission. Only one of the early writers gives any indication of where and from whom information was sought, the Reverend Robert Godfrey (1872–1948). His name is not mentioned above, because it was Xhosa bird names and bird lore that he put together, notably in his 1941 publication *Bird Lore of the Eastern Cape Province*. His name needs to be mentioned here now, though, because Godfrey actually does say where he got his information from. His book is filled with entries such as:

- "A boy of Holy Cross Mission says of the *inkanku* that . . ." (1941: 56).
- "An Emfundisweni scholar states that it feeds its young . . ." (1941: 55).
- "This is shown by the Rev. J.H. Soga's interpretation [that . . ." (1941: 51).
- "A St Cuthbert's pupil allows the young quails somewhat longer time to . . ." (1941: 40).
- "The following wonderful account of *umqunduluthi* comes from Jackson Nteta, Emfundisweni . . ." (1941: 47), and so on.

And in an earlier publication, Godfrey writes "Mr. Fred Madlingozi tells me that at one particular hole, five of these [glossy starlings] were caught in succession . . ." (1922: 47). Here we see Godfrey not only acknowledging a Xhosa source by name, but prefixing it with the honorary title 'Mr'. It can truly be said that Godfrey was a man before his time.

13.3   THE MODERN ZULU BIRD NAMERS

13.3.1   Those involved in the 2013–2017 Zulu bird name workshops

In the previous chapter I described the Zulu bird name workshops held between 2013 and 2017: the aims and goals and the processes followed, the revision of the previously recorded names and the creation of new names and steps taken for their acceptance.

More must now be said about the Zulu-speaking participants whose knowledge of and enthusiasm for birds led to the successful creation of a list of individual Zulu names for each species of bird occurring in KZN.

Part of their knowledge was the oral tradition each one grew up in: including older Zulu bird names and traditional bird lore. But another part was their identification of bird species in the modern Western tradition of birding, and this came from training and working as bird guides, either as self-employed guides or attached to game reserves, safari companies and other similar tourist ventures. Their training was usually a formal course linked to bodies such as BirdLife South Africa and the Wildlife and Environment Association of South Africa (WESSA), coupled with mentoring by experienced bird guides.

In Table 13.1, I have listed the qualifications of all those whose participated in the 2013–2017 Zulu bird name workshops. Most have bird guide qualifications on the National Qualifications Framework (NQF) at various levels. One of them has a doctorate in Zulu bird lore.

There is no space for detailing the experience and qualifications of each and every participant, but here is the full list of those who can be regarded as the 'bird namers of today'.

Table 13.1   Participants in the 2013–2017 Zulu bird name workshops.

Name	Sex	Qualifications	2013	2014	2015	2017(a)	2017(b)
Bukhosi, Theo	M	Not available	X				
Dlamini, Siya	M	Bird Guide Level 1; Dip. in Tourism Management	X				
Dube, Abed-nigo	M	NQF Bird Guide Level 1		X	X		
Gabela, Junior	M	NQF Bird Guide Level 3			X	X	X
Khuzwayo, Thabile	F	Bird Guide Level 1; Dip. in Tourism Management			X	X	X
Mdlalose, Jethro	M	PhD (Zulu tradition-al bird beliefs)				X	
Mhlongo, Bheki	M	NQF Level 2; Dip. in Tourism Management		X			
Mhlongo, Sakhamuzi	M	Nat. Dip. In PR and Bird Guiding	X	X	X	X	X

Table 13.1 *Continued*

Name	Sex	Qualifications	2013	2014	2015	2017(a)	2017(b)
Mthembu, Bongani	M	NQF Bird Guide Level 2		X			
Mthembu, Sakhile	M	NQF Level Biodiversity Conservation		X			
Mthembu, Themba	M	NQF in Nature Conservation	X	X	X	X	X
Mthenjwa, Sakhile	M	Not available	X				
Ngcobo, Daluxolo	M	Not available					X
Ngubane, Benson	M	Bird Guide Level 3; Dip. In Tourism management	X		X	X	
Ngwenya, Vusi	M	Bird Guide Level 1; Dip. In Tourism management		X			
Ntshangase, Phindile	F	Not available				X	
Nyandeni, Bheki	M	NQF Bird Guide Level 2		X	X		
Porter, Roger	M	MSc in Environmental Studies				X	X
Sithole, Bheki	M	NQF Bird Guide Level 2		X	X	X	
Xaba, Nontuthuko	F	Not available				X	
Turner, Noleen (Facilitator)	F	PhD in Zulu linguistics	X	X	X	X	X
Koopman, Adrian (Facilitator)	M	PhD in Zulu linguistics	X	X	X	X	X

The workshop participants came with a range of qualifications including tourism management, bird guiding, (biodiversity) conservation, environmental studies and traditional bird beliefs.

Just as in the past it was mainly the boys in Zulu society who gained the most knowledge about birds and the environment through their herding of cattle, today it is mostly men who carry out bird-guiding activities in KZN. In Table 13.1, of the nineteen Zulu-speaking participants, only three are female, and one of them – Nontuthuko Xaba – was present as a guest from the Women's Leadership and Training programme (WLTP).

As there is not enough space to give full profiles of each of the workshop participants, I have selected just three whom Turner and I regarded as the most knowledgeable about birds and their Zulu names, who contributed most to the discussions about bird naming, and whom we regarded as the 'core' group of the project as a whole.

The three are Sakhamuzi Mhlongo, Themba Mthembu and Junior Gabela.

13.3.2   Sakhamuzi Mhlongo

Sakhamuzi Mhlongo comes from Hlathikhulu village near eShowe in KZN. According to African Conservation: Photodestination:[18]

> Sakhamuzi is a passionate birding guide who is also working for WESSA. He takes the educating of the youth extremely seriously and regularly leads sessions at schools and universities promoting birding and the value of avi-tourism.
>
> He also assists with birding research whenever possible and assists with the Important Bird Area monitoring. He knows his areas intimately, knowing the perfect locations to find the area's specials. He has been a Birdlife SA accredited guide since 2005.
>
> His guiding areas include: Nkandla Forest, Dlinza Forest, Eshowe, Mthunzini, Richards Bay, Ongoye Forest [and] Umlalazi Nature Reserve.

The brief description above does not take into account the fact that 'Sakhi' (or 'Sakkie' in some internet blogs) has been a participant in all five of the Zulu

---

18.   See www.photodestination.co.za/sakhamuzi-mhlongo.html (accessed 6 October 2017).

Sakhamuzi Mhlongo.

bird name workshops held between 2013 and 2017 and has been a stalwart in the process of revising, adapting and extending the list of Zulu names for birds in KZN.

On the last day of the June workshop in Mkhuze in 2017, Mhlongo came in from a tea break holding his mobile phone high in the air in excitement. He had just been notified by BirdLife SA that he had been awarded one of their coveted Owl Awards for conservation. These awards were presented at a BirdLife South Africa function in October 2017 and reported on in *African Birdlife*.

> Sakhamuzi trained as a bird guide in 2002 and has been one of South Africa's top guides since then, operating in southern Zululand. With a reputation for delivering an expert, professional service to local and international birders, he is regularly contracted by birding tour companies.
>
> It's in environmental education, however, that Sakhamuzi makes the most impact. Over the past decade he has encouraged respect for nature and a love for birds in thousands of local learners. Recently he has been conducting weekly environmental classes in Eshowe for the Department of Education and helping to revitalise the Zululand Birding Route, as well as mentoring two prospective bird guides.[19]

19. 'Birdlife South Africa Owl Awards 2017', *African Birdlife* Nov/Dec 2017: 65.

Mhlongo's own thoughts about birds and bird guiding are recorded in the same issue of *African Birdlife* in an editorial on the training given to bird guides by BirdLife South Africa:

> The BirdLife South Africa bird-guide training has opened many doors for me. I have learnt so many things, including an understanding of the importance of conserving birds and their habitats. BirdLife South Africa gave me the hope and courage to use nature to make a living in Nyoni,[20] the small village where I stay, close to Amatikulu Nature Reserve. I was trained in 2000 and, on completion of the course, was presented with a certificate and given binoculars and a bird book to start my own bird-guiding business. BirdLife South Africa has sponsored my attendance at tourism events, including the British Birdfair. Birds are important in my life and in my Zulu culture, and I use my knowledge about bird behaviour to educate people in my community. Birding tourism has become a solution to my financial problems and, believe it or not, I make a living through bird guiding. I wear my khaki uniform with the BirdLife South Africa logo with pride.[21]

13.3.3   Themba Mthembu

A good way to introduce Themba Mthembu is by quoting the African Conservation: Photodestination website:[22]

> Themba has a well-developed guiding profile, whereby up to 98% of his clients come from referrals.
>
> He has unique access to the Nibela Peninsula where the Rosy-Throated Longclaw occurs in good numbers and has found where best to see the Plain-Backed Sunbird. He has developed the Vuvu Khanye Development Programme that focuses on awareness raising in local communities and is a keen participant in the Important Birding Areas monitoring.

---

20. Appropriately, the village where Sakhamuzi Mhlongo lives is named after the Zulu word for 'bird'.
21. 'Community Spirit', *African Birdlife* Nov/Dec 2017: 4.
22. See www.photodestination.co.za/themba-mthembu/html (accessed 6 October 2017).

Areas covered include:

St Lucia, iSimangaliso Wetland Park, False Bay, Nibela peninsula, Mkhuze Game Reserve, Thembe Game Reserve.

A recommendation on Mthembu's website by a birding couple from the United Kingdom states:

Themba is an amazingly knowledgeable and experienced ornithologist. Over two days, he introduced us to dozens of birds to be found in the rich birding area of St Lucia. He and his assistant were also delightful company and we had great conversations: really learning a lot about South Africa today.

I couldn't recommend this more highly. We had a wonderful time over two days: the first day a nearly three hour walk around the town of St Lucia; the second a near 8 hour day in the very rich World Heritage Site of St Lucia National Park.[23]

Themba Mthembu.

In an email to me dated 15 November 2017, Mthembu outlines the development of his career as a bird guide since matriculating in 1997, and it contains some interesting details. As a young boy, he herded cattle on the Nibela Peninsula

23. See https://zulubirding.jimdo.com/references/ (accessed 6 October 2017).

(in the iSimangaliso Wetlands Park) where his uncle was the "tribal chief". Here he also hunted antelope and hunted and trapped birds, giving him the classic background of traditional knowledge about birds. He pursued a career in conservation immediately after leaving school and in 2001 was accredited as a bird guide with BirdLife South Africa. From 2002 to 2004, he worked at Ndumo Game Reserve as an environmental educator for the local schools. During 2006 and 2007, he upgraded his guiding qualifications to NQF4 (the highest level) and also trained as a guide in white water canoeing, following this up in 2007 by working for the St Lucia Kayak Company, where he worked in "dangerous waters with hippos, sharks and crocodiles".

In 2008, Mthembu established his own "zulubirding and ecotours" company[24] and also became an adviser to other birding and ecotour companies. In the same year he also was accredited as an assessor for the ETDP-SETA (Education, Training, Development Practices Sector Education and Training Authority) and was contracted by the Tourism World Academy for tourist-guide training.

Backed with all these extra qualifications, licences and contracts, he started doing contract work for Heritage Tours and similar companies.

Since 2010, Mthembu has been training local bird guides at the Tembe Elephant Park as well as training guides for the Tourism World Academy; this work was recognised in 2015 by a National Best Tourist Guide of the Year Award – a "dedicated Premier Tourism Award".

He has also established his own environmental education centre opposite the gates of Tembe Elephant Park, where there are four part-time trainer guides who take out students and tourists. His intention is for this to be fully open in December 2017, and, once it is, he expects to spend 50 per cent of his time there.

Mthembu estimates that at present he spends half of his time guiding birders and half of his time in education and training of birders and tour guides, and so, like Sakhamuzi Mhlongo, fusing a love of birds with a strong desire to educate youth and tourists in environmental and conservation issues.

Mthembu has been present at all five of the 2013–2017 workshops, has been one of the major contributors to discussions about Zulu bird names, and was certainly the most vocal of all the contributors over the whole project.

---

24. See www.zulubirding.jimdo.com (accessed 3 October 2017).

13.3.4   Junior Gabela

Nicky Forbes begins her article about Junior Gabela in the birding magazine *African Birdlife* as follows:

> Siphamandla Junior Gabela, or Junior as he is known to the many birders who have had the privilege of being guided by him, was born and grew up on the banks of the aMatikulu estuary in KwaZulu-Natal. It was here that, as a young boy, he became interested in birds and was inspired by the identification skills and teaching of the district conservation officer working at the aMatikulu Nature Reserve at the time (Forbes 2017: 66).

The African Conservation: Photodestination website provides further information about this 'core' member of the Zulu bird name project:

> Junior is based at Amatikulu Nature Reserve where the Swamp Nightjar is [a] special bird. He is an expert with all the small "Little Brown Jobs" and also knows the best locations for the localised Green Barbet.
>
> He works for WESSA undertaking research and participating in school outreach programmes. He has a quiet passion for birding and an uncanny knack of locating even the most hard to find species. He and Sakhamuzi Mhlongo work well as a pair providing cross referrals during busy periods.[25]

Junior Gabela.

---

25. See www.photodestination.co.za/junior-gabela.html (accessed 3 October 2017).

We can see here that, as is the case with both Sakhamuzi Mhlongo and Themba Mthembu, Junior Gabela combines bird guiding with environmentally related education. In an email to me dated 12 November 2017, Gabela describes himself as a "passionate bird watcher, enthusiastic environmental educator and having strong entrepreneurial skills". Besides being a bird guide for over ten years, and being strongly involved in environmental education, he assists with bird surveys conducted by BirdLife South Africa, marine estuarine research, works with students from various tertiary institutions in nature conservation, and with tourism students from both the University of South Africa as well as the University of Pretoria and the University of KwaZulu-Natal (UKZN), particularly in their practical modules.

He is an accredited bird guide with BirdLife South Africa, and is working towards his Field Guides Association of Southern Africa (FGASA) level 2 qualifications.

He has established his own business – Birding KZN with Junior Gabela – which offers bird watching, canoeing, environmental education, bird identification courses and team-building. This business was given a Best Business in the Ilembe District Award by the Ilembe Chamber of Commerce, awarding Gabela third place in the 2017 Entrepreneur Competition.

Gabela estimates that at the time of writing he spends approximately 60 per cent of his time in bird guiding, about 30 per cent leading educational tours and about 10 per cent in research projects.

He only joined the Zulu bird names workshop programme in 2015, but very swiftly became a major contributor to the discussions and is regarded as one of the three 'core' contributors.

13.4   BIRDS AND EDUCATION

As can be seen from the brief profiles of Mhlongo, Mthembu and Gabela, bird guiding and environmental education go hand in hand, within the dual contexts of (i) an exponential rise in birding and avitourism in South Africa, and (ii) greater emphasis in the past decades on 'green' issues and the importance of conservation education. It is difficult to separate bird-related conservation education from birding and bird guiding, but for organisational reasons only I divide these as follows:

- Conservation and environmental education in general and about birds in particular.

- The more formal education and qualifications required of those who wish to enter employment as bird guides or more generally as tour guides.
- Birding and avitourism, more specifically under 13.4.4.

13.4.1   Birds and conservation or environmental education

It would take far more than a section to explain in detail all the environmental educational programmes available and I can only give a brief taste of the different sorts of activities where birds, nature and education come together. Given that Sakhamuzi Mhlongo is strongly involved with WESSA, this seems a good place to start.

WESSA has three educational centres in KZN: their headquarters at uMngeni Valley Nature Reserve near Howick, their Twinstreams Environmental Education Centre at uMthunzini on the KZN north coast[26] and a third centre at Treasure Beach on the Bluff in eThekwini. Information about these three KZN educational centres is available on WESSA's website.[27]

Twinstreams runs financially entirely on its school educational programmes and the Umlalazi Municipality puts R150 000 per annum into ecological education for needy schools in the area. Mhlongo is not only involved in the WESSA educational programmes, but also runs courses at the eShowe Environmental Education Centre, a project of the Department of Higher Education and Training.

In the box below are some comments from Mhlongo about birds, education and conservation, which he emailed me in response to some questions I posted to him, Mthembu and Gabela.[28] I have edited Mhlongo's response.

As regards partnership with existing schools, or indeed, universities, it is notable that Mhlongo, as reported on the last morning of the June 2017 Zulu bird name workshop, works in partnership with Professor Craig of the University of Pretoria and his biology students. Craig teaches the theory, while Mhlongo runs the practical aspects of the programme (see text box below). This puts me in mind of another teaching partnership programme, one run by the University of Vermont in the United States, and one that could, I think, be successfully copied here in South Africa. [Text continues on page 413.]

---

26. See www.twinstreams.org.
27. See www.wessa.org.
28. By email 27 November 2017.

I probably spend about 70% of my time guiding and 30% on education. The schools are particularly interested in exposing their learners to the environmental and birding programme. Their major problem is with the costs of such outings: entrance to Dlinza Forest, transport, food, guiding fees, and so on. Recreational birders who pay for my 'birding beginners' courses can far more easily pay for such costs. I count these as part of my educational time.

When the school stays at the Eshowe Educational Centre, the Centre pays me directly; when a school wants to book my services during weekends and holidays they pay me directly.

As to whether I have a set lecture for school visits, it will depend entirely on what grade of learners is involved, and what subject is being focused on. Although all school visits focus directly on environmental and conservation issues, the science teachers may wish a stronger focus on food chains, parasitism and the like, geography teachers may want to know more about migration routes, and language teachers may want to know more about the bird names and the traditional beliefs. As a general rule, when conducting birding education for schools, you need to link the experience to the school curriculum, and make it a joint learning experience with the teacher and the learners.

The Zulu-speaking teachers and learners are very interested in knowing the Zulu names of birds, and the stories behind the names. They and their teachers always ask about cultural beliefs, such as beliefs about owls, the hamerkop and other 'birds of omen'. This gives me, as their bird guide and educator, a chance to play a major role to educate and change mind-sets about birds under threat, so as to help and to protect some of our endangered species.

The educational project with Pretoria University with Professor Craig was a rewarding annual project which introduced the practical aspects of birding and education to tertiary-level students. Unfortunately, the project had to be halted because of budget constraints.

The programme is called 'Environmental Studies 295: Sustainable Education: Birding to Change the World' and it is the brainchild of Dr Trish O'Kane of the Department of Environmental Studies at the University of Vermont (UVM).[29] Briefly, in this programme second-year university students studying Environmental Studies partner with learners from the nearby Flynn Elementary School. Both school and university are adjacent to Lake Champlain, the Winooski River that feeds into it, and extensive woods, forests and gardens. At the university students attend lectures on environmental justice, pedagogy, social justice issues, and public education, sustainability and nature study. At the beginning of the course the students learn how to identify 35 Vermont birds by sight and sound. Once they have these skills, they meet once a week for an outdoor session, with each UVM student pairing up with a "co-explorer" – a pupil from the elementary school. These two then become "bird-buddies" for the remainder of the course, with each student attempting to teach the younger learner the essential elements of nature study, conservation principles, environmental issues, and, of course, the basics of birding. The university student then reports back each week to his or her tutor or lecturer on how things went, and what problems may have arisen.

Each of the weeks has a named theme, for example Week One "Birds as a Portal" and Week Six "Winter World". Of particular interest to the South African situation are the themes of Weeks Eleven and Fourteen, respectively "Birding While Black" and "Flocking: How Birding Creates Community".

13.4.2   Bird clubs and the Women's Leadership and Training Programme

It was July 2015 when I first heard about bird clubs in KZN, involving Zulu-speaking youth. Marilyn Aitken, the biodiversity coordinator of the Women's Leadership and Training Programme (WLTP) came to see me at my home, bringing with her Nomusa Mkhungo, Nontuthuko Xaba and Buyelaphi Sibiya. In discussion with the four I learnt that the WLTP had (at that time) four bird clubs working with Zulu-speaking children aged nine to young adults of 25 years. The bird clubs are part of a number of leadership projects such as

---

29. I am extremely grateful to Nancy Jacobs, author of the 2016 *Birders of Africa*, for drawing my attention to this course, and sending me a copy of the syllabus and programme. www.uvm.edu/uvmnews/news/birding-change-world. For more information about this course see https://youtu.beRyzTkooqPXQ.

the Gender Project (that looks at women abuse, the *ukuthwala* custom and other issues affecting women). The WLTP Birding Project aims to train young women for leadership, conserve threatened species and heighten awareness of the Zulu cultural significance of birds and their cultural and economic potential. In the KZN Midlands (in the Sisonke Municipality) they work in association with BirdLife South Africa; and they are also linked to a birding club in Soweto. One of the three young women, Buyelaphi Sibiya, works with the Mabula Ground Hornbill Project (MGP).[30]

Interested in hearing more about the WLTP and their bird clubs, Turner and I invited the WLTP to send a representative to attend the 5–7 June 2017 Zulu bird name workshop and take part in the discussions. Nontuthuko Xaba attended and proved a lively contributor to the discussions about traditional Zulu beliefs about birds. When talks centred on educational and social issues linking bird with Zulu-speaking children, Xaba shared the following information on the WLTP.

This NGO focuses on women working in rural areas (e.g. Nhlokozi, Drakensberg, but there is also one in KwaMashu). They focus on gender issues and cultural and environmental issues. (For example, girls in rural areas often don't finish matric because they are forced into an early marriage because their father wants cows. In this way girl = cow.)

The WLTP started in Gauteng in 2007, based on the Grail International Women's Movement, (started by the Dutch Jesuit Father Jacques van Ginneken), which is still functioning in Johannesburg. One of the modules that WLTP follows is a programme for young girls in rural areas.

Under the environment module, which deals with deforestation, pollution of wetlands, overharvesting for *muti* purposes and a number of related issues, there are basically two sub-groups: (i) the sustainable vegetable farming group, and (ii) the birding group.

There are four areas with birding groups: Hlokozi (with two clubs), Centocow, Underberg and KwaMashu.

Nontuthuko is involved in the Hlokozi group(s), where 50 or so teenagers meet once a month at a weekend and go for birding outings. Nontuthuko is an area co-ordinator and a bird guide. She uses birding apps on cell phones, using *Roberts'* bird-related quizzes and games. For field trips they create bird lists for use on cell phones and laptops.

---

30. See http://ground-hornbill.org.za/. See also 'The Power of the Thunderbird', *African Birdlife* Sept/Oct 2018: 12–13.

Mhlongo noted that for the last seventeen years Richards Bay Minerals (RBM) had held birding weekends, and school and other groups had attended these and gone on walks and 'mini-tours'.

13.4.3 Writing of children's readers

In her doctoral thesis about children's readers in the Zulu language (and in all the other African languages in South Africa) my former colleague Dr Phindile Dlamini from UKZN explains that as so few suitable readers have been produced in the 'indigenous' languages, the various provincial education departments have resorted to translating them from English. Dlamini has herself written and published several children's readers in Zulu, but the 2007 reader *Ngiphe Izinyo Elisha* is the only one incorporating a bird and a traditional belief about a bird. This reader is about the role of the Yellow-billed Kite as a 'tooth fairy' in traditional Zulu bird lore, which I will discuss further in section 13.6.2.

There seems no reason why Zulu traditional bird lore, and the explanation of bird names as they relate to this lore, should not become a staple content of readers written for children in Zulu. The same would apply to readers written in other South African languages. It is government policy to promote traditional African heritage and, if such children's readers became accepted as textbooks, this would be extremely worthwhile financially for the author.

The genre of children's basic readers seems suitable for the preservation of Zulu bird lore, but other publications, such as school zoology textbooks, also come to mind where the names of various species of birds need to be included. UKZN is one of several universities in South Africa that is encouraging the use of African languages in lectures and tutorials and, if this is applied to ornithological teaching at tertiary level, then the species-specific names developed in the 2013–2017 Zulu bird name workshops could become essential for class notes and other written teaching material.

13.4.4 Bird guide training

In 2009 the Department of Trade and Industry (DTI) commissioned a report, 'Niche Tourism Markets: Avitourism in South Africa'. It is available on the BirdLife South Africa website,[31] and I shall be referring to it in the section on

---

31. See www.birdlife.org.za/images/Birding_Routes/Doc/dti-information_booklet.pdf (accessed 19 October 2017).

avitourism below. On the issue of licences and permits for bird guides, the DTI Report comments as follows:

> Bird Guiding Requirements: In South Africa it is illegal to conduct guided tours without the proper qualification. In order to practice as a local bird guide in South Africa, guides need to be licensed and accredited by the relevant provincial tourism authorities as well as by Tourism, Hospitality, Sport Education and Training Authority (THETA).

It goes on to talk of Professional Driving Permits and other requirements of bird guides and tour guides operating professionally.

During the final day of the June 2017 Zulu bird name workshop, when the discussion centred around professional training for bird guides, I took notes, which on inspection a few days later seemed to be an incomprehensible jumble of acronyms: CATHSSETA, NQF, SAQA, FGASA, SETAs, SKS and more. SETA stands for Sector Education and Training Authority and it is part of the acronym CATHSSETA: the Culture, Arts, Tourism, Hospitality and Sports Sector Education and Training Authority.[32] FGASA, as mentioned earlier, is the Field Guide Association of South Africa,[33] NQF stands for the National Qualifications Framework, and SAQA is the South African Qualifications Authority[34]. SKS stands for Special Knowledge and Skills and this qualification can be held in different 'subjects', for example SKS (Birding); SKS (Dangerous Game [DG]) or SKS (Wildflowers). All of these different authorities are involved in one way or another in controlling the profession of bird guiding.

I do not wish to go into the different courses and qualifications offered by WESSA, by the Wakkerstroom Tourism and Education Centre, or other institutions offering training for bird guides and general tour guides. But I will give below details of the qualifications offered by the Field Guide Association of South Africa (FGASA), if for no other reason than to give an idea of how demanding the requirements are for professional certification.

---

32. See https://cathsseta.org.za/.
33. See https://www.fgasa.co.za/.
34. See http://www.saqa.org.za/, which includes details of the National Qualifications Framework (NQF).

Data from the FGASA website about the qualifications they offer

FGASA offers four qualifications:

a) The Local Bird Guide Certificate
b) The Regional Bird Guide Certificate
c) The National Bird Guide Certificate
d) The SKS (Birding) Qualification

Initial requirements

For all of these qualifications, an aspirant bird guide needs to have attained a FGASA Field Guide (NQF2) or an Advanced Field Guide (NQF4) or the FGASA Specialist Field Guide qualification. Details of these four qualifications can be found on the FGASA website.[35] A valid, up-to-date and recognised First-Aid certificate is also needed for all four qualifications listed below.

a) The Local Bird Guide Certificate: If one has all of the above, then one applies for a practical assessment, where the candidate takes a registered FGASA birding assessor on a guided birding experience. Should he or she pass, then he or she is qualified to operate in specific sites.

b) The Regional Bird Guide Certificate: If one has all the initial requirements, one selects a particular region or one (or more) of the following biomes:

Savannah Biome	Montane Grassland
Karoo	Forest
Arid habitat	Marine
Fynbos	Albany Thicket

One then (i) completes a regional Bird Guide Workbook, which is submitted for assessment; (ii) must pass a slide (visuals) and sound identification of 80 core species of bird found all over the country; (iii) must pass a slide (visuals) and sound identification of between 20 and 80 species depending on the region or biome chosen; and (iv) take a registered FGASA birding assessor on a guided birding experience. Should he or she pass all these requirements, he or she may practice as a guide anywhere in the region or in any biome selected for the testing.

---

35. See https://www.fgasa.co.za/images/17.pdf.

c) The National Bird Guide Certificate: One follows the same process as for the regional Bird Guide Certificate except that after the core testing the candidate must pass slide (visual) and sound (call) for all eight of the biomes listed under (b) above. If the candidate has already passed a practical assessment in the field for the Provincial Certificate, he or she does not need to be tested again.

d) The SKS (Birding) Qualification: First, the candidate must have attained the FGASA Specialist Field Guide Qualification and be declared competent in three core areas: 1) the birding theoretical assessment; 2) the relevant birding slide identification assessment; and 3) the bird guiding skills practical assessment.[36] The candidate must then pass the slide and sound assessment for a chosen region or biome (and also complete a regional level workbook). Then comes the major difference:

The candidate must pass theory examinations in several subject areas including:

Bird history, classification and conservation
Bird anatomy and physiology   Breeding
Defence and survival   Food and feeding methods
Bird behaviour   General birding knowledge.

Once all these assessments and examinations have been passed, the candidate will "be deemed an SKS (Birding) National Guide".

The complex sub-divisions and layers of tests and assessments (theoretical and practical) and the examination only lead to qualifications as a bird guide. Should one wish also, or instead, to be a tour guide, then none of the above are relevant and a whole new authority comes into place, the Culture, Arts, Tourism, Hospitality and Sports Sector Education and Training Authority (CATHSSETA).

13.4.5 Community bird guides

BirdLife South Africa provides the information shown in the box below on the notion of community bird guides.[37]

---

36. The details of this can be found at https://www.fgasa.co.za/images/6.pdf.
37. See www.birdlife.org.za/gobirding/community-bird-guides (accessed 27 October 2017).

An exciting initiative along all the birding routes and in areas adjacent to birding hotspots and IBAs [Important Bird and Biodiversity Areas] is the availability of Community Bird Guides. The guides provide improved security and valuable information on where elusive and special bird species may be found, and in some cases can gain you access to otherwise restricted locations. Although we have detailed below suggested rates for the use of these guides, they are self-employed and free to set their own rates but they are affordable and offer very competitive rates within the guiding industry.

BirdLife South Africa recognised the important role that local communities can play in conservation and as a result became involved in bird guide training and skills development. BirdLife South Africa launched the Community Bird Guide Training Programme in 2001. The programme has relaunched in 2016 following a short break due to funding and personnel shortages.

The programme is designed to provide local community members with an opportunity to participate in the ever-growing avi-tourism sector, as well as to create an awareness of the economic benefits of birds and their habitats to local communities.

The programme works together with communities in each area to identify potential Community Bird Guides. Funding is then sourced to assist these community members in attending an accredited bird guide training course. An extensive route mentorship and support programme is also in place to help guides to continue to grow and develop once training is completed.

A list of community guides is available on the same Birdlife South Africa webpage. They list community guides for the Eastern Cape (no names; one contact number), the "Kruger to Canyons" area (nine guides listed), Limpopo (four guides listed), Mpumalanga (two guides listed) and KZN (31 guides listed). The remaining provinces and/or regions appear not to have any community guides. KZN has so many more guides than the other provinces that these are listed separately under Southern KZN and Zululand, returning us to the days when the Woodward brothers travelled and birded in the two separate locations of Natal and Zululand.

Looking at the names of the KZN birding guides one can see which separate areas of the province they cover (as well as their contact details) and as can be seen, there is no corner left unguided. There is a significant gender imbalance, with only seven women (22.5%) among the 31 bird guides listed for KZN.

Seven of the guides listed participated in at least one of the 2013–2017 Zulu bird name workshops:

Gabela, Junior: Amatigulu, Mlalazi, Ongoye, Dlinza.
♀ **Khuzwayo, Thabile**: Dlinza, Lake Nhlabane, Nkandla, Ongoye.
Mhlongo, Sakhamuzi: Dlinza, Richards Bay Harbour, Mlalazi, Nkandla, Ongoye Forest.
Mthembu, Themba: Ndumo, Thembe, Tshanini, St. Lucia, Mkhuze.
Ngcobo, Daluxolo: Ntsikeni Nature Reserve.
♀ **Ntshangase, Phindile**: no information available
Nyandeni, Bheki: Lake Sibaya and Lower Mkhuze.

It is not clear why this province should have so many more community bird guides than the other provinces and regions listed. This does mean, though, that Zulu names for birds get a greater airing (or potential for airing) than names in the other African languages of South Africa.

13.5   BIRDING AND AVITOURISM

In section 13.4.4, I mentioned the 2010 report commissioned by the Department of Trade and Industry (DTI), which is available on the BirdLife South Africa website. I give a brief summary of some of the salient features to give some of the context of Zulu bird names today.

As pointed out earlier, the major providers of Zulu names for birds today are the knowledgeable people who work as bird guides and who have undergone rigorous training. That they have undergone this training, and as a result can now earn their living by bird guiding, is precisely because of avitourism, both domestic and international. In the June 2017 Zulu bird name workshop, Porter noted that globally, in the last 30 or so years, birding has grown exponentially, and South Africa has become a top spot for avitourism, because of its diversity of bird habitats and the high proportion of endemic species. There is much competition among birders to get as many birds as possible on their personal 'Life List'. To assist these avid birders, the tour guides who work for most tour companies work together with a local bird guide, creating a demand for their skills, resulting in many Zulu-speaking community bird guides being listed.

These Zulu-speaking community bird guides are in a two-way relationship with the Zulu names of birds: on the one hand, it is their combined knowledge of the different individual species of birds and their command of their own mother tongue that has enabled them to participate in the workshops that have ensured unique Zulu names for each species of bird found in KZN; on the other hand, having these unique names now available in Zulu means they can communicate these names to those avitourists who may wish to know them, but also, perhaps more importantly, feed these new names back into their own communities, especially among the Zulu-speaking schoolchildren who are taking part in the environmental education programme with which these bird guides are involved.

Much of the detail of what follows is taken from the DIT report. The term 'avitourism' refers to people travelling within their own country or travelling to other countries for the express purpose of birding. Birding is primarily a <u>social</u> activity, but one with enormous potential for economic development, and the main thrust of the DTI report on avitourism is on the economic prospects. The introduction to the report states (duplicating Porter's thoughts above):

> Birding is one of the fastest growing nature-based tourism activities world-wide and is experiencing similar growth in interest and popularity in South Africa. It has also been recognised that avitourism is an important part of the global growth in nature-based tourism. South Africa is a premier destination for avitourism, due to its large diversity of birds and endemic species, as well as a full complement of major bird habitats in Africa.[38]

---

38. See https://www.birdlife.org.za/images/Birding_Routes/Doc/dti-information_booklet.pdf.

When the research was done, South Africa's annual number of avitourists was between 21 000 and 40 000 annually, of which between 13 000 and 24 000 were domestic birders.[39] At that time the total annual spend of avitourists was calculated to be between R927 million and R1 725 billion, with domestic birders contributing between R482 million and R890 million annually. Clearly birders contribute a significant portion of South Africa's tourism income.

Of these amounts, research suggested that the domestic and international avitourists spent about R47 million annually on tour guides. The research is not very clear on this point: the implication here is that 'tour guides' means specifically 'bird guides', but that might not be the case. However, the report goes on to say that avitourists showed a preference for birding in small groups of between one and four persons:

> This preference lends itself particularly well to the use of small tour operators and community guides, rather than larger tour operators. Use of community guides is desirable as they have been proven to be effective environmental stewards . . .[40]

The report further notes that avitourism generally has positive environmental and conservation impacts in that 'eco-tourists' or 'nature-tourists' (which includes birders) are generally the more environmentally conscious among tourists generally, and that this provides environmental benefits to local communities as it (i) helps to educate local peoples about the value of biodiversity, and (ii) incentivises the protection and conservation of natural areas.

South Africa has a competitive advantage as an avitourism destination for a number of reasons, including, but not limited to:
- A high species diversity, a number of which are endemic.
- A variety of different ecological biomes in close proximity to each other.
- Ease of viewing large numbers of birds quickly: avitourists can expect as many as 400 species in a three-week trip.
- Complementary wildlife attractions in a number of game reserves.

---

39. Figures calculated in 2009 would have changed significantly by 2017. Also, figures do not include 'hunting tourism', for e.g. 'wingshooting'.
40. See https://www.birdlife.org.za/images/Birding_Routes/Doc/dti-information_booklet.pdf.

- Advanced birding specific tourism infrastructure, which includes Important Bird Areas (IBAs), birding hotspots and birding routes.
- Availability of trained community guides.

Under the section 'Opportunities in Avitourism', on how to cater for avitourists, the DTI report offers advice under the headings 'Opportunities for Existing Tourism Enterprises' as well as 'New Business Opportunities in Avitourism', which lists suggestions of how local communities can benefit economically from the considerable number of domestic and international birders. To the established enterprises it suggests:

> If you live in proximity to a birding route or in an area with natural birding assets, you may be able to turn a hobby into a business.

It then goes on to suggest how the community bird guides (and communities) and the established businesses (bed and breakfasts, holiday lodges, etc.) can link up to attract avitourists to benefit them all.

13.6 THE EVOLUTION OF MEMES: CHANGING DYNAMICS IN TRADITIONAL BELIEFS

I have not used meme theory previously in this book, but the notion of memes is useful in a diachronic study of culture. Memes have been described as units, or entities, of idea, thought or image. They may consist of a thought, or a picture, or a certain phrasing (as in a song or an advertisement) – anything notional, in fact, that can be copied and passed on. The whole point of the notion of memes is that they provide a theoretical framework for the passing on of 'bits of culture'. Like genes, memes change through time, and any one of four changes may take place: (i) a meme may die out completely over time; (ii) a meme may survive unchanged over time; (iii) a meme may change certain features of its nature while remaining unchanged in its basic elements; and (iv) a meme may be born as new, distinct and original. In this section we will consider all four of these options, illustrating them with traditional Zulu beliefs about certain birds. We will then consider a fifth option: deliberately attempting to force change on a long-standing meme, giving two examples of this, also illustrating with existing memes of traditional beliefs about birds.

13.6.1   The dying out of a meme

I have mentioned the traditional Zulu children's game '*Bhula 'Ntsentse Bo!*' several times.[41] I brought this traditional game to the attention of the fourth Zulu bird names workshop, in June 2017, when the discussion was about bringing birds into the education of younger Zulu-speaking children in a way that would interest and amuse them. No one at the workshop had heard of the game. Before that workshop and since I have spoken to a number of Zulu-speaking people, old and young, particularly those interested in birds and those interested in education, and not one person has ever heard of the game. I do assume it existed in the first place, even though it is only the two Samuelson siblings who have ever written about it. Perhaps it only existed in the Samuelson household – a game thought up by Robert and Letitia Samuelson to play with their Zulu friends. But whatever the case, this game appears totally unknown today and must be described as a meme that has died out.

13.6.2   The (virtually) unchanged survival of a meme

When I first read about the Yellow-billed Kite in the role of a 'tooth fairy' (see section 11.3.3, page 337) in Samuelson (1929) and in Godfrey (1941), I thought that this belief could not possibly be in existence today. I was therefore greatly surprised when Marilyn Aitken, of the WLTP, brought three young Zulu-speaking women to meet with me in 2015, that, on leaving, one of the young women – Nomusa Mkhungo (of BirdLife Port Natal) – asked me casually, "You do know the story about the Yellow-billed Kite bringing new teeth, don't you?"

I was curious to find out at the June 2017 Zulu bird names workshop, when cultural beliefs were specifically discussed, whether or not anyone else knew about this belief. I was astonished to find that everyone present, young and old, was perfectly familiar with this role played by **unhloyile**, down to the details of the refrain asking the kite to replace the old tooth with the new, to throwing the old tooth backwards between the legs, and running home without looking back.

Here is a meme, then, that has survived untouched since earlier days. I say 'virtually unchanged' in the heading above, however, because of one very small change brought to my notice late in 2017. The tale of the Yellow-billed Kite as a provider of new teeth came up when I was discussing other issues

---

41. See section 9.3, 'Birds in riddles and children's games'.

with Phindile Dlamini. It was not unexpected, after the June 2017 workshop, to discover that Dlamini knew about the kite's role as a 'tooth fairy', but I did not know until that time that she had published a children's reader in Zulu in 2007 about the Yellow-billed Kite and its dental role.[42] Nor did she know that **unhloyile** was a bird: she had always thought that uNhloyile was a beautiful woman living in the sky, handing out new teeth in exchange for old, and had indeed written as such in her book.

13.6.3   The adaptation of a meme: the Southern Ground Hornbill

In Chapter 10, section 10.3.1, we considered the Southern Ground Hornbill (**insingizi**) as a bird of thunder and rain. We noted how the bird was feared as well as revered for its power, not just for bringing welcome rain, but for bringing storms and destructive, torrential downpours. And in Chapter 9, section 9.3, on riddles, we noted the Zulu riddle of the priest with the red collar, the answer being the Southern Ground Hornbill with its red wattles and neck. The various aspects of the hornbill were revisited at the June 2017 workshop when two members of the group spoke of what they knew about the hornbill.

Older beliefs, recorded in Layard as far back as 1867, were that a Southern Ground Hornbill should be drowned in a river. If this was done in relatively shallow water, rain would come; if in deeper water, torrential downpours would occur. At the workshop, Nontuthuko Xaba spoke of the belief held in her community that if one wants to know whether it is going to rain or not, tying an **insingizi** feather to a branch at the very edge of a river will give the answer: if the feather is swept away by the water, it will rain soon, if not, then it won't.

Themba Mthembu came with a different slant on the **insingizi**: the **insingizi** is, as everybody knows, a bird of thunder. Boys are told when on their way to school that if there are dark clouds on the horizon and they happened to meet on the way a person wearing dark clothing but with a red scarf or tie at the throat, not to greet this person (as would normally be the correct polite behaviour) but to be silent and not even make eye contact. Mthembu remembers his mother reminding him of this every time there was a suspicion of thunder ahead on a particular day.

---

42. See section 13.4.3 (page 415).

13.6.4   The birth of a new meme: the Southern Ground Hornbill (again)

Mthembu's take on this hornbill is an interesting and novel combination of older beliefs about thunder birds and the red 'scarf' or 'collar' around the bird's neck. Sakhamuzi Mhlongo, on the other hand, gives a completely new and quite different story about this species. His tale is not one that was told at the June 2017 workshop to discuss traditional Zulu beliefs about birds, but is rather recorded by Mark Cocker, in his 2013 *Birds and People*, with Cocker giving Mhlongo as the source:

> Among many Zulu village people a common form of insult for a person thought not to be very bright is to call them *Ingududu*, which is our name for the ground hornbill. The birds are always walking in small groups and since these are assumed to be of the same sex, they are thought also to be gay or lesbian. Among some Zulu communities, such as the Nyoni people, if one of the birds approaches you it is a cause of concern, because it's a sign that you too are gay (Cocker 2013: 332).

There is much about this quote that needs to be pursued (what is the link between not being bright, and being gay or lesbian, for instance, and why specifically the Nyoni (bird) clan?)[43] but this will have to be done in later research. The cognitive processes that underlie Mhlongo's statement here are interesting:

> Southern Ground Hornbills walk in small groups (ornithological <u>fact</u>)
> → these are all the same sex (<u>assumption</u>, with no logic behind it) +
> birds and animals have gay and lesbian members in the same way that human societies do (<u>assumption</u>, based on no logic)
> → Southern Ground Hornbills are therefore gay or lesbian (<u>false logic</u>).

What is more interesting is that discussion of homosexuality in Zulu society is relatively new. Many senior leaders of society struggle to come to terms with homosexuality. Yet here is Mhlongo, presumably in about 2012 or thereabouts,[44]

---

43. There is a possibility here that it is not the Nyoni clan which is being referred to here, but the people of the Nyoni village where Mhlongo is based. See his profile in section 13.3.2 where he states: "in Nyoni, the small village where I stay, close to Amatikulu Nature Reserve."

44. Given that Cocker published his book in 2013.

discussing the topic in terms of traditional beliefs relating to birds, and the species he chooses is a top 'bird of omen' in traditional Zulu society, as we have seen several times in the course of this book.

It remains to be seen whether or not this new meme will reduplicate and spawn exact copies (i.e. Southern Ground Hornbills are gays or lesbian in nature) or reduplicate in adapted forms (for example, Green Wood Hoopoes and the Chestnut-fronted Helmetshrikes travel in small groups,[45] therefore they must be gay or lesbian as well). Mhlongo's meme is new, so only time will tell.

13.6.5 The deliberate forced reversal of memes: owls and vultures

There are certain traditional belief about birds that are seen in the twenty-first century to be undesirable in current contexts. In chapters 10 and 11 we looked at traditional beliefs relating to owls and vultures respectively: the owl is seen as one of the familiars of witches, as it operates at night, and is therefore feared and hated; vultures can see things from a great distance, and can 'therefore' see even further, even into the future.

Owls

As regards owls, today there are on-going 'owl box projects' run by different organisations. Jim Taylor, the education officer at the WESSA headquarters at Umgeni Valley, is running an owl box project in the urban townships of Pietermaritzburg and Durban. This not so much to conserve owls as a population, but to control the rapid increase in the rat population, itself a result of uncollected garbage and general filth. The July/August 2016 issue of *African Birdlife* carries a story called 'Encouraging efforts to protect owls in communities along the N3 Toll Route'. The story quotes Tammy Caine of the Raptor Rescue Rehabilitation Centre and project manager of the Owl Box Project as saying:

> With N3TC's [N3 Toll Consortium] support we can now reach far more people and educate more communities on the vulnerability of owls and the need to protect this species. We particularly focus on

---

45. Indeed the Zulu names of the helmetshrike – **abayeni** and **uthimbakazane** – reflect this.

educating children in schools in rural communities. Education is key to the success of any conservation project. With our education outreach programme we hope to instil an awareness of nature and help to develop environmentally conscious adults.

In reference to the still well-established meme of owls being linked to evil and bad luck, Caine continues:

In rural areas fears and superstitions around owls are still rife. We find that working with children in these communities is proving successful to allay these fears and superstition. Children are far more open to adopt new belief systems.[46]

It is the last line that is telling here.

Vultures

The general vulture population of South Africa is under severe threat. The website of the African Bird of Prey Sanctuary near Pietermaritzburg has the following to say about vultures (as paraphrased):[47]

Among various changes in the environment (e.g. "over-grazing causes bush-encroachment that makes it more difficult for the vultures to find carcases"), vultures are at risk from colliding with high-voltage power cables, and being poisoned indirectly by feeding off carcases left out for stock killers. Vultures are also subject to two myths: one is that they are themselves stock killers, which is not true at all. The other is more of a cultural meme: vultures can see very far, therefore they can see into the future; therefore *muti* made of vulture parts will allow the purchaser to see into the future as well and thus gamble successfully at horse races, or when playing Lotto and any other games of chance. As the raptor centre website puts it, "there is a cultural belief that vultures locate their food with a clairvoyant ability".

To this end, the raptor centre has been trying to raise R55 000 to purchase a 'binocular viewing machine' that will enable children from underprivileged

---

46. See http://www.n3tc.co.za/press-media/125-encouraging-efforts-to-protect-owls-in-communities-along-the-n3-toll-route (accessed 3 October 2017).
47. See https://www:africanraptor.co.za (accessed 3 October 2017).

communities, who, as they point out, may never have used a pair of binoculars, to understand magnified vision and to experience themselves how a vulture sees about eight times better than humans. This, like the owl box project, can be understood as an attempt to bring about long-term meme change.

Educational programmes that concentrate on conservation, environmental issues, and the role of birds in both, aimed specifically at the youth, may eventually help change these memes around. If, however, what schoolchildren are being told at school about the importance of protecting and conserving birds, such as owls and vultures, is being negated by what their parents and grandparents are telling them at home, then it is going to be a very long time before these deeply entrenched traditional beliefs are changed.

13.7 BIRD NAMES: CHALLENGES FOR THE FUTURE

In this last section I look at what other scholars and authors are doing with bird names in African languages, other than Zulu and Afrikaans. We look first at work currently being done in South Africa itself among the nine official Bantu-origin languages, and then go on to look at work done in the late 1990s on bird names in Swahili, the lingua franca of East Africa and parts of Central Africa. We end with a brief look at bird names in Seychelles Creole: not strictly part of Africa, but interesting nonetheless as an example of bird naming, and the recording in writing of such naming, of this minority language.

13.7.1 Bird names in South African Bantu languages

On 1 December 2016 a meeting was held at the BirdLife South Africa head office in Johannesburg to discuss 'African bird names' and the cultural significance of South African birds. According to the minutes of the meeting it was attended by the following people:[48]

- Derek Engelbrecht (Professor of Zoology at the University of Limpopo);
- Peter Mokumo (postgraduate student at the University of Limpopo who works in close association with Professor Engelbrecht);
- Mark Anderson (CEO of BirdLife South Africa),
- Johan Meyer (Pretoria-based educator, passionate about bird names);
- Rick Nuttall (Bloemfontein-based ornithologist, with a keen interest in bird names and the cultural significance of birds to people);

---

48. Forwarded to me by email by Roger Porter in February 2017.

- Ingrid Weiersbye (bird artist and co-author of the 2016 second edition of *Roberts Field Guide*);
- Roy Cowgill (the late ornithologist and birder who contributed greatly to the training of African bird guides);[49] and
- Roger Porter (retired ecological scientist and birder, and participant in the 2017 Zulu bird name workshops).

The first item in the minutes is relevant:

1. AFRICAN BIRD NAMES

It is important (and an urgent priority especially for African names of all Red Data listed birds) to have a unique name for each bird species in the various African languages of the southern African region. A group, 'African Bird Name Group', has been formed to facilitate, encourage and coordinate people doing research on bird names. The group has as its objective that each bird species in southern Africa is to have a species-specific name in all the languages of the region.

Action: Johan [Meyer], Derek [Engelbrecht] and Peter [Mokumo] to define a research project. Also to liaise with Adrian Koopman (and others) with regard to producing annotated lists of bird names in the various African languages. Important and essential to have these African names ready to be included in the future Roberts VIII Handbook of SA birds (Derek).

The African Bird Name Group was formed in November 2016.[50] Meyer is interested in bird names in all southern African languages, not just those of Bantu origin. He has done considerable research among Khoi and San languages that have a recorded lexicography of bird names. As a speaker of Northern Sotho, he has concentrated on this language among the Bantu group with a strong focus on the other languages and dialects in the Sotho-Tswana cluster.

Using Northern Sotho as an example, he identifies the same 'Maclean problem' that was discussed extensively in Chapter 2 (page 26ff.):

---

49. See his obituary in the March/April 2017 issue of *African Birdlife*, page 64.
50. See https://sites.google.com/site/africanbirdname/ and contact email birdnamesafrica@gmail.com.

After initial research in most African languages four things usually became clear. Using Northern Sotho as an example, they are:

1. One name is used for many species, e.g. "lepidibidi" is used for all ducks and "ntšhu" is used for all larger birds of prey.
2. A single bird species has many names, e.g. the Western Cattle Egret (*Bubulcus ibis*): Modišane, Modišadikgômo, Ledišadikgômo, Kgogonokane, Kgogobadimo.
3. As standard Northern Sotho is a collection of various different dialects, variations in spelling of names occurs or the names are in different noun classes.
   a. Spelling variation, e.g. the Southern Ground Hornbill (*Bucorvus leadbeateri*): legotutu vs. lehututu.
   b. Different noun classes, e.g. falcon: sepekwa (class 7) vs. pekwa (class 9)
4. A certain species or group of birds have no known name, e.g. storm petrels. This is true for almost all seabirds as Northern Sotho is not spoken at the coast.[51]

While Meyer has been running the African Bird Name Group, doing extensive literature searches for recorded bird names and fieldwork on Northern Sotho bird names, Professor Engelbrecht has been co-ordinating field research into Venda bird names, using postgraduate students. He has also, according to a blog post on the Londolozi website, joined forces with the South African College for Tourism Tracker Academy to record bird alarm calls. Alex van den Heever, a game guide at Londolozi, writes:

For centuries bird alarm calls have been a vital source of information for indigenous trackers as they hunted and gathered food in the bush. A bird will make a different call if it encounters potential threat, such as a leopard, eagle, snake or human. This is called an alarm or distress call. The purpose of the call is to alert other birds of the threat as well as to indicate to the predator that the surprise element is gone. Essentially it's saying "Don't try, I see you!" What is interesting is that the majority of these calls are not recorded or published in any

51. See https://sites.google.com/site/africanbirdname/introduction .

scientific research paper, according to Professor Derek Engelbrecht of the University of Limpopo.[52]

A follow-up meeting of the African Bird Name Group, held at the same venue on 12 October 2017, was originally planned for exactly the same participants, with the addition of myself. However, the majority of participants were unable to attend and eventually the meeting was held with Fanie Du Plessis of BirdLife South Africa (standing in for CEO Mark Anderson), Roger Porter and myself. A brief Skype call with Derek Engelbrecht produced no tangible results in the way of an update on research into Venda bird names, apart from Engelbrecht saying his researchers had found that villagers had little knowledge of birds, were very suspicious of the researchers and their aims and goals, and were mainly interested in knowing what was in it for them. From my own point of view, I was able to report back on the conclusion of the 2013–2017 Zulu bird name workshops and that Turner and I now had a list of species-specific Zulu names for all the birds in our region.

In the absence of further evidence to the contrary, as of the end of 2017, there is still considerable work to be done in the other African languages of South Africa.

In another part of Africa, however, namely Tanzania, an attempt has been made to create species-specific names in Swahili, the national language of Tanzania and a lingua franca of much of central-eastern Africa.

13.7.2   Species-specific names in Swahili

In Chapter 2, in section 2.7, I made reference to and quoted from an article written by Charles Mlingwa on the coining of species-specific names for birds in Swahili.

Mlingwa is an academic teaching in the department of Zoology and Marine Biology at the University of Dar-es-Salaam in Tanzania. In 1997 he published a provisional list of species-specific bird names in Swahili for the birds of Tanzania.

---

52. See https://blog.londolozi.com/2014/09/13/a-new-study-recording-bird-alarm-calls/ (accessed 19 September 2017).

In his introduction, he states:

If every bird species will be assigned a distinct Kiswahili name, there would be effective communication with local communities as regards conservation of birds and their habitats. Moreover, having Kiswahili names for birds of Tanzania (and Africa as a whole) will not only advance awareness of birds, but will also facilitate the usage of Kiswahili language even [in] national institution[s]of higher learning such as universities (Mlingwa 1997: 74).

In Chapter 12, in detailing the entire process of revising Zulu avian nomenclature, of which the 2013–2017 Zulu bird name workshops were a major part, I mentioned that prior to the workshops a thorough literature research of all recorded Zulu bird names had been carried out, and that this list was available at each workshop. Moreover, it was an important principle, adopted at the first workshop, that a previously recorded name would not be discarded for a newly coined name unless there was a very good reason.

It is instructive therefore to compare Mlingwa's provisional list with the bird names recorded in an earlier dictionary to see if he incorporated previously recorded names or not. In writing this book, I did not have access to a comprehensive and modern Swahili dictionary, and had to rely on *The Standard English-Swahili Dictionary* compiled by the Inter-Territorial Language Committee of the East African Dependencies under the 'direction' of Frederick Johnson, published in 1939.[53] The dictionary contains mainly generic words for bird genera and families, but now and then offers a species-specific name within a generic group. There is no space to do an exhaustive analysis of either the dictionary or of Mlingwa's provisional list, but to illustrate the degree of correspondence between previously recorded Swahili names and Mlingwa's coined names, I have chosen to present case studies of two clusters of birds – kingfishers and doves. These two clusters have been chosen because similar case studies of Zulu names for birds in these clusters have been given in this book – the doves at the end of Chapter 2 and the kingfishers at the end of Chapter 12.

---

53. Johnson (1939).

Case Studies: kingfishers and doves

In each case study Mlingwa's species-specific names are in the column on the left and previously recorded names from the dictionary are on the right.

Table 13.1 Case study: kingfishers.

Mlingwa	Mlingwa	Dictionary	Dictionary
Kingfisher (generic)	*kichi, zumbulu*	Kingfisher (generic)	*mdiria*
Pied Kingfisher	*kichi mtilili*	Pied Kingfisher	*dete, detepwani*
Pygmy Kingfisher	*kichi mdogo* (small)	Natal Pygmy	*kisharifu, msharifu*
Mangrove Kingfisher	*kichi mikoko* (*mkoko* = mangrove)	Mangrove Kingfisher	*kijimbimsitu*
Giant Kingfisher	*Zumbulu*		
Malachite Kingfisher	*kichi kishungibluu*		
Brown-hooded Kingfisher	*kichi tumbojeupe* (white-bellied)		
Striped Kingfisher	*kichi michirizi*		
Chestnut-bellied Kingfisher	*kichi tumbojekundu* (red-bellied)		
Blue-breasted Kingfisher	*kichi kufuabluu* (blue-breasted)		
Woodland Kingfisher	*kichi bluu* (blue kingfisher)		

Analysis

Johnson's dictionary gives a generic, and specific names for only three species of kingfisher. It is remarkable that not one of the names given in the dictionary are used by Mlingwa, unless one takes the initial element of *kijimbimsitu* (Mangrove Kingfisher) as being a dialectal variation of Mlingwa's generic *kichi*. Not even Johnson's and Mlingwa's generic names are the same. On the limited evidence available, I would say Mlingwa has either not researched existing, previously recorded names, or that he has chosen to ignore them.

Mlingwa has used the name *zumbulu*, which he gives as one of two generics, as the specific name of the Giant Kingfisher. For all his other nine kingfishers, the generic *kichi* is qualified by an extension. I was not able to translate every one of them, but in the table above one can see that the name for the Mangrove Kingfisher is the generic qualified by the plural of the noun *mkoko* (mangrove), and the Blue-breasted Kingfisher is the equivalent in Swahili: the generic *kichi* qualified by *kufua* (breast, cf. Zulu *isifuba*) and the adopted adjective *bluu*.

In comparing this table of Swahili names of kingfishers with the appendix at the end of Chapter 12, which gives a similar outline of previously recorded names of kingfisher names in Zulu, with the final list approved in the 2013–2017 Zulu bird name workshops, it will be seen that the Zulu workshop-based process of assigning species-specific names has not adopted Mlingwa's approach at all. Where he has used the generic *kichi* qualified by an extension for all species (except the Giant Kingfisher), Zulu has not gone for a generic + qualifier approach at all. In fact, it was only once names had been assigned to all nine kingfishers in the KZN region, that a name was sought to serve as a generic.

Table 13.2   Case study: doves and pigeons.

Mlingwa	Mlingwa	Dictionary	Dictionary
Doves and pigeons generic:	*tetere, pugi, njiwa*	Pigeon (generic)	*njiwa*
Lemon Dove	*kipura* (*pugi kipura*)		
Olive Pigeon	*njiwa kisogocheuupe*		
Bronze-naped Pigeon	*njiwa mweusi* (black pigeon)		
Speckled Pigeon	*njiwa madoa* (< *doa* 'speck')		
Rameron Pigeon	*njiwa domonjano* (yellow-beaked pigeon)		
Feral Pigeon	*njiwa manga* (*hua*)		
Afep Pigeon	*hua mwekundu* (red dove)		
Namaqua Dove	*pugi kombamwiko*[†]		
Ring-necked Dove	*tetere mdogo* (small *tetere*)	Ring-necked Dove	*tetere*
Mourning Dove	*kuyu jichonjano* (yellow-eyed kuyu)		

Table 13.2   *Continued*

Mlingwa	Mlingwa	Dictionary	Dictionary
Dusky Turtle Dove	*kuyu kifua rangipinki* (pink-breasted kuyu)		
Red-eyed Dove	*tetere, jichojekundu* (Red-eyed)	Lesser Red-eyed Dove [seen as generic]	*hua, njiwa*
Laughing Dove	*fumvu, (songoro kanturi)*		
Blue-spotted Wood Dove	*pugi wanda*	Wood Dove	*pugi, pugi wanda*
Emerald-spotted Wood Dove	*pugi kitugu*		
Tambourine Dove	*pugi kikombe (kituku pori)*	Tambourine Dove	*pugi kikombe*
Green Pigeon	*Ninga*	Deland's Green Pigeon	*ninga*
Pemba Green Pigeon	*ninga pemba*		
		'Large Dove'	*mwigo*

† Could this possibly be *ki* + *omba* (ask for) + *mwiko* (wooden spoon), in reference to this dove's long tail?

Analysis

Clearly, as with Zulu, Swahili has different words for different groups of doves and pigeons. This is reflected in Mlingwa's list as well as in Johnson's dictionary. Three words are used generically: *tetere, pugi* and *njiwa*, with *njiwa* used for all the pigeons, and *tetere* and *pugi* used for different groups of doves. The word *hua*, which I take to be cognate with Zulu **ijuba**, does not have the same generic nature of the Zulu word, but is restricted to the 'Afep Pigeon' in Mlingwa's list and the 'Lesser Red-eyed Dove' in the dictionary.

Mlingwa has used qualificative extensions like *mweusi* (black) and *jichonjano* (yellow-eyed), but with the seven different bases *kipura, njiwa, hua, kuyu, tetere, pugi* and *ninga*, so there is not the same overall pattern as with the kingfishers, with all species except the Giant Kingfisher using the same generic.

It is useful to compare this table of Swahili names of pigeons and doves with section 2.6.3 at the end of Chapter 2, which gives a similar outline of previously recorded names of pigeons and doves in Zulu with the final list of names approved in the 2013–2017 Zulu bird name workshops. The Swahili list is much closer to the Zulu list in its general semantic profile, based as both are on previously existing nomenclatural distinctions between species.

Mlingwa clearly sees his 1997 list as work in progress:

> Following this provisional list of Kiswahili names, more people would be encouraged, I hope, to explore more about birdlife in Tanzania. I will, at the same time, very much appreciate criticism which would contribute to producing a comprehensive list of bird names (1997: 74).

It would be interesting to find out whether or not this list has been updated.

13.7.3   Names of birds in Seychelles Creole

There are three official languages in the Seychelles: English, French and Creole.[54] Species-specific names for birds have long been established in English and French, but the 2003 book *Zwazo Sesel: The Names of Seychelles Birds and Their Meanings*[55] documents a much more recent attempt to coin names for all the birds in the Seychelles archipelago. The first part of the book title immediately gives a sense of this creole – *Zwazo* is derived from French *oiseaux* (birds) and *Sesel* is derived from *Seychelles*.

The majority of the breeding species and the more familiar migrants have long had Creole names, developed over 200 years of human settlement on the islands. These have presumably mostly been in oral form, as there is no mention of previously recorded lists. However, 259 bird species have been recorded in total for the Seychelles and over 75 per cent have had no known Creole name. D'Offay and Lionnet's 1982 Creole–French dictionary was the first to record Creole bird names in print. Skerrett, Matyot and Rocamora give the figures of the actual birds on the islands compared to the names in the dictionary as follows (2003:7):

---

54. A 'creole' is a pidginised and simplified version of another language. Unlike pidgin, a highly simplified language used for trade and other forms of communication between two groups of people who are mother-tongue speakers of other languages, a creole is a mother-tongue language (i.e. spoken from birth), and may be the only language spoken by the inhabitants of a region or country. Seychelles Creole is French-based.
55. Skerrett, Matyot and Rocamora (2003).

	Total number of birds recorded for Seychelles	Creole names in the 1982 dictionary
Breeding species	66	57
Extinct species	7	3
Annual migrants	22	7
Vagrants	139	8
Uncertain status	25	nil
TOTALS	**259**	**74**

These figures, then, represent the avian nomenclatural situation for Creole in 1982, before the intervention of Skerrett, Matyot and Rocamora. Noteworthy here is that Creole names existed for 57 of the recorded 66 breeding species at that time. On the other hand, vagrants to the island, meticulously recorded by birders, had not captured the attention of the Creole-speaking locals, with only eight names for 139 recorded vagrants.

By the time of the 2003 publication of *Zwazo Sesel*, the project to find Creole names for all species had been completed, and there are now 264 Creole names for the 259 recorded species on the Seychelles, indicating that one or two species have more than one Creole name. In their introduction to the book, the three authors explain the situation pre- and post-project, and discuss a slow 'evolutionary' emergence of the original Creole names and how their much quicker intervention helped the Creole names to reach parity with the bird names in the other two official languages on the islands:

> Relatively few of the species that have been reported in Seychelles have Creole names. Their evolution has been left to the chance inventiveness of the inhabitants of Seychelles or sometimes, visitors to Seychelles. Every bird name is invented at some point of time, whether as an original name or one adopted from some other language. However, there is no reason why it should be left for a name to emerge slowly. This would be a sure way to guarantee most species never receive any Creole name whatsoever.
>
> In French and English, every species has a unique official name (some having two or more). If a new species is discovered, it will

instantly receive a French name and an English name, as well as a scientific name. Names no longer emerge through the old process of evolution. They are thought out and put down on paper. There is no reason for Creole names not to be created in a similar manner. Their survival thereafter is up to the birdwatchers that use them (Skerrett, Matyot and Rocamora 2003: 3).

There is no space to give more than just a few examples from the book of 264 Creole names for birds, and the eight examples below have been selected partially randomly and partially because they have interesting tales to tell. For each species, I begin with the Creole name, then give the English, scientific and French names, followed by an explanation of the Creole name. We begin with the national bird of the Seychelles: the Seychelles Black Parrot:

- *Kato Nwar*: the Seychelles Black Parrot (*Coracopsis* [*nigra*] *barklyi*). The current official French name is *perroquet noir* (previously *cateau noir*). The French *cateau*, itself derived from the earlier *kakatou* (cockatoo), has been Creolised as *kato* and *noir* as *nwar*.
- *Tayvan Zean-d-Sid*: Southern Giant Petrel (*Macronectes giganteus*). Fr. *fulmar géant, pétrel géant.* As for 'Tayvan':

> Tayvan was first cited by Bory (1804) as used in Réunion for a species of breeding petrel, later identified as Barau's Petrel . . . The name is also vaguely applied in Mauritius to other seabirds when seen at sea. The name is derived from the French *taille-vent*, meaning wind-cutter, evocative of the flight of petrels over the ocean and similar to the English name *shearwater*. It is applied here as the generic name for petrels . . . (Skerrett, Matyot and Rocamora 2003: 12).

> The specific *Zean-d-Sid* is a Creolised version of the French *géant du sud* (giant of the south).
- *Manzer Zwit*: Eurasian Oystercatcher (*Haematopus ostralegus*). Fr. *huîtier-pie.* The Creole name is derived from the French *manger* 'eat' and the French *huître* 'oyster'. We see here a name similar in meaning to older Zulu names like **isixulamasele** (what snatches frogs: the Spoonbill) and newly coined Zulu names like **usikhothaphela** (what catches cockroaches: the Icterine Warbler).

- *Katiti Lapat Rouz Was*: Red-footed Falcon (*Falco vespertinus*). Fr. *faucon kobez*. The Creole generic *katiti* is said to be onomatopoeic in origin. *Lapat Rouz* is derived from French *la pied rouge* (the red foot) and *Was* from the French *ouest* 'west'.
- *Kannar Labek Kwiyer*: Northern Shoveler (*Anas clypeata*). Fr. *canard souchet* (see Plate 53). The Creole generic name *kannar* is derived from French *canard* (duck) and the qualifying phrase (the specific epithet) is from French *la bec* (the beak) and *couiller* or *couillier* (spoon). Thus the 'spoon-billed duck'. Skerrett, Matyot and Rocamora say "*Souchet* may come from the Latin *soccus*, meaning a shoe and later used to describe something comic, like the large and spoon-shaped bill of this duck" (2003: 31). Thus 'Shoe-bill', as well! The coined Zulu name for the similar Cape Shoveler, as we have already seen in this book, is **unofosholo**, derived from *ifosholo* (shovel), itself an adoptive from English.
- *Floranten Mov*: Purple Heron (*Ardea purpurea*). Fr. *héron pourpré*. As for the origin of the Creole generic for herons '*Floranten*':

> This long-established Creole name was first cited . . . in 1892 . . . [it has been suggested that the name] was from the French *florentin*, a piece of equipment used in distillation . . . [which is] . . . a glass container with a long neck (Skerrett, Matyot and Rocamora 2003: 22).

The specific '*mov*', clearly, is from French *mauve* (purple, mauve).
- *Payanke*: generic for Tropicbird (genus *Phaethon*): derived from Fr. *paille-en-queue* (straw-in-tail). This name originally belonged to the Long-tailed Skua, but was redirected to the Tropicbird. Before being 'gentrified' the name originally given by sailors to the Long-tailed Skua was *paille-en-cul* (straw-in-arse).
- *Syer*: Seychelles Scops-owl (*Otus insularis*). Fr. *petit-duc scieur*. The Creole name 'Syer' is derived from English sawyer,[56] in reference to the "remarkable call of this bird, similar to the sound of wood being sawn" (Skerrett, Matyot and Rocamora 2003: 67).

---

56. Depending on how *syer* is pronounced, it may also be derived from the French *scieur* (sawyer).

As this book began with Sibree's Madagascan names for White-eyes it seems fitting to end this list,[57] and the book, with two Creole names originating from Madagascar, and one of them indeed refers to a White-eye:

- *Papang Nwanr*: Black Kite (*Milvus migrans*). Fr. *Milan noir*. The Creole generic *papang* is derived from the Malagasy name for the Yellow-billed Kite; *nwanr* is from the French word *noir* 'black'.

- *Zwazo Linet Malgas*: Madagascar White-eye (*Zosterops maderaspatanus*).[58] Fr. *zostérops malgache* (or *oiseau lunettes*). The generic *zwazo linet* is a creolised version of the alternate French name *oiseau lunettes* (the spectacled bird) and the specific *malgas* is a creolised version of 'Malagasy', the language of Madagascar. It has nothing to do with the Afrikaans name *malgas* for the Cape Gannet.

A last word from Skerrett, Matyot and Rocamora, whose thoughts apply to birds and their names anywhere:

All those who take an interest in the birds they see around them enjoy being able to put a name to them. It is one of the great joys of bird watching and often the first step in becoming interested in their lives and their conservation. Ideally, these names should be in the language most familiar to the individual. Without names to refer to in their mother tongue, it is possible an interest in birds will never take root in the hearts of many (2003: 3).

---

57. See section 1.1, page 1.
58. Skerrett, Matyot and Rocamora say that this specific epithet refers to Madras in India, and was given by mistake by Linnaeus "who probably intended to write *Madagascariensis*" (2003: 82).

Bibliography

Ambrose, D. 2005. *Birds, Including Annotated Species Checklist: Lesotho Annotated Bibliography*, Section 167. Roma: National University of Lesotho.

Ambrose, D. and D.H. Maphisa. 1999. *Guide to the Birds of the Roma Campus, National University of Lesotho*. Roma: National University of Lesotho.

Armstrong, E.A. 1958. *The Folklore of Birds*. London: Collins.

Berglund, A. 1976. *Zulu Thought-Patterns and Symbolism*. Cape Town: David Philip.

Berlin, B. 1992. *Ethnobiological Classification: Principles of Categorization of Plants and Animals in Traditional Societies*. New Jersey: Princeton University Press.

Berlin, B., D.E. Breedlove and P.H. Raven. 1973. 'General Principles of Classification and Nomenclature in Folk Biology'. *American Anthropology* 75: 214–42.

Biyela, Sr. N.G.I. (F.S.F.). 2009. 'Popular Predictor Birds in Zulu Culture'. *Alternation* 16 (2): 35–52.

Boon, R. 2010. *Pooley's Trees of Eastern South Africa*. Durban: Flora and Fauna Publications Trust.

Botha, T.J.R. 1977. *Watername in Natal: 'n Inleiding tot die Studie van Zoeloeplekname*. Pretoria: Raad vir Geesteswetenskaplike Navorsing.

Branford, J. 1980. *A Dictionary of South African English*. Cape Town: Oxford University Press.

Brownlee, C.P. [1896] 1977. *Reminiscences of Kaffir Life and History and other Papers*. Pietermaritzburg: University of Natal Press and Durban: Killie Campbell Africana Library. Originally published in 1896 by Lovedale Mission Press.

Bryant, A.T. 1905. *Zulu-English Dictionary*. Pinetown: Mariannhill Mission Press.

———. 1929. *Olden Times in Zululand and Natal*. London: Longmans, Green and Co.

———. [1949] 1967. *The Zulu People: As They Were before the White Man Came*. Pietermaritzburg: Shuter and Shooter.

———. [1906] 1970. *Zulu Medicine and Medicine Men*. Cape Town: C. Struik. Originally published in 1906 in *Annals of the Natal Museum* II (1).

Bulpin, T.V. 1969. *Natal and the Zulu Country*. Cape Town: Books of Africa.

Chadwick, J.M.K. 1947. 'Zulu Names for Birds'. *Ostrich* 18 (2): 179–82.

Chittenden, H. 2007. *Roberts Bird Guide*. First edition. Cape Town: John Voelcker Bird Book Fund.

Chittenden, H., G. Davies and I. Weiersbye. 2016. *Roberts Bird Guide*. Second edition. Cape Town: John Voelcker Bird Book Fund.

Clancey, P.A. 1964. *The Birds of Natal and Zululand*. Edinburgh: Oliver and Boyd.

Clinning, C. 1989. *Southern African Bird Names Explained*. Johannesburg: Southern African Ornithological Society.

Coates Palgrave, K. 1977. *Trees of Southern Africa*. Cape Town: C. Struik.

Cockburn, J.J., B. Khumalo-Seegelken and M.H. Villet. 2014. 'Izinambuzane: IsiZulu Names for Insects'. *South African Journal of Science* 110 (9/10): 1–13.

Cocker, M. 2013. *Birds and People*. London: Jonathan Cape.

Coetzee, R. 1982. *Funa – Food from Africa: Roots of Traditional African Food Culture*. Durban: Butterworths Publishers (Pty) Ltd.

Cole, D. 1984. 'The Specific Epithet of *Turdus litsitsirupa* (Smith)'. *Bokmakierie* 36 (1): 11–12.

Cole-Beuchat, P.D. 1957. 'Riddles in Bantu'. *African Studies* 16 (3): 133–49.

Colenso, J.W. 1884. *Zulu-English Dictionary*. Third edition. Pietermaritzburg and Durban: P. Davis and Sons.

Cope, A.T. 1968. *Izibongo: Zulu Praise Poetry*. Oxford: Oxford University Press.

Dean, W.R.J. 2017. *Warriors, Dilettantes & Businessmen: Bird Collectors During the Mid-19th to Mid-20th Centuries in Southern Africa*. Cape Town: John Voelcker Bird Book Fund.

Deikumah, J.P., V.A. Konadu and R. Kwafo. 2015. 'Bird Naming Systems by Akan People in Ghana Follow Scientific Nomenclature with Potentials for Conservation Monitoring'. *Journal of Ethnobiology and Ethnomedicine* 11 (75): 1–13.

Delegorgue, A. [1847] 1990. *Travels in Southern Africa*. Volume I. (Trans. Fleur Webb; Introduced and Indexed by Stephanie Alexander and Colin Webb). Durban: Killie Campbell Africana Library and Pietermaritzburg: University of Natal Press. Originally published in 1847 as *Voyage dans l'Afrique Australe* in Paris.

———. [1847] 1997. *Travels in Southern Africa*. Volume II. (Trans. Fleur Webb; Introduced and indexed by Stephanie Alexander and Bill Guest). Durban: Killie Campbell Africana Library and Pietermaritzburg: University of Natal Press. Originally published in 1847 as *Voyage dans l'Afrique Australe* in Paris.

Dlamini, P.D. 2007. *Ngiphe Izinyo Elisha*. Pietermaritzburg: Nutrend.

D'Offay, D. and G. Lionnet. 1982. *Diksyonner Kreol-Franse*. Hamburg: Helmut Buske Verlag.

Doke, C.M. and B.W. Vilakazi. 1958. *Zulu-English Dictionary*. Johannesburg: Wits University Press.

Dunning, R.G. 1946. *Two Hundred and Sixty-four Zulu Proverbs, Idioms, etc., and the Cries of Thirty-Seven Birds, Fully Translated.* Durban: Knox Printing and Publishing Co.

Durban: Includes Towns of KwaZulu-Natal (Street Guide series). 2005. Cape Town: Map Studio.

Faye, C. 1923. *Zulu References.* Pietermaritzburg: City Printing Works.

Finnegan, R. 1976. *Oral Literature in Africa.* Nairobi: Oxford University Press.

Forbes, N. 2017. 'Junior Gabela'. *African Birdlife* Nov/Dec: 66.

Fraser, G.M. 2000. *Flashman and the Tiger.* London: Harper Collins.

Fynn, H.F. 1969. *The Diary of Henry Francis Fynn*, edited by J. Stuart and D. McK. Malcolm. Pietermaritzburg: Shuter and Shooter.

Gcumisa, M. and M. Ntombela. 1993. *Isilulu Solwazi Lwemvelo – Umqulu I.* Pietermaritzburg: Shuter and Shooter.

Godfrey, R. 1919. 'The Birds of the Buffalo Basin, Cape Province' (II). *South African Journal of Natural History* 1 (2): 195–209.

———. 1922. 'The Birds of the Buffalo Basin, Cape Province'. *South African Journal of Natural History* 3 (2): 37–49.

———. 1941. *Bird-Lore of the Eastern Cape Province.* Johannesburg: Wits University Press.

Greenoak, F. 1997. *British Birds: Their Folklore, Names and Literature.* London: Christopher Helm.

Gunner, L. and M. Gwala. 1994. *Musho: Zulu Popular Praises.* Johannesburg: Wits University Press.

Haagner, A. and R.H. Ivy. 1923. *Sketches of South African Bird Life.* Third edition. Cape Town: Maskew Miller.

Haggard, H. Rider. [1885] 1962. *King Solomon's Mines.* London: Cassell & Co.

———. 2000. *Diary of an African Journey: The Return of Rider Haggard*, edited by S. Coan. Pietermaritzburg: University of KwaZulu-Natal Press. First publication of this diary by the Dominions Royal Commission in 1914.

Hammond-Tooke, D. 1960. *Bhaca Society.* Cape Town: Oxford University Press.

Hare, C.E. 1952. *Bird Lore.* London: Country Life.

Hendrickx, F.L. 1944. 'Some Kivu Birds and their Native Names'. *Ostrich* 15 (3): 194–212.

Hockey, P.A.R., W.R.J. Dean and P.G. Ryan. 2005. *Roberts – Birds of Southern Africa.* Seventh edition. Cape Town: John Voelcker Bird Book Fund.

Ichikawa, M. 1998. 'The Birds as Indicators of the Invisible World: Ethno-ornithology of the Mbuti Hunter-gathers'. *African Study Monographs* Suppl. 25: 101–21.

Isaacs, N. [1836] 1970. *Travels and Adventures in Eastern Africa: Description of the Zoolus, their Manners, Customs, with a Sketch of Natal.* Cape Town: Struik.

Jacobs, N.J. 2015. 'Herding Birds, Interspecific Communication and Translations'. *Critical African Studies* 8 (2): 136–45.

————. 2016. *Birders of Africa: History of a Network*. New Haven and London: Yale University Press.

Jacqt-Guillarmod, C. 1932. 'Some Notes on Birds in Basutoland'. *Ostrich* 3 (2): 35–40.

Jenkins, E. 2012. 'Barbara Tyrell's Caravan Verse'. *Natalia* 42: 1–8.

Johnson, F. (ed.) 1939. *A Standard English-Swahili Dictionary*. London: Humphrey Milford.

Kemp, L., N. Mkhungo and N. Monama. 2018. 'The Power of the Thunderbird'. *African Birdlife* Sept/Oct: 12–13.

Khumalo, J.S.M. 1974. 'Zulu Riddles'. *African Studies* 33 (4): 193–226.

Kidd, D. 1904. *The Essential Kafir*. London: Adam & Charles Black.

Koopman, A. 1987. 'The Praises of Young Zulu Men'. *Theoria* 70: 41–54.

————. 1990. 'Onomatopoeia: Song Reference in English, Afrikaans and Zulu Bird Names'. *Nomina Africana* 4 (1): 67– 87.

————. 2001. 'Yebo Gogo: "Formula" or "Catch-phrase"?' In *African Oral Literature: Functions in Contemporary Context*, edited by R.H. Kaschula. Cape Town: New Africa Books, pp. 142–55.

————. 2002. *Zulu Names*. Pietermaritzburg: University of Natal Press.

————. 2005. 'Unpacking Jamludi: An Exercise in Interdisciplinary Onomastics'. *Nomina Africana* 19 (2): 159–84.

————. 2011a. 'Lightning Birds and Thunder Trees'. *Natalia* 41: 40–60.

————. 2011b. 'Using "isiZulu" in English Discourse'. *Natalia* 41: 96–8.

————. 2013. 'The Interface between Magic, Plants and Language'. *Southern African Humanities* 25: 87–103.

————. 2014a. 'The Naming Imperative: Naming Wild Animals'. *Nomina Africana* 28 (2): 17–42.

————. 2014b. 'UMahlekehlathini, uMehlomane and uMbokodo: Zulu Names for Whites in Colonial Natal'. *Natalia* 44: 48–69.

————. 2014c. 'Review of *Dr Henrik Greve Blessing: South African Medicinal Plants from KwaZulu-Natal: Described 1903–1904*'. *Natalia* 44: 94–8.

————. 2015. *Zulu Plant Names*. Pietermaritzburg: University of KwaZulu-Natal Press.

————. 2016a. 'Of Birds and Bombs'. *Natalia* 46: 119–21.

————. 2016b. 'What is a "Turkey Buzzard"?'. *Natalia* 46: 126–8.

————. 2017a. 'Henry Francis Fynn: The Long-tailed Finch that Came from Pondoland?'. *Natalia* 47: 39–42.

————. 2017b. '*Isithwalandwe*: The Wearing of the Crane Feather'. *Natalia* 47: 43–6.

————. 2017c. 'Surname Dynamics in Avian Nomenclature'. *Nomina Africana* 31 (2): 141–52.

————. 2017d. 'The Zululand Journeys of the Woodward Brothers 1894–1899'. *Natalia* 47: 13–27.

————. 2018a. 'From Vocalisation to Verbalisation: Strategies for Turning Bird Calls into Language'. *Language Matters* 49(2): 3–22.

————. 2018b. 'Zulu Bird Names: A Progression over the Decades [Part One: The First Hundred Years, from Delegorgue to Samuelson]'. *South African Journal of African Languages* 38 (3): 261–7.

————. 2019. 'Zulu Bird Names: A Progression over the Decades [Part Two: The Second Hundred Years, with Roberts, and Doke and Vilakazi]'. *South African Journal of African Languages* 39 (1): 9–15.

Koopman, A. and A. Davey. 2000. 'Adulphe Delegorgue's *Vocabulaire de la Langue Zoulouse*'. *South African Journal of African Languages* 20 (2): 134–47.

Koopman, A. and N.S. Turner. 2018. 'Terminology Development in Zulu Avian Nomenclature'. *Nomina Africana* 32(1): 11–21.

————. 2019. 'The Morphology of Zulu Bird Names: Old and New'. *South African Journal of African Languages* 39(1): 1–8.

Krige, E.J. 1950. *The Social System of the Zulus*. Pietermaritzburg: Shuter and Shooter.

Layard, E.L. 1867. *Birds of South Africa*. Revised and augmented by R. Bowdler Sharpe. Four volumes, 1875–1884. London: Bernard Quaritch.

Louwrens, L.J. 2004. 'On the Generic Nature of Common Northern Sotho Bird Names: A Probe into the Cognitive Systematization of Indigenous Knowledge'. *South African Journal of African Languages* 24 (2): 95–117.

Lugg, H.C. 1970. *A Natal Family Looks Back*. Durban: T.W. Griggs and Co.

————. 1975. *Life under a Zulu Shield.* Pietermaritzburg: Shuter and Shooter.

Maclean, G. 1984. *Roberts' Birds of Southern Africa.* Fifth edition. Cape Town: John Voelcker Bird Book Fund.

Malcolm, D. McK. (ed.) 1949. *Zulu Proverbs and Popular Sayings, with Translations* (collected by James Stuart). Durban, T.W. Griggs and Co.

Matjila, D.S. 2015. 'Birds as Subjects in Setswana Folklore: Depiction of their Relationship to Man'. *South African Journal of African Languages* 35 (1): 105–11.

McLachlan, G.R. and R. Liversidge. 1957. *Roberts Birds of South Africa.* Second edition. Cape Town: South African Bird Book Fund.

————. 1970. *Roberts Birds of South Africa.* Third edition. Cape Town: John Voelcker Bird Book Fund.

————. 1978. *Roberts Birds of South Africa.* Fourth edition. Cape Town: John Voelcker Bird Book Fund.

Mlingwa, C.O.F. 1997. 'Birds of Tanzania: A Provisional List of Bird Names in KiSwahili'. *African Study Monographs* 18 (2): 71–120.

Moffett, R. 2010. *SeSotho Plant & Animal Names and Plants used by the Basotho.* Bloemfontein: Sun Press.

Moreau, R.E. 1940. 'Bird Names used in Coastal North-Eastern Tanganyika Territory'. *Tanganyika Notes and Records* 10: 47–72.

————. 1942. 'Bird-nomenclature in an East African Area'. *Bulletin of the School of Oriental and African Studies* 10 (4): 998–1006.

Muir, J. 1940. 'Afrikaans Bird-names in Riversdale, C.P.'. *Ostrich* 11 (1): 1–19.

Mynott, J. 2009. *Birdscapes*. New Jersey: Princeton University Press.

Newman, K. 2002. *Birds of Southern Africa*. Eighth edition. Cape Town: Struik.

Nyembezi, C.L.S. 1958. *Izibongo Zamakhosi*. Pietermaritzburg: Shuter and Shooter.

———. 1974. *Zulu Proverbs*. Johannesburg: Wits University Press.

Odendal, F.F., P.C. Schoonees, C.J. Swanepoel, S.J. du Toit and C.M. Booysen (eds). 1979. *Verklarende Handwoordeboek van die Afrikaanse Taal*. Second edition. Johannesburg and Cape Town: Perskor-Uitgewery.

Oliver, J. 1980. *A Beginner's Guide to Our Birds*. First edition. Durban: Wildlife Society of Southern Africa (WESSA).

———. 1989. *Izinyoni Ezingamashumi Amane Natathu ZakwelaKwa Zulu*. Translated from Oliver 1980 by M. Mchunu and B.G. Nkwanyana. Place unknown: uPhiko LweNgcebo Yemvelo KwaZulu.

———. 2003. *Ngezinyoni Zethu*. Translated from Oliver 1980 by D. Kumalo and N. Turner. Durban: WESSA and BirdLife South Africa.

Paton, A. 1948. *Cry, The Beloved Country*. New York: Scribner.

Paulsen, B.S., H. Ekell, Q. Johnson and K.R. Norum (eds). 2012. *Dr Henrik Greve Blessing: South African Medicinal Plants from KwaZulu-Natal: Described 1903–1904*. Norway: Unipub.

Poland, M., D. Hammond-Tooke and L. Voigt. 2003. *The Abundant Herds*. Vlaeberg: Fernwood Press.

Prozesky, O.P.M. 1974. *A Field Guide to the Birds of Southern Africa*. London: Collins.

Read, H. and B. Dobrée. 1952. *The London Book of English Verse*. London: Eyre and Spottiswoode.

Roberts, A. [1940] 1951. *The Birds of South Africa*. London: H.F. & G. Witherby Ltd.

Robinson, F.N. (ed.) 1957. *The Works of Geoffrey Chaucer*. Second edition. London: Oxford University Press.

Rycroft, D.K. and A.B. Ngcobo (eds). 1988. *The Praises of Dingana: Izibongo zikaDingana*. Durban: Killie Campbell Africana Library and Pietermaritzburg: University of Natal Press.

Samuelson, L.H. 1974. *Zululand: Its Traditions, Legends, Customs and Folk-Lore*. Durban: T.W Griggs and Co. Originally published by Mariannhill Mission Press, Pinetown, no date.

Samuelson, R.C.A. 1923. *The King Cetywayo Zulu Dictionary*. Durban: Commercial Printing Co.

———. 1929. *Long, Long Ago*. Durban: Knox Printing and Publishing Co.

Scott, D.C. [1892] 1929. *Dictionary of the Nyanja Language*. London: Lutterworth Press.

Sibree, J. 1891a. 'On the Birds of Madagascar and their Connection with Native Folk-lore, Proverbs and Superstitions'. *Ibis* 33 (3): 416–43.

———. 1891b. 'On the Birds of Madagascar and their Connection with Native Folk-lore, Proverbs and Superstitions'. *Ibis* 33 (4): 557–65.

————. 1892. 'On the Birds of Madagascar and their Connection with Native Folklore, Proverbs and Superstitions'. *Ibis* 34 (1): 103–19.

Sithole, E.T. 1982. *Izithakazelo Nezibongo ZakwaZulu*. Pinetown: Mariannhill Mission Press.

Skead, C.J. (ed.) with J. Vincent, C. Niven, G.G. McLachlan and J.M. Winterbottom. 1960. *The Canaries, Seedeaters and Buntings of Southern Africa*. Cape Town: South African Bird Book Fund.

Skerrett, A., P. Matyot and G. Rocamora. 2003. *Zwazo Sesel: The Name of Seychelles Birds and Their Meanings*. Victoria, Seychelles: Island Conservation Society.

Stark, A.C. and W.L. Sclater. 1900, 1901, 1903, 1906. *The Birds of South Africa*. Volumes I to IV. London: R.H. Porter.

Torrend, J. 1931. *An English-Vernacular Dictionary of the Bantu-Botatwe Dialects of Northern Rhodesia*. London: Kegan Paul, Trench, Trubner and Co.

Turner, N.S. 1997. 'Onomastic Caricatures: Names Given to Employers and Co-workers by Black Employees'. *Nomina Africana* 11 (1): 50–66.

Tyler, J. [1891] 1971. *40 Years Among the Zulus*. Cape Town: C. Struik Pty Ltd. Originally published in Boston: Congregational Sunday-School and Chicago: Publishing Society.

Tyrrell, B. 1945. *A Medley of Caravan Verse: Written and Illustrated by Barbara Tyrell*. Durban: Knox.

Vilakazi, B.W. 1935. *Inkondlo kaZulu*. Johannesburg: Wits University Press.

————. 1945. *Amal'Ezulu*. Johannesburg: Wits University Press.

Wainwright, A.T. 1983. 'Bird Names and Xhosa Oral Poetry'. In *Names 1983: The Proceedings of the Second Southern African Names Congress*, edited by P.E. Raper. Pretoria: Human Sciences Research Council.

Waring, P. 1978. *A Dictionary of Omens and Superstitions*. London: Souvenir Press.

Webb, C. de B. and J. Wright (eds). 1976, 1979, 1982, 1986, 2001, 2014. *The James Stuart Archive of Recorded Oral Evidence Relating to the History of the Zulu and Neighbouring Peoples*. Volumes I to VI. Pietermaritzburg: University of Natal Press/University of KwaZulu-Natal Press.

Wilder, B.T., C. O'Meara, L. Monti and G.P. Nabhan. 2016. 'The Importance of Indigenous Knowledge in Curbing the Loss of Language and Biodiversity'. *BioScience* 66 (6): 499–509.

Wilson, J. 2011. 'Vernacular Names of Malawi's Birds'. Part 1. *Society of Malawi Journal* 64 (2): 36–51.

————. 2012a. 'Vernacular Names of Malawi's Birds'. Part 2. *Society of Malawi Journal* 65 (1): 46–63.

————. 2012b. 'Vernacular Names of Malawi's Birds'. Part 3. *Society of Malawi Journal* 65 (2): 38–55.

Woodward, R.B. and J.D.S. Woodward. 1897. 'Description of our Journeys in Zululand, with Notes on its Birds'. *Ibis* 39 (3): 400–22.

———. 1898. 'Further Notes on the Birds of Zululand'. *Ibis* 40 (2): 216–28.

———. 1899. *Natal Birds*. Pietermaritzburg: P. Davis and Sons.

———. 1900. 'On the Birds of St. Lucia Lake in Zululand'. *Ibis* 42 (3): 517–25.

Websites:

http://www.s2a3.org.za/bio/Biograph_final.php?serial=215 (accessed 26 April 2017).

https://www.biodiversityexplorer.org/people/bell-Marley-hw.htm (accessed 30 April 2017).

https://icosweb.net/drupal/terminology (accessed 19 September 2017).

www.pansalb.org (accessed 19 September 2017).

http://www.salanguages.com/unesco/isizulu.htm (accessed 19 September 2017).

https://www:africanraptor.co.za (accessed 3 October 2017).

https://www.fgasa.co.za/images/17.pdf (3 October 2017).

https://www.fgasa.co.za/images/6.pdf (3 October 2017).

http://www.n3tc.co.za/press-media/125-encouraging-efforts-to-protect-owls-in-communities-along-the-n3-toll-route (accessed 3 October 2017).

www.photodestination.co.za/junior-gabela.html (accessed 3 October 2017).

https://cathsseta.org.za/ (accessed 4 October 2017).

https://www.fgasa.co.za/ (accessed 4 October 2017).

http://www.saqa.org.za/ (accessed 4 October 2017).

www.twinstreams.org (accessed 4 October 2017).

http://www.wessa.org.za/documents/July%202011.pdf (accessed 4 October 2017).

www.photodestination.co.za/sakhamuzi-mhlongo.html (accessed 6 October 2017).

www.photodestination.co.za/themba-mthembu.html (accessed 6 October 2017).

www.zulubirding.jimdo.com (accessed 6 October 2017).

www.birdlife.org.za/images/Birding_Routes/Doc/dti-information_booklet.pdf (accessed 9 October 2017).

www.africanbirdclub.org/bookreviews/zazo-sesel-names-seychelles-birds-and-their-meanings (accessed 14 October 2017).

https://www.islandconservationseychelles.com/books--charts.html (accessed 14 October 2017).

www.birdlife.org.za/gobirding/community-bird-guides (accessed 27 October 2017).

www.uvm.edu/uvmnews/news/birding-change-world (accessed 27 October 2017).

https://blog.londolozi.com/2014/09/13/a-new-study-recording-bird-alarm-calls/ (accessed 20 November 2017).

https://sites.google.com/site/africanbirdname/ (accessed 20 November 2017).

General index

African Bird Name Group 430–2
African Bird of Prey Sanctuary 428
African National Congress (ANC) 316
Aitken, Marilyn 413, 424
Akan people of Ghana 344
Amatikulu Nature Reserve 409
Anderson, Mark 429, 432
Antananarivo Journal 1
appellative(s) 4
artist's pigments 117, 120
avitourism 415, 419, 420–3

Bainbridge, Bill 347
Bell-Marley, Harold 399, 400
Bennie, John 304
Berglund, Axel-Ivar 342
Bhula 'Ntsentse Bo! 236, 266
Bhunu 213
biltong 337
bird calls *see* birdsong
bird guide certification 417
birding, rise of interest among Zulu
    speakers 11
BirdLife South Africa 348–50, 402, 404–6,
    408, 410, 414, 415, 418, 420, 429
birdsong
    description of in names 63, 159, 163
    imitation of 159, 160
    interpretation of 175–84
    metaphor in names referring to 64, 159

onomatopoeia in names referring to 55,
    165–74
bird traps 240, 241, 331
birds killed for food 251
Blessing, Dr Henrik Greve 393, 394
Boers 315, 320
Bokmakierie magazine 15
botonomasticon 10
brand names and logos 130
Bryant, Father Alfred T. 2, 177, 272, 341,
    398
Buckley, T.E. 319
Bukhosi, Theo 40
Bulpin, T.V. 177
Buthelezi, Mangosuthu 188, 194, 202,
    209, 211, 222, 324
Buthelezi, Mbongiseni 238, 239

Caine, Tammy 427
Canute (Danish king) 105, 353
cattle, colour terms 106, 112–16
Cetshwayo kaMpande 197, 198, 202, 204,
    208, 213, 219, 314, 324, 326
amaChunu people 324
clan praises (*izithakazelo*)
    Dlomo 220
    Duma clan 192
    Khumalo 192
    Malinga 223, 228
    Mlotshwa 192

Mnguni 192
Msomi 221
Mthiyane 211
Mthombeni 192
Nkosi 193
Ntuli 192
Thwala 212
Zungu 237
clan totems 290, 291
Cole, Desmond 378
Coleridge, Samuel Taylor 242, 289
colour terminology in Zulu 117, *see also*
    artist's pigments; cattle, colour terms
Colenso, Bishop John 2, 231, 396
Colenso, Harriette 396
core images 189
courting ritual(s) 320
Cowgill, Roy 430
Cry, the Beloved Country 168
Culture, Arts, Tourism, Hospitality and
    Sports Sector Education and Training
    Authority 416, 418

Dadoo, Dr Yusuf 317
Deadman's Tree *see Synadenium cupulare*
Delegorgue, Adulphe 198, 199, 395, 396
derivational grammar 350
derivational system 154
diminutives, formation of 142
diminutive noun classes 143
Dingane kaSenzangakhona 188, 191, 193,
    194, 205, 209, 323
Dingiswayo kaJobe 188, 196, 216, 314,
    327
Dinuzulu kaCetshwayo 188, 202,
    203fn.26, 212, 214, 221, 293fn.28
discontinuities 35
Dlamini, Phindile 45, 425
Dlamini, Siya 402
Downs, Colleen 15
Dube, Abednigo 402

Du Plessis, Fanie 432
Dunning, R.G. 177

ecotourism 422
Education, Training, Development
    Practices Sector Education and
    Training Authority (ETDP-SETA)
    408
Elephantorrhiza elephantina shrub 239
Engelbrecht, Derek 429–32
English Romantic poets 242, 244
ergonym 131
Eshowe Environmental Education Centre
    411, 412
ethnobiological classification 20
ethnobiology 24
ethno-ornithology 24

Feast of First Fruits 294
Field Guides Association of Southern
    Africa (FGASA) 410, 416–18
fixed phrases 189
Flashman, Harry 316
folk taxonomic systems *see* taxonomy,
    folk
Forestry, Department of 347
formative(s) 140
formulas, formulaic phrases 189
frog (*donder-padda*, 'thunder-frog') 288
Fynn, Henry Francis 178, 208, 214

Gabela, Junior 402, 404, 409–11, 420
Gerstner, Father Jacob 2, 400
Gladiolus ludwigii (*isidwa* bulb) 254
Godfrey, Rev. Robert 2, 177, 401
iziGqoza faction 198, 326fn.15
Grail International Women's Movement
    414

Higher Education and Training,
    Department of 411

uHlakanyana (folktale character) 276
hlonipha language 292
homonyms 107
Huddleston, Father Trevor 317
Hughes, George 124
hunting and trapping of birds 248–51,
    330–2
hydronyms 125, 126

Ibandla lamaNazaretha 188
Ibis (*International Journal of Avian
    Science*) 1, 15
*idumbe likanhloyile see Scadoxus
    puniceus*
imikhovu (zombies) 276
Important Bird Areas (IBAs) 404, 419, 423
impundulu (lightning bird) 295, 296
indigenous knowledge systems (IKS)
    342fn.23
inganekwane (folk tale) 261
inhlamvu, meanings of 111
inkotha, meanings of 112
Inkatha Freedom Party 188
inkwazi, as brand name 130
iNyonikayiphumuli 254
isagila (throwing stick) 249
iSandlwana 213fn.37
isidwa bulb *see Gladiolus ludwigii*
isiluba, meanings of 112
iSimangaliso Wetland Park 349, 407, 408
isiqhova, meanings of 112
Isitwalandwe/Seaparankwe Award 317
izembe (axe, cleansing medicine) 110
izibongo 372
izimbongi 372
izithakazelo see clan praises

Jacobs, Nancy 413fn.29
Jama kaNdaba 316
James Stuart Archive (*JSA*) 230
uJamludi (ox name) 239

Jan Blom *see* frog
Jeremiah kaMtekelezi 215, 216
John Voelcker Bird Book Fund 380

uKhahlamba (Drakensberg) 108, 109
Khuzwayo, Thabile 402, 420
Killie Campbell Africana Library 230
King, Dick 202
Konjwayo of the Embo clan 315
Kosi Palm 100
Kumalo, Doris 348

Langeni clan 220
The Lark Ascending 159fn.3
Layard, Edgar Leopold 4
lightning bird 295, 296, *see also*
    impundulu
liminality 276
linguistic strategy 355–7
The Lion King (Disney film) 105, 371
little brown jobs (LBJs) 35, 409
love charms 293
Lubombo Mountains 80fn.24
Lugg, Harry 177
Lunguza kaMukane of the abaThembu
    326
Luthuli, Chief Albert 199, 200, 317

Mabula Ground Hornbill Project (MGP)
    414
Maclean, Gordon 28, 346, 347, 367,
    382
'Maclean problem' 26, 37, 38, 41, 351,
    430
Mageba 200, 220
Magemegeme of the Dube clan 206, 222
Magogo, Princess Constance 188
MaHlalise of the Mkhwanazi clan 215
Mahlokohloko of the Mhlongo clan 208,
    214
Mandela, Nelson 329

Manqaba of the Mbonambi clan 219

Mantantashiya kaMpande 218

Maphitha of the Mandlakazi clan 203

Marakwet people 320

Matabele 319, 320, 325

Mathole kaTshanibezwe of the Buthelezi
   clan 222

Mbandzeni wasemaSwazini 201, 213

Mbeki, Govan 206

Mbuyazi (Henry Francis Fynn) 208

Mbuyazi (Mbulazi or Mbuyazwe)
   kaMpande 218, 219, 326fn.15

meaning
   associative and symbolic 54
   connotative 52
   lexical 51
   notions of 49

meme theory 19, 423

Mdlalose, Jethro 402

Meredith, George 244

Meyer, Johan 55, 429–31

Mgabi 323

Mhlongo, Bheki 402

Mhlongo clan 220fn.47

Mhlongo, Sakhamuzi 349, 372, 402,
   404–6, 409, 411, 420, 426

Mkhungo, Nomusa 413, 424

Mlingwa, Charles 432

uMngeni Valley Nature Reserve 411

Mnkabayi kaJama 316, 322

Mnyamana of the Buthelezi clan 211

Mokumo, Peter 429, 430

Monase, mother of Mbuyazi 198fn.22

monitor lizard 281

months, Zulu names of
   uLuthuli 304
   uMfumfu 306
   uNcwaba 255
   uNhloyile 255

Morning Star (Venus) 216, 217, 264fn.37

Mosilikatse (Mzilikazi) 325

morphology (as linguistic term) 132

Mpande kaSenzangakhona 189, 193, 199,
   212, 217, 315, 326

Mqinisi 219

Msifile (Henry Francis Fynn) 208

Mswati II (Swazi king) 220, 324

Mthembu, Bongani 403

Mthembu, Sakhile 403

Mthembu, Themba 280, 349, 372, 403,
   404, 406–8, 425

Mthenjwa, Sakhile 403

musical instruments in bird names 164,
   165

musical notation 159

Mzilikazi kaMashobane 325

names, birds
   Afrikaans 37
   common problem with terminology 37,
      38
   comparative semantic frequencies of 58
   comparative semantic profiles of 93, 94
   generic 376
   identity and names 20
   individual proper names of 4, 5
   musical instruments reflected in 164,
      165
   nickname 8, 225
   Northern Sotho 32, 35, 37, 42, 43, 48,
      262, 430, 431
   popular name 8
   problematic issues 26
   proposed typology of 9
   rain bird name 298–300
   scientific binomial 6
   Seychelles Creole 437–41
   song reference in 161
   Swahili 432
   typology of 5, 6, 8, 56
   Venda 55, 132, 431, 432
   vernacular 'book' name 7

names, general
  geographical features 125
  magistrates 124, 309
  mountains and hills 125
  Natal Parks Board staff 125
  people 123–5
  rivers and streams 126
  streets and roads 128–30
Nandi, mother of Shaka 222
Natal Parks Board 124, 347
National Qualifications Framework (NQF)
  402, 416
Nazareth Church 193
Ndaba 188, 203fn.26
amaNdebele 320, 325
Ndlela kaSompisi 219
Ndondakusuka (battle of) 197fn.21,
  198fn.22, 219, 326
Ndumo Game Reserve 408
Ngcobo, Daluxolo 403, 420
Ngoqo of the Mbatha clan 209
Ngqengelele of the Buthelezi clan 221,
  222
Ngubane, Benson 349, 403
Nguni languages 15, 25, 31
Ngwenya, Vusi 403
nidification 1
nouns,
  proper and common 3, 5
  primary function of 50
  secondary functions and secondary
    meanings of 51
noun classes 132
noun stems, classification of 134
Nozishada kaMaqhoboza 207, 215
Ntshangase, Phindile 403, 420
Ntshidi kaLindelihle 216
Ntungazelana 200
Nuttall, Rick 429
Nyandeni, Bheki 403, 420

O'Kane, Trish 413
onomasticon 10
onymisation 121, 122
orality 391
ornithology, scientific 24
ornithonomasticon 10
oronyms 125
Ostrich (Journal of African Ornithology)
  15, 37, 55
ostrich feathers, trade in 320
Owl Awards 405
owl boxes 340, 427

Pan South African Language Board
  (PANSALB) 379, 380
partridge in the thumb (story) 273, 338
Paton, Alan 168
pejorative connotations of terminology
  271
Phakathwayo kaKhondlo (Qwabe chief)
  189, 196
uPhiko LweNgcebo Yemvelo KwaZulu
  348
Phinda Private Game Reserve 347, 349
Phunga 200, 220
Pleiades 303
polysemy 16, 107
polytypy 41, 42
Porter, Roger 76, 234, 363, 371–5, 380,
  403, 430, 432
primary functions 2, 50
proto-Nguni 205, 240, 271
puff adder 219, 220

rain, birds presaging 295–8
rain-making 300–2
Raptor Rescue Rehabilitation Centre 427
regiments, traditional insignia of
  Bulawayo 315
  uDhlambedu 322

imDhlenevu 317, 318
Dhloko 317
Dhlokwe 315, 318
Dukuza 315
izinGulube 317, 318
iHlaba 326
Kandempemvu 322
imKulutshane 326
Mbelebele 315
amaMboza 315
iziMpohlo 315, 325
Mxapo 317
uNdabenkulu 327
Ndhlondhlo 315
Ngobamakhosi 317, 318, 321
Ngwegwe 315
iziNyosi 317, 318, 326
Siklebe 315
Tulwana 315, 322
uluVe 324
Richards Bay Minerals (RBM) 415
Rime of the Ancient Mariner 289
Roberts, Austin 315, 395, 399
Royal Air Force 195

sacred bird, wagtail as 288, 290
salience 24, 35, 355
salient features 355
Samuelson, Letitia 237, 424
Samuelson, Robert 424
Saunders King, J. 313
Scadoxus puniceus (Snake Lily, *Z. idumbe likanhloyile*) 113, 303
semantic frequencies, comparative African and Zulu 58
semantic profiles of
    African names 93
    coined Zulu names 94–105
semantic categories 92
Senzangakhona kaJama 188, 193, 203

Shaka kaSenzangakhona 188, 189, 191, 204, 205, 211, 221–3, 313, 314, 323, 325, 327
Shelley, Percy Bysshe 242
Shembe, Isaiah 188, 193, 202, 217
Shepstone, Sir Theophilus 195, 198fn.22, 201
Sibiya, Buyelaphi 413, 414
Sibree, Rev. James 1, 2
Singcofela kaTshungu of the emaBomvini clan 282
Sirayo 323
Sithole, Bheki 403
Snake Lily *see Scadoxus puniceus*
Sobhuza II (Swazi king) 187, 206, 207, 217
Socatshwa kaPapu 230, 234, 277, 284
Somtseu *see* Shepstone, Sir Theophilus
Sotobe kaMpangalala 186, 190, 191
South African College for Tourism Tracker Academy 431
South African Ornithological Society 37
South African Qualifications Authority (SAQA) 416
St Lucia National Park *see* iSimangaliso Wetland Park
Stuart, James 230, 234, 237
superstition, negative connotations of term 272
Synadenium cupulare (Deadman's Tree, *Z. umdlebe* tree) 218

taboos, bird-related 85, 287–93, 332
taxonomy
    folk 20, 24, 25, 32, 36, 37
    scientific 20
Taylor, Jim 240, 427
Tembe Elephant Park 408
terminology, onomastic 10
thefuya speech 196fn.15
abaThembu people 324, 326
tokoloshe 276

tooth-fairy, Yellow-billed Kite as 337,
    415, 424
Tourism, Hospitality, Sport Education and
    Training Authority (THETA) 416
Tourism World Academy 408
Trade and Industry, Department of 415,
    420
traditional ecological knowledge (TEK)
    342
Tshanibezwe kaMnyamana of the
    Buthelezi clan 209
Turner, Noleen 41, 346, 348–51, 361, 363,
    365, 366, 372, 374, 375, 380, 403, 404
Twinstreams Environmental Education
    Centre 411

ujojo finch as 'Everyman' 205
umdlebe tree see Synadenium cupulare
umthothwane plant 239, see also
    Elephantorrhiza elephantina
uncede, as a 'little brown job' 35
University of KwaZulu-Natal 15, 410
University of Pretoria 410–12
University of Vermont 411, 413
uthekwane, use of name as brand 130

Van Ginneken, Father Jacques 414
Venus see Morning Star
Vilakazi, Benedict Wallet 17, 185, 234,
    242–6
Voight, Leigh 116

Wakkerstroom Tourism and Education
    Centre 416
Webb, Colin 229
Weiersbye, Ingrid 430
whistling spirits (abalozi) 339
Wildlife and Environment Society of
    South Africa (WESSA) 340, 348,
    402, 404, 409, 411, 416, 427
Williams, Vaughan 159fn.3

Wilson, G. Walker 309
witchdoctor, inaccurate use of term 272
Women's Leadership and Training
    Programme (WLTP) 345, 404,
    413–15, 424
Woodward, Reverends Richard and John
    2, 3, 13, 392, 398
Wordsworth, William 242
Wright, John 229, 230, 234, 238, 239

Xaba, Nonthuthuko 268, 403, 404, 413,
    425

Youth Bird Clubs 345
Yucatec Maya people 344

Zigqoza faction (of Mbuyazi) 197fn.21,
    198, 326fn.15
Zulu Bird Name Workshop(s) 9, 11, 18,
    24, 41, 43, 90–4, 144, 145, 154, 171,
    172, 174, 232, 256, 268, 280, 309,
    348, 349, 363, 401, 402, 408, 415
isiZulu, non-use of preface xiiifn.1
Zululand Birding Route 405
Zulu Language Board 379
Zulu plant names 39
Zulu royal genealogy 187–8
Zuma, Nongejeni 195
Zwelithini kaBhekuzulu (Zulu king) 324
Zwide of the Ndwandwe clan 211

Bird name index

abayeni 133, 365, 427

Abdim's Stork 75

Abdimia abdimii 75

Acacia Pied Barbet 374

Accipitridae 24, 25

Actitis hypoleucos 353

Actophilornis africanus 83

Afep Pigeon 435, 436

African Black Duck 358, 368

African Black Swift 298

African Broadbill 65, 66, 103, 141, 371

African Cuckoo-Hawk 98, 141

African Darter 376

African Dusky Flycatcher 101

African Finfoot 103, 371

African Firefinch 134

African Fish Eagle 16, 27, 29, 41, 79, 106, 114, 116, 121, 127, 129, 130, 131, 133, 294, 333, 358

African Goshawk 141, 155

African Green Pigeon 43, 45, 46, 82

African Harrier-Hawk 78, 84, 102, 370, 385

African Hawk-Eagle 27, 91, 149

African Hoopoe 61, 68, 69, 73, 79, 153, 304

African Jacana 139, 153, 225

African Marsh Harrier 52, 145

African Olive Pigeon 45, 151, 173

African Openbill 101, 153

African Oystercatcher 97, 346

African Palm Swift 87

African Pied Wagtail 33, 84

African Pygmy Kingfisher 83, 89, 381, 382, 384–8, 434

African Rail 79

African Scops Owl 140

African Snipe 354–7

African Spoonbill 152, 240

African Stonechat 80, 334

African Wattled Lapwing 97, 138

African Wood Owl 183, 294

African Yellow White-eye 77

Afropelia capicola 44, 46

akóhalàhinála 72

akòholàhindràno 72

akòholàhundràno 299

Alauda hamgazy 199

albatross 288, 351

Alcedo semitorquata 382, 383, 386

Alpine Swift 298, 367

amadada 150

amadojeyana 102

amadojeyanabomvu 101

amajuba 125, 133

amalanda 265

amanqe 246, 257, 264

amapopo 291

Anas clypeata 440

andóvy 86

andriesdak 310

Andropadus flaviventris 6

Andropadus flaviventris flaviventris 6, 7

Andy-the-dawn 310

Angola Kingfisher 382, 383, 384, 388

Angolan Swallow 91

angolaswael 91

angòly 79

Anhinga rufa 376

a-nini-a 60

Ant-eating Chat 101, 153

Anthus 83

Anthus brachyurus 4, 5

Anthus richardi 309

Anthus similis 309

Ant-Thrush 75

Anvil Bird 163

Aplopelia larvata 8, 43, 45, 46

Apus affinis 298

Apus barbatus 298

Aquila 22, 23, 27, 28

Aquila verreauxii 22, 23

Arctic Skua 102

Arctic Tern 99, 154, 366

Ardea purpurea 365, 440

Ardeotis kori 325

Arenaria interpres 352

Arrow-marked Babbler 96, 157, 162

Ashy Flycatcher 148

Au-to-dor 179

Aves 25

Avocet 352, 374

Ayres's Hawk-Eagle 27, 369

babasi 79

babbler 158

babewatoto 86

Baillon's Crake 144

Bald Ibis 135, 172

Baltimore Oriole 97

Banded Harrier Eagle 166

Barau's Petrel 439

barbet 170, 374

Barn Swallow 91

Barratt's Warbler 100

Bartailed Godwit 99, 352, 354, 356

Bar-throated Apalis 140

bata mchikichi 76

bateleur 9, 14, 22, 23, 29, 31, 41, 225,
    226, 274, 277, 278, 281, 289, 297, 301,
    302, 327, 329, 343, 358

Bateleur Eagle 24, 25, 27, 30, 54, 83, 109,
    124, 135, 201, 202, 203, 204, 219, 282,
    283, 287, 295

Bearded Vulture 27, 28, 146, 358

Bearded Woodpecker 146, 151

bee-eater 71, 75, 78, 85, 86, 113, 144

berghaan 22, 28

bibing 78

bird of the sky 295

bishop 25, 115

Bishop bird 210

Bittern 74

black and grey cuckoo 299

Black Boubou Shrike 60, 166

Black Crake 73, 144, 215, 365

Black Crested-cuckoo 307

Black Crow 265, 275, 287

Black Cuckoo 62, 152, 182, 225, 227

Black Eagle 22, 23, 27, 91, 116

Black Egret 79

Black Harrier 97

Black Heron 172

black ibis 179

Black Kite 441

Black or Common Ibis 179

Black Parrot 439

Black Stork 364

Black Swift 360, 364, 367

Black Widowfinch 144

Black-and-white Cuckoo 85

Black-and-white Flycatcher 60

Black-backed Puffback 70, 71, 86, 95, 225, 226, 369

Black-bellied Bustard 329, 373

Black-bellied Starling 33

Black-breasted Snake Eagle 27

Black-collared Barbet 170, 360

Black-crowned Tchagra 79, 87, 88, 116

Black-eyed Bulbul 9, 69, 173

Black-headed Canary 151

Black-headed Heron 139, 172

Black-headed Oriole 9, 31, 135, 136, 358, 369

Black-headed Paradise Flycatcher 286, 292

Black-rumped Buttonquail 255

Black-shouldered Kite 77, 81, 82, 340

Blacksmith Lapwing 65

black-tailed finch 113

Black-winged Kite 134, 143, 340

Black-winged Lapwing 168

Black-winged Pratincole 75

Bleating Warbler 64, 163, 173

blokbek 362

Blouswael 91

Blue Crane 54, 92, 121, 130, 135, 227, 229, 289, 313, 314, 316, 317, 322, 326, 328, 329, 358

Blue Heron 172

Blue Jay 325

Blue Kingfisher 383

Blue Korhaan 143

Blue Swallow 91, 147

Blue Waxbill 101

Blue-billed Firefinch 134

Blue-breasted Kingfisher 434, 435

Blue-headed Coucal 79, 88, 143, 166

Blue-spotted Wood Dove 45, 436

Bokmakierie 74, 300, 356

bokmakierie 61

boleseboku 70

Booted Eagle 27, 149

bosduif 6, 8

bosmusikant 164

bosveld-tjeriktik 160

Boubou Shrike 60

Brimstone Canary 362

Broad-billed Roller 369

Broad-tailed Warbler 376

bromvogel 288, 301

Bronze Mannikin 75

Bronze-naped Pigeon 435

Bronze-winged Courser 150

Brown Fly-catcher 73

Brown Skua 102, 147, 370

Brown Snake Eagle 27, 101

Brown Tambourine Dove 44, 46

Brown-headed Parrot 78, 151

Brown-hooded Kingfisher 80, 145, 377, 381, 382, 384–8, 434

Brown-throated Martin 368

Brubru (Shrike) 105

Bubulcus ibis 289, 431

Buceros buccinator 167

Bucorvus leadbeateri 431

bud-bud 61

Buff-spotted Flufftail 7

Bugeranus carunculatus 3, 4

bulbul 84, 247, 330

Bully Canary 143

Bully Seedeater 362

bunyiro 81

Buphagus africanus 76

Buphagus erythrorhynchus 76

Burchell's Coucal 98, 133, 139, 142, 289, 299, 329

Burchell's Courser 88

Burnt-necked Eremomela 147

Bush Blackcap 96

Bush Musician 164

bushshrike 95, 126, 260, 261

bush-pheasants 308

bustard 63, 135, 151, 161, 162, 325, 329, 373

Butcher-bird 126, 260
Button Quail 64, 77, 165, 209, 383
buttonquail 255
buzzard 155, 204, 205
bvuo 76
bvuobvuo 76
bvuwe 76
bvuwo 76
bwabwalala 84
Bycanistes bucinator 168

Cabanis's Bunting 74
Calandrella cinerea 309
calaos 167
Calidris alba 352, 354
Calidris canutus 105, 352, 353
Calidris ferruginea 352, 354
Calidris minuta 352, 354
Camaroptera brachyura 108
Camaroptera olivacea 108
canard 440
canard souchet 440
canary 151, 257
Cape Batis 105, 364
Cape Bittern 74
Cape Canary 95, 97, 155, 163
Cape Dikkop 87
Cape Dove 293
Cape Eagle-Owl 51, 70, 148
Cape Gannet 103, 151, 153, 370, 441
Cape Grassbird 311, 376
Cape Gull 99
Cape Kite 287
Cape Laughing Dove 44, 46
Cape Longclaw 234, 239, 240, 245, 246
Cape Parrot 78, 226, 358
Cape Raven 83
Cape Robin 280, 282
Cape Rock Thrush 74, 100, 154
Cape Shoveler 91, 97, 369, 440
Cape Sparrow 87

Cape Teal 376
Cape Turtle Dove 8, 42, 44, 46
Cape Wagtail 33
Cape White-eye 138, 145, 394
Capped Wheatear 80, 334
Cardinal Woodpecker 170
Carmine Bee-eater 73
Caspian Tern 95, 156, 171, 367
cateau 439
cateau noir 439
cattle egret 80, 85, 130, 254, 289, 334
Cecropis abyssinica 151
Cercomela familiaris 84
Ceryle maxima 383
Ceryle rudis 382, 383, 386
chapuluka 60
Charadrius tricollaris 365
chekiuwa 60
chemalango 162
chemilunda 68
Chera progne 322
Chestnutbacked Sparrow-Lark 97, 155
Chestnut-bellied Kingfisher 434
Chestnut-fronted Helmetshrikes 426
Chestnut-vented Tit-babbler 96, 160
chickadee 165
chicken 147, 250
chidazi 71
chiffchaff 165
chihaha 168
chikhunguo 363
chikupe 82
chinkwemaula 78
chinsoso 305
Chinspot Batis 364, 365
chiombankhanga 78
chipiyo 60
chithathale 78
chitotola 78
chivundikile 79
chiwalewale 87

chokiro 60

Chrysococcyx cupreus 97, 114

Ciconia 71, 72, 75

Ciconia abdimii 75

Ciconia ciconia 75

Cinnamon Dove 8, 43, 45, 46, 151, 360

Cinnyris afra 33, 111

Cinnyris chalybea 33

Circaetus 27, 28, 29

Cisticola textrix 137

cisticola 35, 60, 147, 161, 239, 256

Clamator 85

Clamator jacobinus 230, 306

Clamator levaillantii 306

Cloud Cisticola 137, 377

Coccystes cafer 114, 231, 306

Coccystes jacobinus 114, 306

cock 252

cockatoo 439

Colius capensis 273

Colius striatus 323

Collared Pratincole 151

Collared Sunbird 111, 143

Collared Turtle Dove 43, 46

Columba 45

Columba arquatrix 42, 44, 45, 46, 47

Columba delegorguei 45

Columba guinea 6, 8, 42, 43, 46, 47

Columba larvata 45

Columba livia 45

Columba palumbus 8

Columba phoeonata 43, 46

Columba tympanistria 44, 46

Columbidae 45

Columbiformes 45

coly 297

Common Buzzard 141

Common Cuckoo 95, 139

Common Greenshank 165, 353, 356, 357

Common Moorhen 73, 99, 147

Common Myna 102, 141, 371

Common Ostrich 34, 317

Common Quail 82

Common Ringed Plover 98

Common Rock Pigeon 44, 46, 47

Common Sandpiper 61, 74, 83, 85, 98, 166, 352, 353, 356

Common Scimitarbill 364

Common Waxbill 3, 32, 98, 130, 142, 143, 366

Common Widow-bird 113

Cooscot 8

Coqui Francolin 78, 161, 339, 373

Coracias caudatus 325

Coracias garrulus 33

Coracopsis barklyi 439

Coracopsis nigra 439

cormorant 76, 351

Corn Crake 137

Corvidae 276

Corvus albicollis 363

Corvus capensis 126

Corythornis cristata 382, 386

Corythornis cyanostigma 383

Cossypha heuglini 311

coucal 339

coucou 95

courser 150

crakes 144

crane 298

crane 303, 314, 315, 318, 325

Creatophora carunculata 75

Creatophora cinerea 75

Crested Barbet 65, 88, 97, 114, 138, 144

Crested Eagle 282

Crested Francolin 170, 373

Crested Guineafowl 265

Crested Ibis 72

crimson-wing 70

Crithagra sulphurata 362

Croaking Cisticola 147, 161

Crossley's Warbler 82

crow 2, 255, 276, 290, 307, 328, 363
Crowned Crane 3, 61, 71, 72, 133, 139, 288, 369
Crowned Eagle 22, 27, 29, 133
Crowned Guinea-Fowl 111
Crowned Hawk-Eagle 76, 112
Crowned Hornbill 143, 371
Crowned Plover 116, 146
Cuckoo Finch 104
cuckoo 98, 114, 142, 165, 300, 304, 305, 308, 310, 344
Cuckooshrike 80, 334
cuculo 95
Cuculus solitarius 49, 50
curlew 165, 351
Curlew Sandpiper 166, 352, 354, 356
Cut-throat Lark 245
Cypsiurus parvus 87

D'Arnaud's Barbet 62
Dabchick 83
dagbreker 84, 310
dangòromainty 66
dangòrovana 75
Dark Chanting Goshawk 7
Dark-backed Weaver 164
Dark-capped Bulbul 69, 128, 183
Dark-capped Warbler 103
dassievanger 22, 23, 76
Deland's Green Pigeon 436
Delegorgue's Pigeon 45
Demoiselle Crane 315
Dendropicos namaquus 151
Desert Cisticola 64, 377
dete 434
detepwani 434
Dialiptila phaeonota 44, 46, 47
dicky bird 35
Diederik Cuckoo 63, 298
dikbeksysie 362

Dikkop 87, 374
Dilophus carunculatus 114
dix-huit 59
doholantalambo 78
domestic chickens 252
domestic cockerel 308
domestic fowl 147, 247, 263
domestic hen 210, 211
dondolamswa 75
dove 8, 15, 33, 40, 41, 43, 46, 48, 67, 69, 77, 85, 130, 133, 134, 151, 178, 210, 215, 248, 251, 259, 280, 281, 313, 326, 332, 433, 435, 436
Draadstertswael 91
drake 226
Drakensberg Rockjumper 80, 100, 334
Drakensberg Wailing Warbler 311
drongo 133, 276, 333, 344
dubbele geelsysie 362
duck 7, 13, 14, 33, 36, 37, 38, 92, 96, 99, 150, 165, 226, 289, 291, 355, 357, 358, 368, 431, 440
Dusky Indigobird 144
Dusky Turtle Dove 436
dzanjamphako 78

eagle 7, 14, 21, 22, 23, 25, 26, 28, 29, 35, 36, 40, 41, 91, 92, 109, 112, 124, 130, 143, 147, 149, 185, 188, 201, 202, 203, 254, 259, 261, 294, 313, 326, 335, 357, 358, 369
Eastern Bronze-naped Pigeon 45, 151
Eastern Long-billed Lark 155
Eastern Nicator 79, 102, 291, 334
egret 69, 80, 210, 214, 324, 395
Egyptian Goose 358
Egyptian Vulture 143
Emarginata familiaris 310
emba 436
Emberiza longicauda 322
Emerald Cuckoo 138, 179, 184, 395

Emerald-spotted Wood Dove 17, 44, 45,
   51, 181, 361, 436
emu 287
Estrilda astrilda 3
Ethiopian Snipe 60, 352, 355
Euplectes progne 321
Eurasian Golden Oriole 97
Eurasian Hobby 376
Eurasian Oystercatcher 439
Eurasian Wren 37
European Bee-eater 101, 153
European Cuckoo 95, 139
European Golden Oriole 97
European Honey Buzzard 101, 371
European Roller 369
European Swallow 32, 91
Europese Swael 91

Fairy Flycatcher 74, 97, 150
Falco biarmicus 155
Falco vespertinus 440
falcon 25, 155, 201, 339, 340, 376, 431
Falconiformes 25
Familiar Chat 84, 310
fandiafásika 74
fangàlatróvy 78
fangàlimótivòay 81
Fan-tailed Warbler 376
Fan-tailed Widowbird 140, 150, 207, 210
fariki 78
faucon kobez 440
Feral Pigeon 45, 435
fiàndrivòditàtatra 74
Fieldfare 162
Fiery-necked Nightjar 7, 62, 179
finch(es) 25, 67, 68, 97, 104, 135, 148,
   150, 155, 205, 207, 208, 209, 210, 251,
   322
firefinches 70
firìoka 82
firìotsàndro 82

Fiscal Flycatcher 60, 98, 142
Fiscal Shrike 9, 77, 98, 114, 116, 133, 142,
   225, 307
fitatra 162
fitìlibèngy 80, 334
fitosívy 88
flamingo 68, 145
Flappet Lark 66, 84, 170
floranten 440
Floranten Mov 440
flufftail 150
flycatcher 148, 276, 329
Forest Buzzard 135
Forest Weaver 164
Fork-tailed Drongo 86, 121, 333
fòtsièlatra 67
fowl 253, 289
francolin 133, 161, 254, 373, 396
Francolinus natalensis 310
Freckled Nightjar 100, 139, 145
fulangombe 73
fulmar géant 439
Fulvous Whistling Duck 368, 369
fumvu 436

gàdragàdra 162
Gallinago nigripennis 352, 354
gamefowl 127
gannet 351
geeldikbeksysie 362
geelgat 62, 69
Geoffrey's Plover 73, 83
Geronticus hagedash 167
gewone tortelduif 8
Giant Eagle Owl 148, 183
Giant Kingfisher 195, 381, 382, 384,
   386–8, 434–6
Giessvogel 300
Glareola melanoptera 75
Glareola pratincola 75

Glaucidium perlatum 112

Glossy Ibis 88, 99

glossy starling 133, 137, 210, 216, 217, 264, 401

Gnat 105

Gnet 105

Goàika 2

goat-sucker 179, 308, 309, 336

Goaway Bird 160

golden yellow weavers 74

Golden-green Cuckoo 97, 114

Goliath Heron 51, 152, 225

gongofutu 71

goose 276

Gorgeous Bushshrike 142, 163, 170, 369

goshawk 78, 155, 376

Grass Owl 360

grass-birds 308

Great African Kingfisher 383

great blue turaco 291

Great Reed Warbler 100

Great Spotted Cuckoo 98, 142, 174

Great White Pelican 149

Greater Flamingo 149

Greater Honeyguide 63, 149, 152, 163, 360

Greater Kestrel 149

Greater Sand Plover 163

Greater Striped Swallow 149

Green Barbet 100, 139, 144, 370, 409

Green Coucal 63, 65, 94, 136, 137, 162, 171, 369

Green Malkoha 94, 136, 137, 171, 369

green parrot 114

Green Pigeon 43, 45, 291, 436

Green Sandpiper 143

Green Starling 33

Green Twinspot 366

Green Wood Hoopoe 66, 153, 158, 162, 364, 394, 426

Green Woodpecker 162, 300

Green-backed Camaroptera 64, 108, 127, 143, 163, 173

greenbul 330

Greenshank 95, 352

Green-spotted Wood Dove 44, 46, 51, 361

Green-winged Pytilia 98, 142, 366

Grey Crowned Crane 169, 297

Grey Cuckooshrike 143

Grey Go-away-bird 60, 69, 98, 135, 136, 142, 160, 174

Grey Heron 139, 172

Grey Penduline Tit 171, 369

Grey Plover 146, 373

Grey Sunbird 150

Grey Waxbill 97, 153, 201, 371

Grey-backed Bush Warbler 289, 329

Grey-backed Camaroptera 329

Grey-headed Bushshrike 95, 137, 163, 370

Grey-rumped Swallow 91

Grey-winged Francolin 373

Groenpiet 310

Groot Sprinkaanvoël 75

Grosbeak Weaver 87

Ground Hornbill 175, 245, 275, 276, 278, 296, 414, 426, *see also* Southern Ground Hornbill

Groundscraper Thrush 64, 367, 378

Grus paradisea 313

Gryskruisswael 91

guguk 95

guineafowl 35, 124, 130, 248, 250, 311

gull 84

gymnogene *see* African Harrier-Hawk

Gypaaetus 27, 28

haan 168

Hadada (Hadeda) Ibis 4, 59, 63, 85, 126, 143, 162, 166, 167, 168, 172, 179, 283, 292, 392

Haematopus moquini 346

Haematopus ostralegus 439

haha 59, 168
Halcyon albiventris 382, 383, 386
Halcyon chelicuti 382, 386
Halcyon cyanoleucus 383
Halcyon irroratus 383
Halcyon senegalensis 382, 383, 386
Halcyon senegaloides 382, 386
Half-collared Kingfisher 382–4, 386–8
Haliaeetus vocifer 106, 114
Haliaeetus 27, 28
Hamerkop 9, 16, 34, 63, 73, 86, 112, 123–5, 129, 130, 179–81, 210, 212, 220, 227, 228, 234, 262, 274–8, 281, 284, 285, 287, 295, 296, 300, 342
hammer-head 179, 234, 284, 285, 288, 296
hammerhead crane 179
hammer-kop 297, 328
Harlequin Quail 150
harrier 155, 376
Haw-di-das 179
hawk 7, 36, 87, 133, 143, 155, 201, 204, 255, 335, 337, 339, 376
Helmeted Guineafowl 111, 254, 358
helmetshrike 427
Helotarsus ecaudatus 295
hen 263
héron pourpré 440
heron 67, 71, 73, 74, 76, 81, 172, 179, 351, 365, 440
Heuglin's Robin 75, 84, 162, 164, 311, 374
Hieraaetus 27, 28
hilembe-lembe 81
hlekabafazi 394
hoephoep 61
hoètrika 83
Honey Buzzard 145
honeybird 214, 258
honeyguide 63, 111, 144, 149, 152, 163, 241, 242, 258, 262

honeysucker 32, 33, 77, 210, 214
hoopoe 61, 126
hornbill 125, 176, 218, 263, 287, 291, 344, 395
Horus Swift 364
hosétrika 83
Hottentot Teal 150, 368
House Martin 368
House Sparrow 100
houtkapper 162
hua 435, 436
hua matembwe 70
hua mwekundu 435
huîtier-pie 439

i(li)-Hōbe 46
ibhada 237, 241
ibhobobo 43, 46
ibhoyi 289
ibhoyibhoyi 137
ibis 80, 334
Ibis addidas 167
ibomvana 144
ibomvana 67
Icterine Warbler 101, 439
idada 13, 14, 33, 36, 355, 357, 358
idadelimlomophuzi 368
idadelimnyama 99, 368
idadelincane 150, 368
idlantuthwane 101
ifefe 33, 133, 148, 325, 369
ifefebomvu 369
ifefelihle 369
ifefeluhlaza 369
ifefemidwa 148, 369
ifuba 373
ifukezi 148
igedezi 63
igqumusha 126, 260
i-guondwana 73
igwababa 255, 307

igwababane 265, 307
igwalagwala 54, 119, 124, 130, 136, 150, 151, 195, 198, 238, 323, 396
igwalagwala eliluhlaza 150
igwalagwala logu 151
igwayigwayi 137, 169
igwedlamanzi 103, 371
igwigwelimlotha 373
igwigwi 137, 171, 365, 373
igwigwigwi 366
ihahane 59, 168, 172
ihahanemhlophe 172
ihem 61, 169, 297
ihlabahlabane 384–8
ihlabankomo 225
ihlabankomo 83, 360, 364, 367
ihlalankomo 76
ihlalanyathi 76, 121–3, 129, 135
ihlekehle 96, 157, 158
ihlekehleke 96
ihlokohloko 33, 137, 208, 238, 241, 258
ihlolamvula 51, 85, 103, 298, 360, 364, 367
ihlungulu 363
iHobe 33
ihobhe 8, 43, 45, 46, 47, 48, 326
ihubhulu 363
ijankomo 83, 360, 364, 367
ijikanyawo 102, 370
ijip 59
ijiyankomo 83, 367
ijuba 8, 42, 43, 45, 46, 47, 48, 67, 130, 133, 151, 248, 251, 326, 436
ijubalaphansi 151, 360
ijubantendele 43, 46, 47, 67
ijubantondo 42, 43, 45, 46
ijubantonto 43, 46
ijubelintamemhlophe 151
ikholwane 125
ikhwelematsheni 74, 100
ikhwelentabeni 99, 154

ikhwezi 118, 133, 216
ikhwikhwi 137, 169
iklebedwane 143
iklewu 60
iklosi 171, 369
ikoalakoala 396
ikoé 395
ikwinsi 33
ilanda 80, 130, 334, 395
ilindankomo 80, 334
ilongwe 358
ilunga 114, 307, 333
ilunga legwaba 307
imbekle 365
imBucu 350
imbumbazane 43, 46, 47
imbuyelelo 145, 384–8
imbuzana 64, 108, 127
imbuzane 143, 147, 173
imbuzane yomnqawe 147
imbuzaneluhlaza 173
imbuzanephuzi 173
imbuzi yehlathi 64, 163, 173
imi-ngquphane 114
impangele 111, 124, 130, 248, 250, 254, 265, 311, 358
impevu 365
impisiyolwandle 102, 147, 370
impofana 97
impofazana 66, 114, 144
impundulu 295
imvuliyeza 367
imvunduna 88, 97, 114
incumba 44, 46
incuncu 71, 112, 118, 121, 214
incwaba 365
incwincwemphunga 150
incwincwi 32, 33, 71, 112, 150
incwincwincwi 94
indebebomvu 96
Indian Myna *see* Common Myna

Indicator major 111
indlamadoda 9, 22, 23, 30, 31, 109, 111, 226
indlangamandla 102, 154
indlankumbi 75
indlantuthwane 153
indlanyoka 29, 101
indlanyokensundu 27, 101
indlanyokephuzi 27, 150
indlanyokomnyama 27
indlanyoli 30, 31
indlanyoni 30, 31, 225
indlanyula 30, 31
indlazanyoni 30, 31, 297
indlazi 119, 133, 170, 255, 273, 293, 297, 323
indlazinyoni 297
indlovuyenduna 89
indudumela 65
indwa 54, 92, 227, 313
indwazela 80, 377, 382, 383, 385–8
indwe 118, 121, 130, 135, 313, 317, 358
indweza eluhlaza 362
ing'ang'ane 59
ingagalu 126
ingcungcu 71
ingedana 144
ingede 144, 258, 360
ingilinkingci 164
ingongoni 170, 360, 370
ingoyi-ngoyi 352
ingozwana 356
ingqanga 30, 31, 225
ingqangendlela 298
ingqangqamathumba 85
ingqungqulu 9, 22, 23, 24, 27, 29, 30, 31, 41, 109, 124, 135, 202, 203, 226, 274, 277, 279, 281, 282, 294, 295, 297, 358
ingqwayingqwayi 203
ingududu 9, 134, 136, 175, 296, 369, 426
ingwangwa 137, 169

inhlamvu 111
inhlava 63, 111, 152, 163, 258
inhlavebizelayo 63, 111, 152, 163, 360
inhlazazana 111
inhlekabafazi 66, 153, 158, 363
inhlolamanzi 368
inhlolamfula 368
inhlolamvula 85, 152, 368
inhlolamvula yamadawala 368
inhlolamvula yasekhaya 368
inhlolazulu 298, 367
inhlunuyamanzi 83, 382–4, 386–8
inkankane 59, 119, 126, 143, 167, 168, 172, 283, 292
inkankanelunga 173
inkankanemhlophe 172
inkanku 114, 135, 230, 231, 305, 307, 344
inkanku 401
inkashana 293
inkayishana 293
inkenkane 240
inkombazane 43, 46
inkonjane 32, 143, 147, 193, 195, 256, 259, 264
inkonjane emqalomhlophe 149
inkonjane yamawa 149
inkonjanencane 149
inkonjanenkulu 149
inkonjanesifubabomvu 149
inkonjanesiloside 147
inkonjenesibhakabhaka 147
inkosiyezinkozi 27, 358
inkotha 144, 360
inkotha enkulu 149
inkothana 144, 149
inkothanyosi 101, 153
inkovana 144
inkovu 300
inkukhu 147, 247, 250, 252, 264
inkukhuyamanzi 73, 99, 147

inkwababakazana 44, 46
inkwali 127, 133, 161, 169, 170, 255
inkwalimanzi 73
inkwalitwetwe 80
inkwambakazana 44, 46
inkwazi 14, 16, 27, 41, 106–8, 121–3,
    127, 129, 133, 294, 333, 358
inongqwatshi 304
inqe 33, 100, 147, 148, 197, 254, 281, 326
inqelendlovu 68, 147, 148
inqemvuma 100, 148
inqilo 240
inqomfi 17, 121, 130, 135, 185, 229, 234,
    236, 237, 239, 240–4, 246, 284, 331,
    343
inqondanqonda 170
inqwathane 143
insansa 367
insigees 301
insindaphi 178
insingizi 9, 129, 135, 175, 218, 219, 274,
    281, 287, 296–8, 425
insukakude 99, 154, 366
inswempe 161, 338, 339, 373
inswinswi 137, 155, 367
intaka (Xhosa = bird) 25, 205
intaka (Zulu = finch) 67, 68, 97, 104, 135,
    148, 150, 205–7, 215
intaka mkhosi 30, 31
intaka yamadoda 30, 31, 225
intaka yamathamsanqa 336
intaka yeenkomo 336
intaka yempi 30, 31, 225
intaka yemvula 296
intaka yotshaba 30, 31, 225
intakansinsi 67, 148, 210
intakanzwili 97, 155
intakemahlombabomvu 150
intakembila 225, 226
intendele 43, 67, 130, 217, 218, 250, 286,
    311

intengu 117, 121, 133, 333
intenjane 117, 118
intinga 377
intingamafu 377
intinginono 121, 135, 263, 358
intiyane 32, 130, 366
intiyaneluhlaza 366
intlaba mkhosi 225
intonjana 88
intshe 92, 133, 135, 317, 358, 395
intsingisi 293, 296
intsomi 118, 121
intungunono 298, 358
intuntwane 143
inyoni 205
inyoni yezulu 295
inyoninyoka 376
inzwece 136, 324
inzwinzwebomvu 369
inzwinzwi 94, 171, 366, 368, 369
inzwinzwinzwi 94, 171, 369
iphemvu 365
iphothwe 119
iqabathule 232, 391
iqhude 126
iqola 114, 117, 133
isadube 70, 98
isagqukwe 8, 43, 46
isagundwane 98
isagwaca 135, 141, 150
isagwacesibomvu 150
isakabuli 133, 134, 321, 322
isambatha 365
isandondwane 360
isangulube 142, 163, 170, 369
isankawu 96, 142, 163, 370
isantinti 365
isanxa 141
isanzwili 396
isaqgukwe 360
isaqola 98, 142

iselantaka 104
iselayolwandle 102
iseme 135, 151
iseme lasentshonalanga 151
isiBelu 396
isibhelu 43, 45, 46, 182, 360
isibulalambiza 88, 350
isicelankobe 101
isicelankothe 101
isiCelegu 396
isicibamanzi 103, 153, 370
isiCibilili 350
isicivó 95, 369
isicukujeje 366
isigqobhamithi 151
isigqobhamithi saseningizimu 151
isigqobhamithintshebe 146, 147
isigqobhamnenke 101, 153
isigumbamphalo 86
isigwaca 84, 249, 269
isigwe 114, 118, 121, 200, 241
isihlalamahlangeni 74
isihlalamatsheni 74, 99, 154
isihuhwa 27, 28, 29, 112, 133
isikhobothi 135
isikhokhwane 358, 360
isikhombazana 43, 44, 46, 47, 119
isikhombazana sasehlanze 44
isikhombazana sasenkangala 43
isikhombazana sehlathi 44, 46, 47
isikhombazane 43, 151, 181
isikhombazane sasehlanze 46
isikhombazane sasenkangala 45, 46
isikhombazane sehlanze 151
isikhombazane sehlathi 45, 151, 360
isikhotha 148
isikhova 33, 36, 133, 144, 148, 151, 211, 212, 260, 294, 395, 399
isikhova sexhaphozi 151
isikhova sotshani 360

isikhovampondo 148
isikhovanhlanzi 148
isikhukhukazi 211
isikhwehle 133, 161, 170, 310, 373
isikhwehlesiqhova 373
isikhwenene 151, 226, 358
isikhwenene esikhandansundu 151
isikilothi 384–8
isinqonqotho 170
isiphikeleli 145, 382–8
isiphungumangathi 27, 29, 41, 135, 282, 334, 335, 337, 358
isipigileni 382, 388
isipopopo 369
isiqhananazana 215, 365
isiqhanazana 144, 215, 365
isiqonqotho 360
isiquba 382, 383, 386, 388
isiqula 383, 388
isishishi 137, 366
isithandamanzi 73, 153
isivuba 195, 382–4, 386–8
isiwekeweke 137
isiwelewele 63, 137
isixula 382–4, 386–8
isixulamasele 76, 152, 439
isizinzana 144
isizinzi 144
isomi 221, 290
Ispidina natalensis 383
Ispidina picta 382, 386
it(h)ilongo 164
ithendele 247, 249, 254, 373
ithendelomlotha 373
ititihoye 61, 135, 136, 147, 168, 373
ititihoyelimlotha 146, 373
ititihoyenomqhele 146, 147
ivevenyane 165
ivuba 150, 383
ivubelincane 149
ivukuthu 43–7, 117, 151, 170

ivukuthu lehlathi 45, 151, 173
ivuzi 376
ivuzigazi 97, 153, 201, 371
iwabayi 363
iwamba 151
iwamba lasenyakatho 151
iwili 173
izikhova 134, 277, 279
izikova 395
izingududu 263
izinkonjane 191, 192, 194
izinsingizi 263
izintaka 322

jacana 124, 125
Jackdaw 165, 310
Jackdawn 310
jacky hangman 9
Jacobin Cuckoo 85, 135, 230, 298
jandiederik 310
jay 307
jeremane 67
jichojekundu 436
jìjỳ 162
jìjỳ 63
Johnny Blacksmith 163

Kaapse tiptol 69
kabikabutshe 84, 311
kadekere 81, 134
kadima mbuzi 80, 334
kadurha 82
kadzidzi 143
Kaffrarian Grosbeak 322
kafunzi 143
kahene 64, 163
kahuji 69, 134
kahuji mirunda 69
kakatou 439
kakelaar 158, 163
kakkuk 95

kakozi 143
kakukhwe 143
kalyabuhuka 75
kambalazi 73
kambanda 84
kamkubezi 82
kamlenje 79
kamwendomphalo 78
kanchenge 143
kaneelduifie 8
kannar 440
Kannar Labek Kwiyer 440
kansire 143
kanyamarhaza 74, 143
karwhekeru 67
kashuge 143
kasongolela katuri 62
kasuku ndogo 71
katawa 86
katiti 440
Katiti Lapat Rouz Was 440
katlagter 158
Kato Nwar 439
katumbulambewa 77
kaweleweswa 82
Kelp Gull 99
kestrel 25, 82
kgogobadimo 431
kgogomeetse 73, 99
kgogonoka 73, 99
kgogonokane 431
khako 59, 168
khoholira 81
khoiti-mohlaka 74
khungubwi 363
kibarabara 74
kibasu 70
kibikula 84
kibikula lamabwe 84
kibikula nyasi 84
kìboranto 83

kibwenzi 69
kichajatui 70
kichi 434, 435
kichi bluu 434
kichi kishungibluu 434
kichi kufuabluu 434
kichi mdogo 434
kichi michirizi 434
kichi mikoko 434
kichi mtilili 434
kichi tumbojekundu 434
kichi tumbojeupe 434
kidusawarumbi 84
kiebitz 59
kievit 59
kigendagenda 83
kigoma msindo 64
kihunguluwa 77
kijimbimsitu 434
kijogoo 65
kijogoo mburo 73
kijogooshamba 65
kikoti 70
kikwele kwechi 170
kilwa 299
kimakima 76
kimpululu zeze 64 , 165
King of the Red-bills 210
king-finch 210
kingfisher 18, 76, 80, 143, 332, 357, 377, 381, 382, 385–8, 433–5
kinuka 79
kinunerako 77
kinyamhuwi 81
kipulipuli 60, 166
kipupwi 81
kipura 435, 436
kirìondànitra 82
kisharifu 434
kishosoungula 66

kishunde mabuwa 88
kitambi 67
kitàndry 80, 334
kite 225, 233, 287, 303, 337
kitolondo 70
kitolondo kanga 70
kitolondo mzitu 73
kitòry 89
Kittlitz's Plover 103
Kitty-me-Wren 37
kituku pori 436
kivuyu 71
Klaas's Cuckoo 62, 104, 142, 154
klaasperdewagter 80, 334
klaasskaapwagter 310
Klein Sprinkaan Vogel 75
Klein Sprinkaanvoël 75
kluitjiekorrel 69
Knob-billed Duck 92, 97, 369
knoet 105
knorhaan 161
Knot 105, 352
Knysna Turaco 54, 150, 238, 281, 358
Knysna Woodpecker 151
kokoko 166
komakacoka 68
komandugu 162
Koning Roodebec 210
Koning Rooibek 210
konkoit 170
korhaan 161
korhaan 126, 135, 373
Kori Bustard 325
korongo taji 72
kowa 299
kransduif 6, 8
kuifkop 69
kukačka 95
kukkuk 95
kukuku 61
kukulka 95

kumbakima 76
kumbanti 84
kumbo 60
kurukuru 166
kuyu 436
kuyu jichonjano 435
kuyu kifua rangipinki 436
kwale kwechi 170
kwarara 59, 168
kwêvoël 60, 136, 160

Lammergeier 146
landa 395
langavura 85, 298
langazua 85
langòroaomby 80, 334
langórobè 66
langòrovalàfa 74
Lanius 105
Lanius collaris 114
Lanner Falcon 80, 141, 155, 339, 376
Lappet-faced Vulture 78, 102, 154
lapwing 373
Large Dove 436
Large-billed Lark 60
lark 7, 25, 36, 67, 85, 115, 116, 133, 189,
    199, 210, 216, 232, 233, 243, 244, 245,
    246, 249, 251, 256, 277, 284, 291, 304,
    308, 309, 310
Lark-heeled Cuckoo 289
Laughing Dove 42, 45, 62, 436
làvasalàka 69
Lazy Cisticola 102
Le Vaillant's Cuckoo 231
Ledišadikgômo 431
leeba 42
leebamašu 42
leebarui 42
leebarupi 42
legotutu 431
lehehemu 61, 169

lehututu 431
Lemon Dove 45, 84, 151, 360, 435
Lemon-breasted Canary 139
lengangane 59, 168
le-ngao 168
le-nkagata 168
lenkunkuroane 60
lepidibidi 431
Lesser Black-winged Plover 373
Lesser Collared Turtle Dove 43
Lesser Crested Tern 95, 136, 143, 155,
    174, 367
Lesser Double-collared Sunbird 32
Lesser Flamingo 149
Lesser Gallinule 71
Lesser Honeyguide 144
Lesser Kestrel 149
Lesser Pied Kingfisher 84
Lesser Red-eyed Dove 436
Lesser Spotted Eagle 27, 91, 147, 149, 369
Lesser Striped Swallow 149, 151
Lesser Swamp Warbler 149
Lessercollared Turtle Dove 46
letsetseropa 378
letšoanafike 74
Levaillant's Cuckoo 306
lightning bird 295, 296, 375
likuŋguvû 363
Lilac-breasted Roller 325, 369
lily-trotter 83, 124
Limosa lapponica 352, 354
litsitsirupa 378
Little Bee-eater 73, 133
Little Bittern 68
little brown job 35, 409
Little Grebe 83
Little Sparrowhawk 155
Little Stint 254, 352, 356
Little Swift 103, 152, 298, 367
Little Tern 143, 155, 174, 367
Livingstone's Turaco 151

liwundi 264
Lizard Buzzard 78, 203, 376
locustbird 75
loerie/lourie/loury/laurie *see* turaco
lolenzoka 62
Long-billed Pipit 309
longclaw 121, 135, 185, 199, 234, 238, 241, 244, 284
Long-crested Eagle 22, 27, 29, 69, 135, 334, 335, 339, 358, 411
long-tailed black finch 316, 328
Long-tailed Dove 69
long-tailed finch 208, 209
Long-tailed Namaqua Dove 44, 46
Long-tailed Paradise Whydah 251
Long-tailed Skua 440
Long-tailed Whydah 18
Long-tailed Widow 7
Long-tailed Widowbird 133, 134, 208, 316, 321, 332
Lophaetus 27, 28
Lophoaetus occipitalis 334
lovebird 71
Loxia sulphurata 362
lubaka 84
lubondo 71
lubozi 79
Ludwig's Bustard 151
lufite 88
lunyuwa mazi 68
luvungabwasi 86
luvuzi 76
luzila 73
lyrebird 164

machema 79
Macronectes giganteus 439
Macronyx 199, 237
Macronyx ameliae 199
Macronyx capensis 234
Macronyx croceus 234

Madagascar Swallow 82
Madagascar White-eye 441
magpie 300, 310
mahem 288, 297
mahem 61, 169, 369
mahemkraanvoël 169
Malachite Kingfisher 382–4, 386–8
Malachite Sunbird 66, 77, 88, 144
malgas 441
'*malioache* 65
'*Mamolangoane* 263
manàboandràno 85
man'an'ani 59, 168
manàranòsy 80, 334
manbuekendu 286
Mangrove Kingfisher 377, 382–8, 434, 435
manjamphako 78
Manzer Zwit 439
Marabou Stork 147, 148
Marsh Owl 151
Marsh Sandpiper 103, 352–4, 356
Martial Eagle 22, 27, 28, 133, 149, 358
Martial Hawk-Eagle 112
martin 85, 368
matumandago 77
mazanamulungu 85
mbadule 72
mbeŋguni 262
mbilili 69
mchimbamchanga 86
mchunga-mguruve 87
mdiria 434
Megaceryle maxima 382, 386
meidjie 62
Melba Finch 98, 366
Melodious Lark 95, 155, 161, 163
mguna 63, 162
mhokeuta 62
mhupupu 61
Milan noir 441

Milvus aegyptius 113, 233
Milvus migrans 441
Mirafra africana 231, 232, 299
mkaribisha mgeni 86
mkatasanda 86
mkoka 85
mkubwamkubwa 60, 166
mlangilambago 63, 162
mlele 178
mmabolesana 70
mmaliwatjhe 65
mmamalianoka 262
mmamasionoke 73
mmamokete 68
mnanandjutshi 75
mng'ang'a 168
Modišadikgômo 431
Modišane 431
mojalipela 76
mokoroane 71
mokotatsie 75
molisalipela 80, 334
molomo-khaba 71
Montagu's Harrier 104
mo-otla-tshepe 65
Morning Warbler 87
Morus capensis 151
Mosilikatze's Roller 325
Mosilikatzi's Bird 325
Motacilla aguimp 84
Motacilla clara 84
motintinyane 60
Mountain Buzzard 340
Mountain Chat 99
Mountain Wagtail 84, 148
Mountain Wheatear 74, 154
Mourning Collared Dove 45
Mourning Dove 435
mousebird 78, 123, 125, 170, 214, 255,
    273, 293, 297, 323
mpirahira 82

mposanji 66
mpuji mbago 73
msharifu 434
mshongogolo 63
msilimba 70
msoka 81
Mud Lark 179
mugeke 79
muhangali 71
mukokafodya 71
mukugwhe 88
mu-kwe wa Leza 299
mulyambasi 75
mungo 60
mununi 77
mushumbiza-ngoma 64, 165
muvuzi 76
mvulawe 85
mwadonta 85
mwalabu 68
mwalala 168
mwanana 63, 162, 168
mwanawawa 168
mwendophikalo 78
mwigo 436
Myrmecocichla formicivora 101

Namaqua Bush Dove 182
Namaqua Dove 43, 44, 45, 46, 69, 145,
    249, 435
namasupuni 71
nankapakapa 82
Narina Trogon 68, 135
Natal Bush Partridge 310
Natal Francolin 161, 310
Natal Grassbird 376
Natal Kingfisher 83, 382–4
Natal Pygmy Kingfisher 434
Natal Robin 95, 140, 165, 371
Natal Spurfowl 161, 169, 310
ncedetje 239fn.16
ncedetjie 239fn.16

ndege chai 72
ndege lubozi 79
ndege mpunda 70
ndiyembili 79
nduku 67
ndyabusogya 77
Neddicky 239, 256, 281, 292, 377
Neddicky lark 235
neruvuta 88
Newton's Warbler 73
ng'ang'ane 168
ngechechangu 62
ngilikilajako 62
ngolankuchange 63
uNgqwatshi 232
nguhe 84
nguongo 60
uNhloyile kaGelegele 81, 113, 233
Nicholson's Pipit 309
nightjar 84, 85, 86, 135, 161, 179, 276,
    290, 309, 332, 336
Nilaus afer 105
ninga 436
ninga pemba 436
njiwa 435, 436
njiwa domonjano 435
njiwa kisogocheuupe 435
njiwa madoa 435
njiwa manga 435
njiwa mweusi 435
nkomatsabola 329
nkong'ota 162
nkongorho 71
nkukumadzi 73
nkule 82
nkungamadzi 73
nna-lugurhu 74
nnamba 264
Northern Grey Tit 63
Northern Lapwing 59, 297
Northern Shoveler 440

nshule 69
ntšhu 431
Numenius phaeopus 352, 354
Numida coronata 111
Numida meleagris 111
Numidian Crane 315
nyabanya 79
nyakalibwato 84
nyakanana 63
nyalundshirandshira 69
nyamoto 67
nyamuloba 76
nyamundubike 84
nyange 85, 289
nyawawa 59, 168
nymirhere 69
nyunda 79
nzwiyi 362

Oena capensis 44, 45
oiseau lunettes 441
Olive Bushshrike 174
Olive Pigeon 44, 46, 47, 70, 435
Olive Thrush 155, 367
omkoloani 395
omoucé 395
Orange Ground Thrush 137, 155, 367
Orange-breasted Bushshrike 95, 174, 378
Orange-breasted Rockjumper 100
Orange-shouldered Bunting 322
oriole 126, 260
Oriolus larvatus 31
osprey 25,
ostrich 18, 35, 92, 133, 135, 313, 317,
    318, 320, 326, 358, 395
Otus insularis 440
owl 7, 25, 36, 54, 72, 81, 85, 112, 133,
    134, 143, 144, 148, 151, 165, 183,
    210–12, 257, 258, 260, 264, 275, 276,
    277, 279, 280, 286, 289, 290, 292, 294,
    329, 332, 340, 395, 427, 428, 429

oxpecker 76, 121, 129, 135, 225
oystercatcher 276

paauw 289
paille-en-cul 440
paille-en-queue 440
palalithupa 74
Pale Flycatcher 98
Pallid Flycatcher 98
Palm-nut Vulture 76, 100, 148
papang 441
Papang Nwanr 441
Paradise Flycatcher 68, 133, 135, 136,
    256, 259, 286, 324, 325, 328
Parasitic Jaeger 102
Paríamàso 1, 3
parrot 151
partridge 67, 127, 130, 133, 210, 217, 218,
    247, 249, 250, 251, 254, 255, 273, 286,
    338
passenger pigeon 4
payanke 440
Pearl-breasted Swallow 91
Pearl-spotted Owlet 7, 112, 144
peeweep 297
peewit 59
Pel's Fishing Owl 148
pelican 76, 351, 373, 383, 387
Peliperdix coqui 161, 339
penguin 4
Penthetria albonotata 210
Pêrelborsswael 91
perroquet noir 439
petit-duc scieur 440
petleke 83
pétrel géant 439
petrel 439
Petronia 97
Phaethon 440
phakoe-ea-balisa 80, 339
pheasant 127, 396

Phezukwomkhono 310
Philomachus pugnax 352, 354
pic de la pluie 300
Pied Avocet 353–6, 373
Pied Crow 255
Pied Kingfisher 82, 381–4, 386–8, 434
Pied Starling 137, 169
pied wagtail 290
pietmajol 69
Piet-my-vrou 49, 61, 160, 369
pigeon 8, 15, 35, 40, 41, 43, 46, 48, 67,
    77, 125, 133, 151, 435, 436
Pink-backed Pelican 149
Pink-billed Lark 96
Pink-throated Longclaw 199
Pink-throated Twinspot 98, 370
Pin-tailed Whydah 68, 69, 86, 88, 133,
    152, 210
pipit 83, 121, 123–5, 308
pjemtjete 61
Plain-backed Sunbird 406
plantain-eater 358, 385
pleu-pleu 300
plover 103, 135, 136, 147, 210, 365, 373
Pogoniulus 63
Polemaetus 27, 28
Polemaetus bellicosus 112
popopo 61
poppoo 61
Porphyrio 68
Porphyrula 68
poup 61
pratincole 151
Prinia mistacea graneri 74
Pririt Batis 70
ptarmigan 4
Pternistis natalensis 161, 310
pugi 435, 436
pugi kikombe 436
pugi kipura 435
pugi kitugu 436

pugi kombamwiko 435

pugi kombe 67

pugi kubo 71

pugi wanda 436

pupupu 61

Purple Gallinule 99

Purple Heron 73, 365, 440

Purple Indigobird 151

Purple Roller 369

Purple Water Hen 83

Purple-crested Turaco 54, 136, 238, 323, 359

Pycnonotus barbatus 128

Pycnonotus bulbul 62

Pycnonotus Geelgat 69

Pycnonotus tricolor 128

Pygmy Falcon 376

Pygmy Goose 165

Pyromelana orix 210

quail 64, 73, 84, 127, 135, 141, 150, 165, 248, 249, 251, 254, 255, 268, 291, 336, 401

Quisty 8

rail 73, 78, 79

rainbird 103, 296, 298–300

rain-crow 300

rain-fowl 300

rainpie 300

Rallidae 299

Rallus gularis 79, 299

Ramanjèreky 1

Rameron Pigeon 42, 45, 151, 173, 435

raptor 82, 210

Rattling Cisticola 161

raven 265, 268, 269, 276, 307

Recurvirostra avosetta 352, 354

Red Bishop 70, 121

Red Bishop Bird 114, 188

Red Knot 353, 356, 357

Red-backed Mannikin 101

Red-backed Shrike 371

Red-billed Firefinch 293

Red-billed Oxpecker 76

Red-billed Quelea 64

Red-billed Shrike 85

Red-billed Teal 78

Red-billed Waxbill 3

Red-billed Wood Hoopoe 66, 153, 158, 364

Red-breasted Swallow 149

Red-capped Lark 309

Red-capped Robin-Chat 95, 140, 165, 371

Red-chested Cuckoo 49, 50, 54, 61, 85, 160, 298, 305, 309, 310, 311, 344, 358

Red-chested Flufftail 150

Red-chested Sunbird 77

Red-collared Widowbird 205, 207

Red-crested Korhaan 373

Red-eyed Dove 42, 45, 82, 436

Red-eyed Turtle Dove 64, 165

red-faced coly 297

Red-faced Mousebird 170, 297, 323

Red-footed Falcon 440

Red-fronted Tinker Barbet 63, 163

Red-fronted Tinkerbird 163

Red-headed Finch 67, 97, 139, 145

Red-headed Quelea 96, 138

Red-knobbed Coot 70, 143

Red-legged Pheasant 80, 127

Red-naped Lark 188, 189, 196

Red-necked Spurfowl 80

Redshank 162

Red-shouldered Widowbird 140

Red-throated Wryneck 96, 139

Red-winged Bush-shrike 114

Red-winged Partridge 311

Red-winged Pratincole 151

Red-winged Starling 115, 116, 121, 126, 221, 290

Reed Cormorant 76
reed-hen 68
Reeve 352, 354, 356
regen vogel 300
regenpfeifer 104
regnkrake 300
Ring-necked Dove 42, 45, 60, 435
Riparia riparia 368
road lark 309
robe-re-bese 72
Rock Dove 45
Rock Martin 368
Rock Pigeon 5, 6, 8, 42, 43, 46, 47, 67,
    116, 170
roibek 210
roller 33, 133, 148
roodebek 210
roodebekje 210
rooibekje 3, 7
rooibekkakelaar 158, 163
rook 165, 275, 287
rooster 126
Rose-ringed Parakeet 98, 370
Rosy-throated Longclaw 199, 406
Rudd's Lark 84
Ruddy Turnstone 141, 352, 356
Ruff 98, 144, 352–4, 356
Rufous Chatterer 75, 162
Rufous-breasted Sparrowhawk 155
Rufous-naped Lark 67, 124, 188, 189,
    196, 197, 231, 246, 294, 298, 299, 304,
    360, 390

saccaboola/sakaboola/Sakabula 7, 321
Sacred Ibis 172, 173
Saddle-billed Stork 96
Sakabuli 208
Salute Bird 378
Sand Martin 368
Sand Plover 96
Sandcock 162

Sanderling 103, 140, 352, 354, 356
sandpiper 74, 79, 103, 351
Sandwich Tern 143, 155, 174
sayeikia 62
Scarlet-chested Sunbird 96, 138
Scimitar-billed Woodhoopoe 364
Scopus umbretta 123, 227
seabird 431
Secretarybird 25, 121, 125, 135, 263, 298,
    358
Sedge Warbler 99
sehoeletsane 63
seinoli 76
sekikoko 86
selukungu 87
selunchungi 69
semchocho 85, 298
semphoma 84
Senegal Coucal 291
Senegal Lapwing 373
Sentinel Rock Thrush 74, 100
sepekwa 431
Serinus canary 79
Serinus sulphurata 362
Serinus sulphuratus 362
setoma-mahloane 72
Seychelles Scops-owl 440
shamushule 69
shearwater 439
sheisafu 75
Shelley's Francolin 373
shemakhome 87
shematigili 72
shempule 76
shemsana 65
shishungi 69
Shoe-bill (duck) 440
shorebird 18, 351, 355
Short-tailed Pipit 4, 5
shrike 114
Sickle-winged Chat 104

sidintsidina 83
Singing Bush Lark 155, 163
Sipàromàso 1, 3
skaapwagter 80, 334
skylark 159, 242, 245, 246
Small Locust Bird 75
Small Natal Bush Dove 181
Smithornis 65
snake eagle 29, 101
snipe 300
Sociable Weaver 104, 144
sokelo 79
Sombre Greenbul 155, 169, 173
sonbuoko 335
songoro kanturi 436
songororokanturi 62
Sooty Falcon 155
South African Cliff Swallow 149
South African lark 304
South African skylark 304
Southern Banded Snake Eagle 27, 149, 150
Southern Black Flycatcher 346
Southern Black Tit 137, 366
Southern Boubou 63, 126, 260
Southern Fiscal 9, 77, 98, 116, 133, 142, 225, 260, 307, 333
Southern Giant Petrel 439
Southern Ground Hornbill 7, 9, 116, 125, 129, 134–6, 175, 210, 218–20, 263, 265, 274, 275, 281, 287, 288, 289, 293, 296, 300–2, 329, 341, 369, 425–7, 431, *see also* Ground Hornbill
Southern Pochard 96, 132, 163, 370
Southern Red Bishop 66, 144, 148, 200, 241, 242
Southern Red-billed Hornbill 143, 371
Southern White-faced Owl 183
Southern Yellow-billed Hornbill 105, 371
sparrowhawk 78, 155, 256, 260, 376
speckled bush-pheasant 310

Speckled Mousebird 69, 133, 170, 255, 297, 323
Speckled Pigeon 6, 8, 42, 45, 170, 173, 435
Speckled Rock Pigeon 44, 46, 47
Spectacled Weaver 97
Spermestes cucullatus scutatus 75
Spilopelia 45
Spilopelia senegalensis 45
Spizaetus coronatus 112
spookvoël 95, 138, 163, 370
Spoonbill 71, 76, 439
spoon-billed duck 440
Spotted Dikkop 81, 374
Spotted Eagle-Owl 51
Spotted Flycatcher 101, 141
Spotted Ground Thrush 367
Spotted Thick-knee 81, 265, 374
Spotted-back Weaver-bird 33
Spurfowl 373
Squacco Heron 74, 102, 140
Standard-winged Nightjar 70
Stephanoaetus 27, 28
Stephanoaetus bellicosus 112
Stephanoaetus coronatus 112
Steppe Buzzard 289
Steppe Eagle 27, 91
Stictoenas arquatrix 44, 46, 47
Stierling's Barred Warbler 70, 98
stint 351
stompstertarend 22, 23
Stone Curlew 87
Stonechat 80
stork 7, 71, 75, 96
storm petrel 431
stormbird 218, 219
stormcock 162
Streptopelia 45
Streptopelia capicola 8, 42, 45
Streptopelia decipiens 45
Streptopelia semitorquata 42, 45

Streptopelia senegalensis 42, 45
Striped Crested Cuckoo 85
Striped Flufftail 146, 149
Striped Kingfisher 382, 385–8, 434
Struthius camelus 317
Stumpy Toad 37
sunbird 33, 35, 71, 77, 112, 150
sungurire katuri 62
suwagulamilanzi 87
Swainson's Spurfowl 255
swallow 32, 35, 75, 82, 83, 92, 99, 126,
    143, 147, 149, 185, 190–5, 210, 225,
    256, 259, 264, 290, 332, 339, 344
Swamp Hen 99
swartkopgeelgat 69
Swee Waxbill 150
Sweety-Sweety-Woo 378
swempie 161
swift 83, 85, 103, 152, 225, 290, 298, 368
Swift Tern 99
Syer 440
Syrnium woodfordi 399

Tachmarptis melba 298
tagilamkoko 87
taillevent 439
taji 72
Tambourine Dove 43, 44, 45, 46, 47, 67,
    71, 73, 151, 181, 182, 360, 436
Tauraco porphyreolophus 323
Tawny Eagle 27
Tayvan 439
Tayvan Zean-d-Sid 439
teewhup 59
Telophonus senegalus 114
Telophorus zeylonus 61
Temminck's Courser 139
Terathopius 27, 28
Terathopius ecaudatus 24, 25
Terek Sandpiper 352, 353, 356
tern 94, 99

Terpsiphone perspiculata 328
Terpsiphone viridis 324
tetere 435, 436
tetere mdogo 435
Tetrapteryx paradisea 328
Teuchit 297, 298
thahakhube 67
thick-knee 87, 374
thoboloko 83
Three-banded Plover 171, 365, 373
thrush 133, 155, 395
thunder bird 296, 300, 375
Timmer Doo 8
tinker 163
tinkerbird 63
tinkey bird 277
tinktinkie 60
tinky 330
tiptol 183
tlo-nke-lesoho 61
toitòy 61, 166
tolòkoràno 73
Too-zoo 8
topolamavi 87
toppie 183
toppie 9, 128, 183
touraco 321
towhee 165
Trachyphonus vaillantii 114
Treron 45
Treron calvus 43, 45
Trichophorus flaviventris 6
Tringa glareola 352, 353
Tringa hypoleucos 352
Tringa nebularia 352, 353
Tringa stagnatilis 352, 353
triotrio 166
triotriotsa 166
Troglodytes troglodytes 9, 37
Tropicbird 440
Trumpeter Hornbill 164, 168

tsàramàso 1, 3, 71
tshibiribiri 166
tshihungwe 83
tsiázotonònina 85, 293
tsìkoròvana 84
tsitsimungu 69
tsoroane-lipholile 60
Tufted Umber 179
turaco 18, 67, 124, 130, 150, 151, 197, 198, 200, 314, 323, 324, 326, 358, 385, 396
Turdoides jardineii 157
turkey buzzard 7, 9, 274, 275, 288
turnstone 73, 83, 89, 102, 351, 352, 356
Turtur 45
Turtur afer 45
Turtur capicola 43, 46
Turtur chalcospilos 44, 45, 46
Turtur semitorquatus 43, 46
Turtur tympanistria 45
Twinspot 70
Tympanistria bicolor 43, 46

ubantwanyana 138
ubhaklakliyo 95, 156, 171, 367
ubhamukwe 3, 4, 5, 133
ubhavuzile 150
ubhavuzile omidwayidwa 149
ubhavuzilobomvana 150
ubhavuzilomidwayidwa 146
ubhenqu 105
ubhobhoyi 61
ubhoboni 60
ubholoba 365
ubucubu 118, 134, 255
ubuklekle 60, 352, 354–7
ucijomhlophe 98, 353, 356
ucilo 115, 216, 249, 256, 281
udemezane 143
udlezinye 225
udlihashe 225

udokotela 364
udoye 298
udwetya 311
ufuba 373
ufukwe 133, 289, 299
ufumba 373
ugazini 67, 97, 139, 145, 201
ugilonci 172
ugilonko 172
ugolantethe 75
uhele 204
uheshana 376
uheshane 155, 376
uheshanobomvu 155
uheshanomncane 155
uheshe 36, 133, 155, 376
uheshomlotha 155
uhlalanyathi 225
uhlazalwesiwa 74
uhlazanyana 144
uhlazazana 66, 77
uhuye 232, 304, 394
ujamelihlathi 100
ujamelumhlanga 74, 100
ujamelumhlanga omncane 149
ujikanyawo 78
ujojekhaya 152
ujojo 133, 152, 205, 208, 209, 214, 251
ujolwane 100
ukhandabomvu 96, 138
ukholo 234, 303, 358
ukholumidwayidwa 369
ukholwase 145
ukholwase omkhulu 149
ukholwase omncane 149
ukhonzane 44, 45, 46
ukhozi 22, 23, 28, 29, 36, 40, 91, 124, 130, 133, 143, 147, 188, 254, 294, 326, 357, 358, 369
ukhozilwentshebe 27, 146, 147, 358
ukhozimuhlwa 27, 91

ukhozolumabala 27, 91, 147, 149, 369
ukhozolumabalabala 27
ukhozolumadladla 27, 149
ukhozolumidwayidwa 27, 91, 149
ukhozolumnyama 22, 23, 27, 91
ukhozolunsundu 27, 149, 369
ukhozolusisila 27, 91
uklebe 155, 204, 205, 376
uklebe osankonjane 376
uklebemawa 376
ukliyo 95, 136, 143, 155, 156, 174, 367
uluve 133, 135, 324
umabengwane 399
umabhashinhlayela 95, 174
umabhashinhlayela ohlaza 174
umabhelwane 140
umabhengwane 140, 183, 294
umacibudaka 99
umacutha 102, 140
umadevaphuzi 97, 138
umadevu 97, 138
umadletshana 140
umadolobomvu 96
umafusini 74
umagevuza 96, 163
umagumejana 98, 370
umahube 140, 210
umajikamatshe 352
umakhwifikhwifi 353
umalaleni 139
umalusinkomo 102, 334
umambathingubo 68, 153
umamhlangeni 52, 145, 155, 376
umandubulu 112, 183
umangqwashi 232, 294, 304, 361, 391
umangqwatshi 232, 304
umangube 140
umantuluza 63
umanyosini 101, 371
umanyovini 101, 140, 145, 371

umaphendulamatshe 102, 141, 353, 356,
    357
umaphithizela 103, 140, 354–7
umasengakhoth'idolo 225
umashwili 155, 173
umashwilomidwa 155, 173
umasikulufu 103, 141, 371
umatatazela 103
umathandaluzibu 153
umathantatha 103
umathebethebana 82
umathebethebana omkhulu 149
umathebethebana omncane 149
umathebethebeni 256, 260
umathithibala 371
umatsheni 100, 139, 145
umatshikiza 103
umawube 210
umazalashiye 104, 141, 154
umbalane 151, 257
umbalane okhandampisholo 151
umbangaqhwa 87, 265, 374
umbhicongo 9, 31, 358
umbhukwane 143
umbicini 72, 145
umbuzana 163
umcelu 333
umcwicwicwi 136, 137, 171, 369
umdlambila 76
umdlampuku 76, 225, 340
umehlwana 142
umehlwane 72, 138, 142, 394
umfelokazi 144, 346
umfelokazi omlenzemhlophe 151
umgongolozi 9, 31, 358
umgoqongo 358
umgquphane 119
umhlanebomvu 150
umhlangeni 99
umhloshana 354, 373

umjekejeke 137
umjenenengu 135
umkhololwane 143
umkholomphunga 105, 371
umkholwane 125, 143, 371, 395
umklewu 135, 160, 174
umkliwu 60
umkluwe 60
umlungwana 340
umngcelekeshu 309
umngcelu 36, 121, 123–5, 133, 308, 309
umngqithi 325, 373
umnqube 105, 364
umnquduluthi 352
umntoli 309
umqaqongo 358
umqhalophuzi 96
umqoqongo 9, 31, 135, 358
umshwelele 360
umswelele 280
umtshivovo 170, 297, 323
umtshwelele 280
umunswi 133, 155, 367, 395
umunswi wehlathi 367
umunswili 155, 367
umuntswi 118
umvemve 33, 129, 133, 148, 213, 240, 286, 288, 308, 333, 336
umvemventaba 148
umvuliyeza 103, 152, 298
umvuzi 376
umxhwagele 135, 172
umzwangedwa 80, 81
umzwili 155, 163
umzwilili 95, 97, 155
unceda 239
uncede 35, 147, 239, 256, 377
uncede womhlane 377
uncedoselesele 147
undlunkulu 87

undodosibona 152
ungangomfula 89
ungceda 181, 239, 277, 292
ungcede 239, 249, 256
ungoqo 209, 248, 255
ungoqongo 31
ungozwana 354
ungqangendlela 197, 232, 360, 390
ungqengendlela 232
ungqwashi 67, 195, 196, 231, 232, 304, 390, 392
unhlekwane 113
unhloyile 113, 135, 229, 233, 255, 303, 358, 360, 424, 425
unkombose 43–6, 145, 249
unobathekeli 183
unobulongwe 88, 139, 150
unobulongwonsundu 150
unochibi 99
unochweba 99
unocilongo 95, 164, 374
unocu 98, 370
unodaka 99, 354–7
unofosholo 91, 97, 369, 440
unogandilanga 63, 163
unogilonki 139
unogolantethe 152, 264, 364
unogqabhakazi 98, 144, 353, 354, 356, 357
unogxumetsheni 100
unohhemu 3, 61, 133, 139, 169, 297, 369
unohlohlweni 104
unokhifi 354–7
unokhukhuza 95
unokilonki 172
unokilonki elikhandamnyama 172
unokilonkomnyama 172
unokukhukhuza 139
unolwandle 99
unomanduli 99

unomaswana 226
unomlomophuzi 96, 150
unomntan'ofayo 182, 225, 227
unompempe 95, 165, 353, 356, 357
unomtsheketshe 111, 360
unomtshingo 95, 164, 374
unomunga 374
unomyayi 126
unondwayiza 83, 125, 139, 153
unonengekhanda 63
unongcangiyana 78
unongilobomvu 96, 139
unongobotsha 382, 385, 386, 388
unongoyana 100, 139, 144, 370
unongozolo 382–8
unongozolwane 387, 388
unongqwashi 232, 298, 304, 361, 390
unonkalankala 377, 387, 388
unonklilwane 143, 155, 174, 367
unonkliyo 143, 155, 174
unonkositini 95, 140, 165, 371
unonsundu 97
unonzwili 95, 155, 163
unonzwilobomvu 155
unopheshwana 353, 356, 357
unophuzana 97, 140
unoqand'ilanga 163
unosidlekekazi 104, 144
unosigqokomnyama 96
unosikhutha 376
unosimila 92, 97, 369
unosongo 98
unosungulo 364
unothezane 104
unothwayiza 103, 353, 356, 357
unothwayizana 354, 356, 357
unovilane 102, 140
unovimba 105, 353, 356, 357
unowanga 364
unoxhongo 365
unoxwil'impuku 77, 225, 340

unoyenge 364
unozalashiye 104
unozalizingwenya 225
unozalizingwenya 51
unozalizingwenyana 152
unozila 97, 346
unqilo 240
untilontilo 232
untilontilo 361, 390
u-Ntloyile 228
untloyile ka'Gelegele 228
untshili 170
unukani 79, 145, 364
unununde 352, 354, 355
upapasa 86
uphalane 143
uphendu 352
uphezukomkhono 49, 50, 61, 305, 309, 358, 369
Upupa epops 61
Upupidae 61
uqaqashe 232, 361, 391
uqholompunga 97, 150
uqotshane 178
Urocolius indicus 323
usacingo 105
usafukwe 98, 142
usamdokwe 44, 45, 46
usamklewu 98, 142, 174
usangqwashi 67, 141, 232, 304, 361, 390
usantiyane 98, 142, 366
usibagwebe 200, 360
usibó 9, 31, 136, 358, 369
usifubabomvu 96, 138
usikhothamlotha 148
usikhothanambuzane 101
usikhothaphela 101, 439
usipheshula 355, 356, 374
usiphikeleli 388
usipoki 95, 137, 164, 370
usiqhovana 65, 97, 138, 144

usomheshe 141, 155
usomthende 98, 141
usonambuzane 100, 141
usonkanyezi 98, 141
usothathizwe 102, 141, 371
usoxhaphozi 355
uswenka 360
ut[h]ekwane o'ziluba 227
uthekwane 9, 16, 122–5, 129, 179, 220, 227, 274, 277, 284, 296, 297
uthekwane ka'ziluba 227, 228
uThekwane kaZiluba 9, 112
uthimbakazane 365, 427
uthuthula 352
uve 256, 259
uvuze 311
uvuze 376
uzangozolo 382–4, 386–8
uzavolo 62, 135, 179, 309
uzazu 105, 371
uzibukwana 97

Vanellus vanellus 297
vànofòtsy 66
vànomainty 66
Verreaux's Eagle 22, 23, 27, 76, 91
Verreaux's Eagle-Owl 183
Vidua funerea 144
Vidua macroura 210
Vidua phoenicopterus 322
vìkyvìky 83
Vinago delalandi 43, 46
visvanger 386, 388
vivak 59
vòrombàraràta 73
vòrombéndrana 73
vorompatsa 76
vòrondríaka 73
vòronodábo 77
vóronòsy 73
vórontàniómby 80, 334

vulture 28, 33, 54, 68, 100, 147, 148, 197, 207, 210, 215, 216, 246, 252, 254, 257, 264, 281, 289, 326, 329, 427–9
Vulterine Fish-Eagle 76

wader 18, 351, 355
wagtail 80, 84, 129, 133, 148, 210, 213, 240, 246, 280, 286, 288, 308, 333, 334, 344
Wahlberg's Eagle 22, 27, 91, 149, 369
Wailing Cisticola 161
wanamlungu 85
warbler 63, 69, 96, 147, 162, 235, 249, 256, 308
waterbird 83, 99, 365
waterhen 365
watertrapper 371
Wattled Plover see African Wattled Lapwing
Wattled Starling 66, 75, 114, 144
waxbill 3, 32, 89, 255
weaver 25, 64, 67, 258, 261
Western Cattle Egret 431
Whimbrel 352, 354, 356
Whip-poor-will 179
Whiskered Tern 97, 138
White Bellied Stork 75
white cattle egret 265
White Stork 71, 72, 75, 152, 264, 364
White-backed Duck 94, 171, 366, 368
White-breasted Dove 8, 43, 46, 47, 67, 182
white-browed alethe 291
White-browed Coucal 65
White-browed Robin-Chat 75, 95, 162, 164, 311, 374
White-crested Helmetshrike 133, 365
white-eye 1–3, 71, 441
White-faced Duck 94, 171
White-faced Owlet 166
White-faced Whistling Duck 368

White-fronted Bee-eater 360
White-fronted Dove 182
White-fronted Plover 7, 103
White-necked Jacana 73
White-necked Raven 265, 266, 363
White-starred Robin 98, 141
White-throated Swallow 91, 149
white-winged partridge 338
White-winged Tern 99
whydah 25
widow bird 187, 316, 321, 322, 346
widow 25
willie 169
willietiptol 169
Wire-tailed Swallow 91, 147
Witkeelswael 91
witkruisarend arend 22, 23
Wood Dove 436
Wood Owl 140
Wood Sandpiper 98, 143, 352–4, 356
Wood Pigeon 8
Woodcock 162
Woodford's Owl 294, 399
woodhoopoe 79, 145
Woodland Kingfisher 145, 382–4, 386–8, 434
woodpecker 151, 162, 170, 200
Woodwards' Barbet 370
Woolly-necked Stork 153
Wrannock 37
wren 9

Xenus cinereus 352, 353

Yellow Bishop 70
Yellow Warbler 103
yellow weaver 33, 74, 87, 137, 208, 214, 229, 235, 238, 241, 242, 258
Yellow-bellied Bulbul 5–7
Yellow-bellied Eremomela 173
Yellow-bellied Greenbul 237, 241, 242

yellow-bellied wagtail 74, 166
Yellow-billed Duck 368
Yellow-billed Kite 18, 81, 84, 113, 135, 228, 229, 233, 234, 255, 303, 337, 338, 339, 344, 358, 415, 424, 425, 441
Yellow-billed Oxpecker 76
Yellow-billed Stork 96, 150
Yellow-breasted Bulbul 241
Yellow-fronted Canary 257
Yellow-rumped Tinkerbird 369
Yellow-shouldered Wydah [*sic*] finch 210
Yellow-streaked Greenbul 155, 173
Yellow-throated Longclaw 17, 130, 229, 234, 239, 246
Yellow-throated Petronia 97, 140
Yellow-throated Sparrow 97
Yellow-throated Woodland Warbler 96

Zitting Cisticola 161
Zosterops 1, 2, 71
Zosterops madagascariensis 1, 3
Zosterops maderaspatanus 441
zostérops malgache 441
zumbulu 434
zwazo linet 441
Zwazo Linet Malgas 441